# Earth's Flips of State

# Earth's Flips of State

## *The Co-Evolution of Life and Planet*

JONATHAN COWIE

OXFORD
UNIVERSITY PRESS

OXFORD
UNIVERSITY PRESS

Oxford University Press is a department of the University of Oxford. It furthers the University's objective of excellence in research, scholarship, and education by publishing worldwide. Oxford is a registered trade mark of Oxford University Press in the UK and in certain other countries.

Published in the United States of America by Oxford University Press
198 Madison Avenue, New York, NY 10016, United States of America.



CIP data is on file at the Library of Congress.

ISBN 9780197825372

ISBN 9780197825365 (hbk.)

DOI: 10.1093/9780197825402.001.0001

Paperback printed by Integrated Books International, United States of America

MIX
Paper
FSC FSC® C183721

The manufacturer's authorized representative in the EU for product safety is Oxford University Press España S.A. of Parque Empresarial San Fernando de Henares, Avenida de Castilla, 2 – 28830 Madrid (www.oup.es/en or product.safety@oup.com). OUP España S.A. also acts as importer into Spain of products made by the manufacturer.

*In the memory of*

*Graham Connor*

*(16 January 1957–26 December 2018)*

# Introduction and Acknowledgements

There is an early 20th-century saying, 'may you live in interesting times'. Certainly, we are now living in an interesting time. Never before have we, humanity, had access to so much information at our fingertips. Never before have we threatened so many species. Never before have we faced the challenge of transforming the way we live so profoundly into an environmentally sustainable mode.

Yet our present is shaped by the past. Exactly how is something we rarely consider in any depth, let alone on the grandest of scales: the billions of years of Earth's history.

This book looks at how carbon-based life and our planet have evolved—co-evolved—together to the present rather strange (in the broad context of the bulk of the past half billion years or so) state. The Earth has not always been in its present form: it has existed in other states and 'rapidly' (in the geological sense of time) flipped from one state to another. Each new, *key* state has seen a different form of life dominate, and so a co-evolution of life and planet narrative has developed. Indeed, all these key flips of state have certain characteristics that are not all found with other changes—sometimes profound changes—the planet has experienced. All these characteristics can be seen today, which raises the question as to whether or not we are about to see the Earth system transition (flip) to a new *key* state: one in which the planet's state is determined by a single species and its technological complex.

Another question arises: if the afore co-evolution of life and planet narrative took place here on Earth, could it take place elsewhere? There are implications for life elsewhere in the Solar System and the Galaxy. It might even shed light on the Fermi Paradox as to why we have not detected alien intelligence!

To answer these questions we need to draw on a number of scientific disciplines, from biology and geology to environmental and Earth system sciences. Yet, the way much of science is undertaken is usually in a very specialised way. To speak across many disciplines, and indeed additionally to address a more popular science readership, I have attempted to present matters in as simple a way as possible. (Having said that, I have not stinted on the academic referencing of fundamental papers, so there are plenty of academic avenues and background for those wishing additional exploration.) Further, because issues and questions have been approached from different directions, it can seem from time to time there is some repetition. In actuality, this is because it is the same thing being viewed from different perspectives as well as because a bigger picture is being slowly assembled before coming to the final chapter's conclusions.

Readers should carefully heed this introduction. Because this book draws on so many disciplines, *before* commencing to read the main body of this book's text, it really is worth briefly checking out the book's appendices, as these explain technical nomenclature, abbreviations, and the bio-geological timescale. Just to get an initial feel for what's in them. Many readers should find this a great help!

Before you start, I need to make it clear that not all research provides a clear-cut answer to some of the topics covered. In some instances, there are rival ideas. Indeed, many specialists in varying disciplines may have their own take or at least nuance on some of the ideas and science presented. On occasion, I will remind you of this, and sometimes I will summarise rival hypotheses. As mentioned, this book is academically referenced so that those who wish to can 'search' the internet for the key sources and see competing ideas in the way their proponents presented them. As said, never before have we had access to so much information at our fingertips.

This said, so as not to overwhelm readers with academic references, I have limited these to the pertinent ones (though there are still many here) and I am assuming that most readers will have a reasonable understanding of school-level science. Further, this is not a detailed analysis: this is a big topic and what is being attempted here is to present a broad-brush picture.

The first half of this book (Chapters 1–5) recounts the evolution of life and our changing planet over roughly 4 billion years from our planet's formation to several hundred million years ago. Across this time, the Earth has changed: seen a thicker crust develop, the instigation of plate tectonics, the transformation from having a chemically reductive surface to an oxidised one. Life too has evolved, from simple (prokaryote) cells to more complex (eukaryotes) and then multicellular life.

But these two evolutionary strands—of life and planet—are not independent: life depends on its environment and the environment in turn is transformed by life. What is happening is a co-evolution of life and planet.

This co-evolution was not uniform: there were distinct periods of critical transition from one planetary state to another. As mentioned, all these key, co-evolutionary 'critical transitions' have some characteristics in common.

However, not all of the Earth's critical transitions were due to the co-evolution of life and planet and involved a new fundamental stage of biological evolution: some were more happenstance, such as an asteroid wiping out the dinosaurs. Chapter 6 looks at these other (non-biological-evolution-driven) critical transitions—these 'jokers' and 'wild cards'—that were independent of any major leap in biological evolution. It also teases apart some of the characteristics of these 'joker' critical transitions and those key others driven by a new innovation arising out of biological evolution. Having said that, Chapter 8 contains examples of how critical transitions (due to factors such as climate change, hence environmental change), independent of any major leap in biological evolution, can spur biological evolution with a look at how our own species evolved.

If the narrative covered in this book's first half has any credibility, then we might use it as a lens through which to view the possibility of life arising elsewhere than the Earth. This is explored in Chapter 7.

Chapter 8 looks at the past few million years and how humans have co-evolved with changing environmental conditions that themselves undergone critical transitions. It also develops an additional characteristic of the key, biology-mediated transitions from one planetary state to another. This characteristic, together with *all* the others associated with previous life-mediated planetary transitions, can be seen today. There is now the prospect of an imminent new flip of state, and we may be in the midst of this transition right now!

The last chapter is an unavoidably lengthy one, as it brings together much of the previous chapters and puts into context what has happened across geological deep-time with where we seem to be heading. It includes issues that feature in the news today, such as CoVID, plastic pollution, and the rise of artificial intelligence (AI). All these are symptomatic of current change and the likely crossing of a critical threshold. This chapter emphasises the need to actively manage the forthcoming transition and illustrates some of the economic resources which might be brought to bear that have not had the attention they warrant elsewhere. Included are some of the barriers to this management and safeguards needed. It also explores the implications for life elsewhere (away from the Earth) as well as what the future may have in store for us. Such has been the explosion in AI the past 2 years (during which this book was in peer-review), that if I were to draft this chapter today, the discussion on AI would be different (I might, some other time, explore this further elsewhere). However, I can reassure readers that no AI was used in the writing of this book (other than as an aid to get one piece of data mentioned below), and my publisher assures me that the copy and proof reading was done by humans. Finally, the last chapter proposes some ways in which we might test the scientific merits of this book's narrative hypothesis.

This book has taken a few years to write. Yet, over the past half-decade, new research has been published and help has been kindly provided. That the narrative I have developed is reasonably robust is testified by the new research published, in recent years while writing, simply slotting into this book's narrative line. That some colleagues in the science community have willingly helped has also been encouraging.

First up, thanks must go in general to members of the Earth System Science group (formerly the Gaia Society) of the Geological Society. Over the decades their symposia have always been interesting and some others attending them have tolerated my seemingly naive questions as someone more rooted in the life sciences than the geosciences (much understanding of biological evolution and past climate change comes from the geological record). Special thanks needs to be given to those that provided material as well as copyright permission for some of the diagrams. These include: Buzz Baum, Rob Grant, Robert Hazen, and Hiroyuki Imachi, together with the Met Office and the UN's Intergovernmental Panel on Climate Change. Guidance on photosynthesis came from Simon Geikie.

Hiroyuki Imachi kindly provided more detailed micrographs of the Japan Agency for Marine-Earth Science and Technology's team's Lokiarchaeota isolate, discussed in Chapter 4, and has also has just informed me that in November 2024 that species gained an official designation by the International Committee on Systematics of Prokaryotes. For completeness, and for those in the biosciences wishing to explore this aspect further, the isolate's official nomenclature is: phylum—Promethearchaeota; class—Promethearchaeia; order—Promethearchaeales; family—Promethearchaeaceae; genus—*Promethearchaeum*; and species—*syntrophicum*.(*1*) (The average reader need not worry about this.)

Rob Grant kindly provided an AI-gleaned data point—different AIs were used as a check—know-how to estimate the size of data storage in the technosphere for 2025(2) used in Figure 8.7. Its various references were looked up in case of AI hallucination. Estimates are just that, 'estimates', but such is the current rate of technosphere growth that if this does not relate to the year 2025 it almost certainly does to the mid-to-late 2020s. Rob works in the AI sector. Other than this single use as an aid to literature search, as said, there was no AI employed in writing this text, which was done the old fashioned way (albeit with a hefty supply of builders' tea).

Much of this book was written during the 2020–2021 CoVID-19 lockdowns, which for myself was also a digital lockdown without domestic internet access (nor do I own a smartphone): having retired from '9-to-5' (it was never that) I no longer have work internet access, and during lockdowns learned society Fellows rooms with their cybercafés were closed, as also were local library cybercafés, plus my voluntary ventures ceased too (these too had previously given me ample day-to-day internet access). So I am indebted to Mark, Tracy, Victoria, and William Cowling as well as Julie and Stephen Perry, who became dab hands with Google Scholar over various CoVID lockdowns, looking things up when I phoned and/or downloading papers to memory sticks. Appreciation to Pat Fearnside for the CoVID social bubble and occasional, additional net access. I am additionally grateful to Tony Bailey for the back-up, fall-back access plan (fortunately not required but it was most reassuring to have). Finally, I thank all involved at Oxford University Press: my Commissioning Editor Jeremy Lewis; Project Editor Jodie Keefe; Project Manager Fathima Nisana; Copy Editor Sylvia Cannizzaro; Clare Hofmann did the cover design providing options and those largely associated with the $SF^2$ Concatenation team were polled for the choice as to which to use. Thanks also goes to the OUP marketing team with whom, at the time of writing, I have yet to liaise.

Of course, any errors that may have crept in, or failings in the narrative I portray and any oversimplifications, are my own.

Meanwhile, I hope you enjoy the ride on which you are about to embark.

*Jonathan Cowie*
E. Midlands, Great Britain
www.science-com.concatenation.org
Archived at: www.webarchive.org.uk

## References

1. Imachi, H., Nobu, M. K., Kato, S., *et al.*, Promethearchaeum syntrophicum gen. nov., sp. nov., an anaerobic, obligately syntrophic archaeon, the first isolate of the lineage 'Asgard' archaea, and proposal of the new archaeal phylum Promethearchaeota phyl. nov. and kingdom Promethearchaeati regn. nov. *International Journal of Systematic and Evolutionary Microbiology*, **74**, 006435 (2024).
2. Grant, R., *Technosphere size estimate in bytes for 2025.* Personal communication to Jonathan Cowie, January 2025.

# Contents


1. The Young Earth 1
2. Towards a Flip 35
3. Towards a Snowball 69
4. A Billion Years towards the Second Flip 107
5. The Second Flip and Other Snowballs 135
6. Jokers and Wild Cards 170
7. Exobiological Implications 199
8. A New Flip? 241
9. Implications 292

Appendix 1: Abbreviations and symbols 357
Appendix 2: Nomenclature, Data, and Terms 361

*Nomenclature and data sources for appendices 1 and 2* 376
Appendix 3: Bio-geological Timescale 377

*Index* 379

# 1
# The Young Earth

In one way this book recounts your story, it being the story of the events and circumstances that led to you and your life. As such, it is also very much the story of your offspring's future. Of much less personal importance to you—but arguably at least as fascinating—this narrative hopefully casts some light on the potential for multicelled, carbon-based life forms elsewhere, on other Earthlike planets.

Yet though this is a story, it is not simply made up; it is a story based on scientific evidence. So, to follow this narrative, an understanding of the science in a number of different disciplines is useful: this is not just a story but is as much a *tour de force* journey through the sciences that joins up many dots to make a big picture. Stick with it and it will take you from our earliest origins, as the most primitive of life forms, through to some of the challenges we face today as a modern, technological society. Ultimately, it even speaks to life elsewhere huddling around some of the couple of billion stars in our galaxy. But let's not get ahead of ourselves.

To begin at the journey's 'beginning' you need to know that there is an 'end', as well as the distance between the two, in order to decide whether this sojourn is one worth taking. This story covers the highlights of our world's extensive chronicle of life and planet. It is a story that covers a long time: billions of years. As told here, it begins with the birth of the Solar System and, lasting a few billion years, ends in the near future but with more far-reaching implications.

Now, you will appreciate that our world's history has lasted a long time, but you may not truly appreciate the scale of things and your place within it, let alone when the major flips in planetary state took place. There is though one illustrative way I would like to do this that helps gain a sense of proportion.

Consider the span of the Earth's existence in the perspective of your arm's length, from your shoulder to your fingertip: your shoulder represents the birth of the Earth and the tip of your forefinger, the present. On this scale the few thousands of years of recorded human history (effectively the time since writing was invented) easily nestles within a fingerprint valley at your forefinger's end. Such is this scale, of the roughly 4.6 billion year arm's length of our planet's existence to date, that the couple of centuries of our own post-industrial revolution, technological society is but a fraction of a micrometre: extend your arm and brush your forefinger against an object and the representation of time of our technological

*Earth's Flips of State*. Jonathan Cowie, Oxford University Press. © Oxford University Press (2026).
DOI: 10.1093/9780197825402.003.0001

society is smaller than the size of a bacterium that you have unconsciously just wiped away. In that sense, this puts any concerns you may have about things in general—such as filling in your tax return—into perspective.

## A sense of scale

The afore viewpoint is humbling: it neatly put us in our place in time, if not space; we are really quite insignificant.

Yet this perspective is also illuminating. As this book is being written (in the early 21st century), we are discovering an ever-increasing number of exoplanets—that is to say, planets outside of our Solar System. We are now finding so many that we are now not just looking simply for exoplanets, the hunt is on for exo-Earths; searching for Earthlike planets elsewhere in our own neighbourhood of the Galaxy. Indeed, we have found a handful that are close enough in likeness to Earth that they could have liquid water on their surfaces. Yet, even supposing we do find one very much like Earth (given there are well over 100,000,000,000 stars in our Galaxy—actually, possibly 200 billion stars or even more as indicated by recent surveys [*1,2*]—it is an odds-on bet that there really are a number of Earth near-twins somewhere out there), because of various factors, the chances still are that our newly discovered Earthlike planet will be devoid of a technological civilisation *as* advanced as our own. After all, if we do find an Earthlike planet, then we could discover it at any time in its analogue Earthlike history. And if in the arm's-length scale of our Earth's history, human technological society was less than the size of a bacterium, a randomly selected pin-point along the arm of a similarly aged, alien world counterpart would be unlikely to hit upon that specific bacterial-sized length.

Of course, it is a moot science fictional point as to whether we might find on an older analogue Earth either a very ancient, long-lived, *more* advanced technological civilisation, or one that flourished briefly before dying out leaving its world to live on with just archaeological remains of the techno-civilisation it once sported. There is much to debate. What might be considered optimists may contemplate that once a civilisation establishes science and technology, this knowledge is never fully lost and so the technological civilisation continues for a very—even perhaps a 'geologically'—long time. (We will return to this briefly later.) On the other hand, pessimistically, even if our technological civilisation lasts a thousand years (not even the diameter of a red blood cell on our arm's-length scale) it will be doing well; there are many pitfalls for a technological civilisation to avoid. These range from outright failure (such as nuclear war self-destruction) to failing success (the global society flourishes, growing, eating up the biosphere's resources and disrupting biosphere processes to the extent that the civilisation destroys itself and its intellectual heritage). In-between these two extremes, some alien civilisations may decide to have small, self-sufficient populations and stay at home until after billions of years their sun begins to die and they up sticks to move to a new home around a younger star (3).

But all this is rather fanciful. Whatever Earthlike planets we do find, they will be interesting irrespective of whether or not they sport a technological society. Moreover, as we discover more exo-Earths (especially old ones), so the probability of learning of implications for our own technological species will increase. Again, here we are drifting into an area traditionally occupied by science fiction, though this book's story is firmly rooted in science.

All this raises the question, how truly Earthlike does an Earthlike exoplanet have to be before we can consider it to be properly illustrative of our own planet?

One thing, an Earth system (or a biosphere) scientist will tell you, is that—if a world is truly Earthlike—the Earthlike world will have evolved through time and through 'flipping' states within its history in ways not unlike our planet. That is to say, the world will have on a number of occasions been in one state of condition before 'rapidly' (in the geological sense of time) changing, or 'flipping', to another distinctly different state. (We will return to the question as to how truly Earthlike an Earthlike exoplanet has to be for life habitability in Chapter 7.)

But before we begin, there is one other aspect of our arm's-length-scale sense of perspective that really needs to be noted. Given the aforesaid, the length from shoulder to finger-tip represents 4.6 billion years of Earth's existence and a few thousand years of written history—let alone the couple of centuries of technological civilisation since the industrial revolution being equivalent to less than the size of a bacterium; so in the scheme of things, the entirety of industrial, or even agrarian, human history is colloquially almost no time at all. Therefore, if we are going to talk about the changing states of the Earth's biosphere we need to appreciate that—on an Earth-lifetime geological scale—a biosphere transition, lasting say a thousand years, or even millions (yes, millions), is very fast indeed: so fast that in geological time of the entirety of our planet's life the change of state can be truly called a 'flip'.

Now, this is not a cheat. The two key 'flips' of planetary state we will discuss did take millions of years to complete, but the change from before to after was extremely profound. Some other of the 'flips' the Earth has gone through discussed in this book have been transition periods lasting many centuries to thousands of years. Indeed, there is one of direct relevance to us that covers just a couple of centuries and whose peak transition (the time of greatest change) may be just a century or even less: a time period almost comparable to the current life of a long-lived human.

True, to the average person, a change lasting 200 years is very slow: conversely, to a geologist, or an Earth system scientist, such a change is not just fast but very abrupt. To the average person 200 years is a length of time that they cannot practically conceive of living all the way through; to a geologist it could easily represent less than a few millimetres of strata. And so words like 'abrupt' do mean very different things to different people. Indeed, in the early part of the 21st century there has been quite a bit of discussion in blogs and social network sites on climate change and climate change science as to the meaning of 'abrupt' change. Here, numerous times people have been pulled up by other commentators over their use of terms like 'fast', 'rapid', and 'abrupt'. Even scientists talking across disciplines—such as ecologists to geologists (and *vice versa*)—find that they give different meaning to the same words. It even happens with closely allied disciplines, such as ecologists talking to palaeoecologists (the study of past ecosystems)—ecologists look at species living in front of their eyes but palaeoecologists view ecology through the prism of the geological record. Bear all this in mind as you read this book.

But how did it all begin? What were the key, founding parameters enabling life on Earth? And, importantly for us, what were the parameters for intelligent life to evolve, and to evolve to you as an observer and book reader? How did the Earth form? What was its initial state from which it first flipped?

## The formation of the Earth-Moon-Sun system

The formation of the Earth by itself did not result in a biosphere. There are objects that are not stars, such as brown dwarves and planet-sized bodies, floating between stars with their systems of planets and minor bodies: they cannot support life as complex as that on Earth, if they can support life at all. (They probably can't unless under some rather exotic set of circumstances.) Our biosphere needs energy, and for that it needs a star (a giant, nuclear fusion power plant), which for us is the Sun (aka Sol, hence the term 'Solar System'). The currently prevailing view is that it also needs a stabilising companion just as a high-wire walker needs a stabilising long pole. For life on Earth, this stabilising helps keep the Earth's angle of tilt, hence the seasonality of weather, more constant: this important for complex life on Earth. The Earth's stabiliser is its moon: the Moon (aka Luna). (Having said that, some others disagree.) (*4*).

As it happened, all three—the Sun, Earth, and the Moon—formed more or less in succession and—in stellar evolutionary let alone geological terms—somewhat quickly.

The Earth and the rest of the Solar System began forming some 4.6 billion years ago. We can date this because grains of radioactive elements (such as uranium-235) were separated. But being radioactive, these began to decay. Uranium-235 (the formal symbol of which is $^{235}U$) decays so that half of it decays over 703,800,000 years into thorium-231 ($^{231}$Th) and then this in turn again decays but much faster and ultimately to lead (Pb). Measuring the amount of the decay products relative to that of U-235 enables us to date when that grain of uranium formed. $^{238}$U also decays, with a half-life of about 4.5 billion years, and this too provides a useful tool. Measuring the amount of the decay products relative to that of the original isotopes enables us to date when that mineral formed. There are a number of radioactive elements in nature, and so we can use a variety of such dating methods, which in turn enables us to check one estimate of age against another. Lead isotopes are particularly useful over very long periods (such as the age of the Earth).

Yet there is a problem. Where does one find such primordial rocks on the Earth? Alas, much of the Earth's surface has been reworked—with tectonic plates subducting and volcanoes creating new strata—so eradicating old surface geology. But we can examine meteorites—part of the leftover material from the formation of the Solar System—to find out the ratios of their element composition and so see how old they are, hence the Solar System.

On Earth and other large bodies, every time a mix of rocks melts its constituent elements separate so undermining the radiological record. This means that the ratio of radioactive elements and their decay products is determined by the length of time only since the rock solidified after it last melted. The cycle of plate tectonic spread and subduction means that the Earth's surface gets reworked. Yet on Earth there are still a very few places with really ancient strata on or near the surface. Using radiometric dating on rocks from these areas, let alone on meteorites (which never experienced plate-tectonic-driven reworking), it is estimated that the Solar System (hence the Earth) formed roughly about 4.57 billion years ago, give or take a hundred million years or so.

The Earth—or to be more exact, the proto-Earth (sometimes known as Tellus)—had originally 'condensed' out of a primordial swirl of gas and dust at the heart of which formed the Sun. As this gas and dust moved, so some parts became more condensed and some less so. Gravity made the more condensed parts attract even more gas and dust, and so the process of accretion accelerated.

How long did it take for the protoplanets—the large asteroid-sized bodies—to form? Well, just as it is possible to use radioactive elements to date the Earth, it is also possible to use them to date the period of protoplanetary formation; after all, some protoplanets would form first and some last. Today's asteroids are the leftover remnants of this process. And asteroids will collide from time to time and so chunks of asteroids will break away. Some of these will fall to Earth as iron meteorites (actually iron-rich meteorites containing many metals). One recent illumination as to what happened during the Solar System's formation was reported in 2014, when a European and US team looked at the variations in the concentrations of the tungsten-182 isotope ($^{182}$W) in such meteorites. Looking at the variations in isotopic concentrations (which reflect the time between the first and last protoplanets formed) they concluded that the molten core of these protoplanets formed over a period of roughly 1 million years: a very short time in the history of the Solar System. Having said that, the estimate is only rough; they estimate that the likely time could be anywhere between 0.6 and 2 million years (*5*). Other earlier estimates suggested that the process took a few tens of millions years (6) and even so, in astronomical-scale terms, this is still very rapid. Indeed, parts of the process of condensing dust into at least small lumps of rock may even have been so fast as to have taken place within just a couple of years!

Amazing as this may seem, we have actually seen key stages of a star's planetary system's birth in real time. In 2012, astronomers reported observing a Sun-like star called TYC 824126521 some 456 light years from Earth. When they first looked at it, it was surrounded by a gaseous and dusty circumstellar envelope. This envelope was being warmed by the early star, and the envelope in turn emitted infrared (heat) radiation. But, less than 2 years later the infrared flux had decreased drastically, by a factor of about 30 as the dust disappeared: some stages of planet formation from condensing dust clouds seem to happen very fast indeed (*7*), even if other parts of the process takes millions of years.

But how did the Moon form?

The very earliest days of the Solar System were not quiet ones. There are a number of theories as to how the Moon came into being, but the one that seems to have the most currency was suggested as far back as 1976 and is as follows. At the dawn of the Solar System there were very many small bodies around that coalesced into the planets. By 4.5 billion years ago there were still leftover bodies, and still there are even today, such as those found in the asteroid belt between Mars and Jupiter. But back in the early days of the Solar System there were also a few large bodies bigger than today's Mars-Jovian asteroids. One of these, a Mars-sized body—known as Theia (or Thea)—came close to the proto-Earth (Tellus) and gave it a glancing blow. This caused much of Theia to melt, as well as part of the embryonic Earth around the impact point. Some of the debris from this glancing collision fell back to the proto-Earth, and part (as will be discussed shortly, maybe a tenth to a third of proto-Earth material), along with the molten remnants of Theia, went into a close orbit about the Earth. This coalesced to form the young Moon. Being molten, volatile elements and compounds (such as what on Earth would be gases and liquids) boiled off the Moon which, with its lesser gravity, would have difficulty retaining them (*8*). Only in the cold, perpetual shadows of the Moon's polar craters can water ice be found today, and even so, it is quite likely that that ice came to the Moon by comets later.

Since this model was proposed, there has been a fair bit of research into how Theia might have collided with the embryonic Earth. Here, physics helps. One law of physics is the conservation of momentum which itself is part of Isaac

Newton's laws of motion. Looking at the Earth-Moon system in isolation—ignoring the Sun, as the effects are less (there is also a separate, little more complicated story here)—both the Earth and the Moon have angular momentum from their spin, and the Moon has angular momentum about the Earth; the sum of these two momentums must always be conserved. We can measure what the total momentum of the system is today, and so we can work backwards to see what type of collision might be needed to get what we see in the present. This is just what US scientists Robin Canup and Erik Asphaug did in 2012. They reported a class of impacts that would yield an iron-poor Moon, as well as the current masses and angular momentum of the Earth-Moon system. This class of impacts involves a smaller—and so more likely—object than was previously considered, and which is consistent with the Moon forming near the very end of the proto-Earth's own accumulation of material (9).

Until recently there has been a debate as to whether Theia originated from another part of the Solar System or whether it formed close to (in a similar orbit to) the Earth. For the answer we again turn to isotopes. We know from meteorites that the various isotopes of oxygen ($^{18}O$, $^{17}O$, and $^{16}O$), that make up the chemical compounds in the rock, vary depending where the meteorite originated in the Solar System. (More specifically, its distance from the Sun.) Yet lunar rock brought back to Earth by the Apollo missions gave a similar mix of oxygen isotopes to those of the Earth, suggesting that Theia formed in a similar orbit to that of the embryonic Earth: though it should be said, such are the complexities in the science, for some this similar isotope evidence even casts doubt on the Theia impact hypothesis altogether.

Only recently have matters begun to clarify. When the University of Cologne in Germany upgraded its mass spectrometer there was an opportunity for a more detailed analysis. (Mass spectrometers are devices used to measure the mass of atoms and their isotopes.) In 2014, a team led by Daniel Herwartz published the results of their own more refined analysis of the Apollo 11, 12, and 16 lunar samples. They found that that the lunar samples had an $^{17}O$ to $^{16}O$ ratio that was 12 parts *per* million higher than rocks from the Earth's mantle. This suggests that the Moon has no more of a proportion of originally Earth material than between a tenth to a third but, frustratingly, such is the lack of a great number of different sample sites that there is still a small possibility that cannot be completely ignored of a more equal Theia–embryonic Earth mix (*10*). The high oxygen isotope ratio suggests that Theia was principally made of a rare material called 'enstatite chondrite'. Now, the vast majority of meteorites that land on Earth are chondrites, but only about 2% of those are enstatite chondrites. (Not that you need to know the detail, these are chondrites that contain orthopyroxines found in basic igneous rocks that are formed at high temperatures). Having said all that, there is still much more we would like to know about the Earth-Moon formation and, indeed, the formation of the Solar System, as this will tell us a lot about the likelihood of other Earths (exo-Earths) around other stars and, indeed, the likelihood of complex life elsewhere in our galaxy.

But why this discussion of the formation of the Moon? Let us return to the question of the importance of the Moon to the Earth's biosphere. The Moon–Earth ratio of mass is 1 to 81.3. This is much higher a satellite–planet mass ratio than anywhere else in the Solar System, which makes the Earth-Moon system kind of unique! Yet the16th-century astronomer Nicolas Copernicus considered that there was nothing special about the Earth necessitating it to be at the centre of the Solar System as many in Medieval Europe then thought: instead Copernicus theorised that the Sun was at the heart of the Solar System. So *if* we take the Copernican assumption that there is nothing unduly 'special' about our Solar System compared to other planetary systems around other stars, and given that our Moon's broadly similar size relative to its parent planet is unique in the Solar System, could it be that such an Earth-Moon sized combination is rare in the Galaxy—that is to say, that few Earthlike planets elsewhere in our galaxy (or even other galaxies) have as large a companion moon as our Moon?

Equally, it follows, the Copernican principle also suggests—if there is nothing special about the Earth as a world giving rise to a sentient species—that a sentient-life-bearing planet needs a large moon and must be in effect part of a double or binary planetary system.

Actually, this last is unduly pessimistic, there is another example of a low mass ratio double body within our Solar System that was discovered towards the end of the 20th century: the Pluto-Charon double is even closer in mass than the Earth and the Moon. However, Pluto is no longer, according to present astronomical nomenclature, classified as a proper planet but is a 'dwarf planet' whose diameter is roughly equivalent to the north–south length of the United States at approximately 1,430 miles (2,300 km).

If either, let alone both, of these conjectures have any merit then they have profound implications for the formation elsewhere of a biosphere like ours with an abundance of multicelled species living on land.

As mentioned at the beginning of this section, the Moon confers axial stability to the Earth. You see, the Earth wobbles very slowly with time, with the angle of its axis (the angle of tilt) increasing and decreasing. As we will find out much later on, this is important for a type of Earth flip: the flip between cold glacial modes and warmer interglacial modes—like the one we are currently in—within ice ages. Without a large moon like our Moon, this stability of the Earth's axis would be lost and so, over geological time, the swings of its axis would be much greater: this affects the way heat is received by the planet from the Sun. The Earth's tilt varies by only about 2 degrees in timescales of hundreds of thousands of years. By comparison Mars, without a large stabilising moon, sees substantial tilt change—by as much as 60 degrees—on timescales of hundreds of thousands to millions of years.

As it stands, the Earth's angle of tilt is such that there are only two small portions of our Earth at the poles, denoted by the polar circles, within which at mid-winter it is dark 24 hours a day: similarly, at mid-summer it is light 24 hours a day. This day-length effect is most pronounced exactly at the poles themselves and lessens moving away from the poles to the 'polar circles' outside of which, no matter what time of year, there is always some light during the day and always some dark at night, even if these do vary throughout the year.

Yet, consider a world where the axial tilt was greater. This would mean that larger parts of its surface would have either light or dark for a whole day (see Figure 1.1). In fact a world whose axis was not nearly perpendicular to the line of its orbit (as ours is) but at right angles and in the same plane as its orbit, then it would have twice a year the entire half the planet in perpetual light and the other half in dark: its polar circles would be the same as its equator and its seasons part of a year-long single day-night cycle. One could envisage on that world the winters being far colder and the summers exceedingly hot. With such temperature extremes, it would be very hard for multicelled species to cope. On our planet at least, species 'know' where they are (in the metaphorical sense). They either live in high latitudes and are adapted to the cold (such as, say, conifers) or are adapted to hot dry conditions (like cacti): they do not have to cope with both extremes at the same time, as would be the case on a world with a great angle of tilt. So a common view (there are others) is that having a large moon is important to a rich and complex a biosphere such as our own as it stabilises the Earth's tilt hence temperature variation on our planet's surface. Life needs some stability and would therefore seem not to like a system that rapidly 'flips' between markedly different states. Yet the Earth has, a number of key times, flipped from one state to another.

## Evolution of the Earth-Moon system

Standing on the Earth's surface a few hundred million years after the formation of the Moon (to be safe from post-Lunar-formation leftover debris raining down), one would see the Moon as being very large in the sky. This was not because it was bigger than it is today, but because it was a lot closer! Indeed, not only was it a lot closer, and appearing much larger, it also moved across the sky more quickly. This was not only because the Moon's orbital period was shorter but also because the Earth was rotating faster and so the Earth's days were far shorter: less than 10 hours long. As we shall see in forthcoming chapters, this may have been important to the evolution of life on Earth, but explaining this, and then

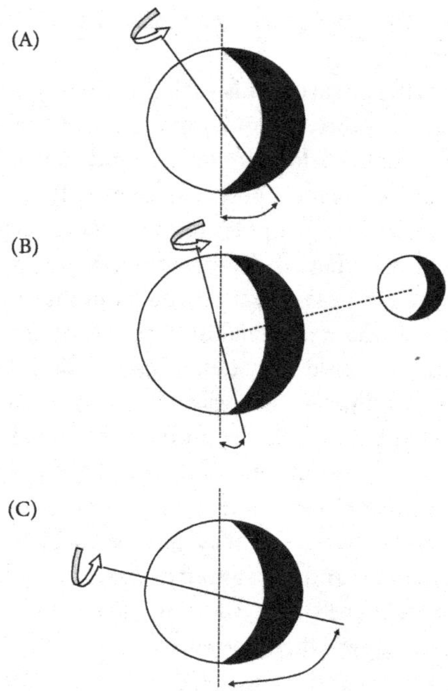


**Figure 1.1** (A) A planet's axis of tilt can increase and decrease over time. (B) But this variability of axial tilt is reduced in the presence of a stabilising moon. (C) If a planet's axial tilt is free to increase considerably, then for half that planet's orbit about its sun half is baked hot while the other half freezes, and the other half of the year the reverse happens. Complex (multicelled) organisms can live in an extreme environment: tropical plants and animals can live in a warm climate, and subpolar plants and animals can live in a cold climate. A planet whose land surfaces are regularly subject to *both* extremes is likely to see the evolution of complex land species impeded.

its evolutionary (as opposed to day-to-day biological) importance, necessitates understanding a chain of phenomena.

We actually have direct evidence of the Moon's orbital period being shorter millions of years ago through biology: bioastronomy. (Bet you never knew you could use biology to do quite a bit of astronomy let alone palaeoastronomy.)

It works like this. Some marine species—those species that release their gametes into the sea to be fertilised—have a reproduction problem: the problem is that, releasing eggs into the sea, how can one be sure that they will get fertilised? Some marine animals can get around this with the female laying the eggs and the male moving quickly in to release sperm. But species that are less mobile, and others, have another way of maximising the chances of fertilisation.

If wide, three-dimensional space—the volume of the seas—is the problem that reduces the odds of fertilisation, then species might be able to bring the odds back up again through time. This is what a number of species do by synchronising timing of their gamete release. However to do this they need some sort of signal to prompt reproduction time. For example, a few species use spring tides, with water pressure being the determining, timing factor. These are large tides that take place near when the Earth, Moon, and the Sun are lined up, which takes place twice a lunar month: twice a rotation (less 'a bit') of the Moon about the Earth. Other species use the full Moon itself with photoreceptors attuned to peak night-time illumination. As full Moons and spring tides only take place when the Moon and the Earth and the Sun are lined up, so they are linked to the Moon's orbit about the Earth.

An example of marine species attuned to these monthly cycles is the annelid polychaeta *Platyneriseis dumerilii* (from the same phylum of species as common garden earthworms) that lives in coastal sands and mud. (If you live by the coast and fish then you may have dug up on beaches similar species to *P. dumerilii* for fishing bait.) *P. dumerilii* can be found mating on the first and last quarters of the lunar month on the French coasts: the Moon's orbit's effect on the tides is the trigger.

In reef-building corals, *Stylophora*'s reproductive stages see it release its larval reproductive stage (planulae) monthly throughout the year (*11*). Again, this is triggered by the Moon's orbit about the Earth.

Today there are 12.4 lunar months (or synodic months) in the year. This corresponds with 12.4 plus 1, or 13.4, orbits of the Moon about the Earth (or sidereal months). (The afore 'plus 1' is due to the Earth-Moon orbital system itself orbiting about the Sun—this extra orbit needs to be taken into account.) This means that those species that rely on the Spring tides for their reproductive signal get their signal 13.4 minus 1 lunar orbits a year. In short, knowing how many reproductive signals such a species gets, we know that these equal the tidal signals and that this relates to the number of orbits the Moon makes.

Of course today throughout the year we can simply count the spring tides. Indeed, we do not even have to do that, we can just look up at the sky to see what the Moon is doing. This is all very well, but it does not tell us what was happening many millions of years ago and how the Moon orbited the Earth! This is where the biology and geology come together and into their bioastronomical own, with biology in the geological record illuminating the astronomy of the Earth-Moon-Sun system in the distant past, so providing an understanding of the palaeoastronomy.

For an example of how this is done, we can look at one species of coral that reproduces using the lunar monthly tidal signal: *Manicina* is a coral whose development is affected by how much energy it can devote to growth.

This energy-for-growth resource, in turn, is affected by a number of factors. Two of the most important are temperature and the energy diverted away from growth instead going towards reproduction. Fortunately, for us, this can easily be seen because *Manicina* lays down growth rings, much as a tree lays down growth rings. However, unlike trees, whose growth rings are associated with the annual cycle of summers and winters, *Manicina* lays down daily growth rings too. It is not an exact business, and confounding factors such as extreme events may mean that a coral occasionally misses a daily growth ring. Outside of the tropics, in the winter these daily rings are laid down closer together and in the summer further apart. It is therefore possible to count how many days there are in the year, especially if one looks at a number of samples to rule out erroneous, or include missing, rings. But there is a further twist: when *Manicina* is diverting energy away from growth into generating planulae (larvae) the daily growth rings narrow (both in summer and winter). In short, it is also possible to see how many lunar months there are in a year in addition to the number of days in a year.

Now, some of you may be wondering, why do we need biology to do this basic astronomy when all we need do is look up at the sky? This is where geology comes in. Find fossils of *Manicina* that are millions of years old and it is possible to examine its growth rings and ascertain the number of both days and months in a year millions of years ago. If we did not have such a palaeo-indicator then we would need a time machine to obtain such hard evidence (Figure 1.2).

Back in 1963, coral expert John Wells noticed that daily growth rings could be seen in these fossil corals within the annual compaction of growth rings and started counting. He was amazed to discover that whereas today corals like *Manicina* have about 360 growth lines a year, fossil corals from the middle of the Devonian (about 380 million years ago) have between 385 and 390 daily rings in an annual sequence (*12*).

The following year Colin Scrutton published his results, building on Wells's work, but he was looking at the monthly constriction of growth rings and not just the annual ones. He concluded that in the middle Devonian, some 380 million years ago, the year consisted of 13 lunar months each of 30.5 days (*13*). This compares with the present time of there being 29.53 days in a lunar month and there being 12.4 lunar months in a year.

But this story of biological astronomy does not end here: there is more. The 16th-century German astronomer and mathematician Johannes Kepler established some of the key mathematical fundamentals of the way orbiting objects move in gravitational fields (such as planets orbiting stars and moons orbiting planets). Importantly, he developed three laws of planetary motion: what came to be known as Kepler's laws. One of these was that the square of the period of a

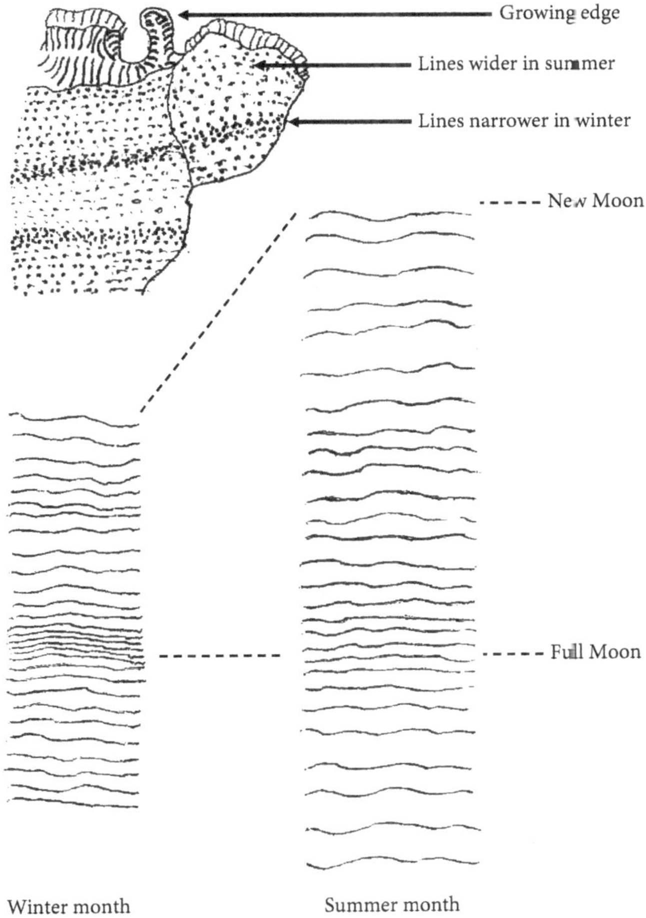


**Figure 1.2** The coral *Manicina* lays down daily rings that bunch together each full Moon as well as, outside of the tropics, in winter.

satellite's orbit (the time it takes to make a single orbit or 'T' in Figure 1.3) about a primary body is proportional to the cube of the distance between the two (the distance between the satellite and the body it is orbiting or 'D' in Figure 1.3).

From Scrutton's work we know the Moon's periodicity in Devonian times in terms of a fraction of a year. We also know from present-day observation

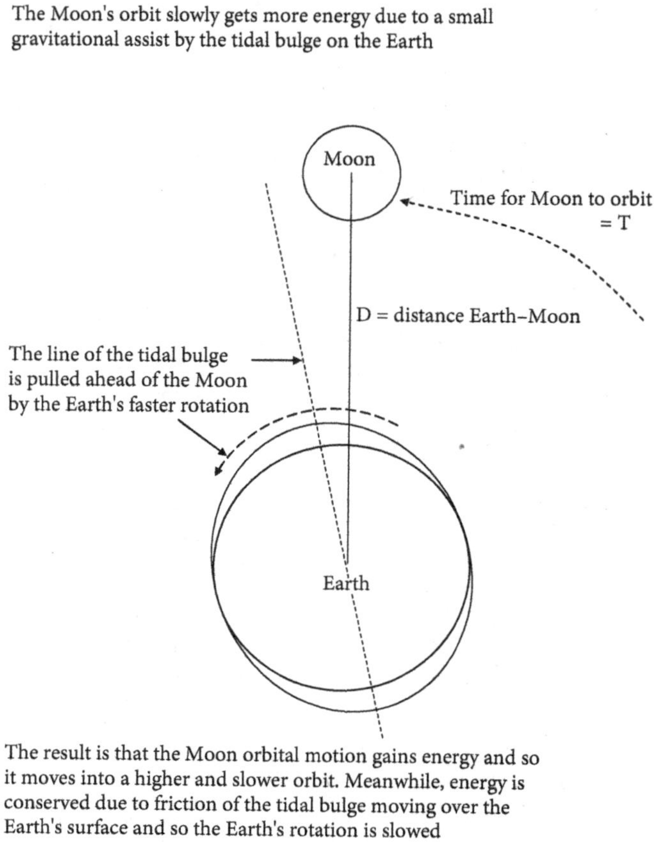


**Figure 1.3** The Moon creates a tidal bulge on Earth's seas that is carried ahead of the Moon's position above the Earth due to the Earth's rotation being faster than the Moon orbits. This tidal bulge gives a weak gravitational assist to the Moon's orbit, and so lunar months increase. Meanwhile, friction of the tidal bulge as it moves over the Earth slows the Earth's rotation. Energy and momentum is conserved, and the Earth's day slowly grows longer. The angle between the Earth-Moon line and the tidal bulge access is exaggerated in the diagram. (In reality it is only about one degree.)

(and not least the Apollo missions and the laser reflectors they left on its surface) the distance from the Earth to the Moon today as being approximately 384,400 km (to take the estimate used back at the end of the 1960s, when this work was first performed). Taking into account the aforementioned 'a bits' and adding one to convert from synodic to sidereal months, and then applying Kepler's law, it is possible to calculate that the distance from the Earth to the Moon in mid-Devonian times was 372,870 km. From then to today, the Moon has increased its distance from the Earth by over 11,000 km in 380 million years! It should be said that some modification to this approximate estimate is required in line with Newtonian mechanics and this gives us a slightly greater figure of a retreat over this time of around 15,000 km (*11*).

So the Moon is retreating from the Earth at a rate of around 4 cm a year: this is a similar order of magnitude to the rate of growth of your fingernails (the latter growth rate being around 3 mm, or 0.12 inches, a month). Will knowing this forever change the way you contemplate trimming your nails?

It is always good to have empirical verification of what was already thought, and this is one such instance, for the mechanism by which this happens was already thought of by Immanuel Kant in 1754 (*14*) and then later mathematically calculated by the astronomer and mathematician George Darwin (geologist and biologist Charles Darwin's second-eldest son and fifth child) in 1897 (*15*). Kant and Darwin's mechanism for the Moon's retreat also explains well why the Earth's rotation has been slowing down, so lengthening its days with time (which Wells verified through his bioastronomy). What is happening is this. There is a tidal bulge of water (where high tides occur) on the Earth's surface beneath the Moon. Yet because the Earth is rotating faster than the Moon is orbiting about the Earth, this tidal bulge takes time to form and is carried forward by the Earth's faster rotation; this inertia results in tidal friction and so the water never manages to move fast enough to be directly beneath the Moon. Another way to look at it is to consider that it takes about 4 minutes for the bulge to re-form from where it was during which time the Earth has rotated by about a degree. Either way, the result is that the tidal bulge is about one degree off of being directly beneath the Moon. This off-centre bulge also has its own, reciprocal (two masses mutually attract each other) gravitational attraction to the Moon. Therefore a consequence of the tidal crest being ahead of the Moon is that the bulge exerts a slight slingshot effect on the Moon, which moves into a slightly higher orbit. But gravity between bodies is a two-way effect. Just as the Moon gets a small slingshot effect, the Moon also exerts to drag the Earth's tidal bulge, and friction—of this water against the rest of the planet—has a slowing effect on the Earth's rotation. This two-way effect also ensures (as Newton calls for) the conservation of the Earth system's (angular) momentum. George Darwin predicted that this process of slowing the Earth's rotation and increasing the size of its orbit will continue until sometime in the future the Earth's rotation will equal the Moon's rotation about the Earth. At this future time the Earth will be tidally locked with the Moon and have one face permanently towards it. Then the day and lunar month will be the same as each other and be roughly the equal to 47 of our present days. There is another chapter to this story in which in the far future the Moon then starts to close in on the Earth as the Sun's tidal friction slows the Earth, placing the Moon ahead of the tidal crest, but I leave it to you to read George Darwin's work for yourself as, the year following George Darwin's calculating the tidal consequences of the Earth-Moon system, he had a review published by the Royal Society which is currently downloadable from the net (*16*). Having said that, the Earth-Moon system is unlikely to ever exist long enough for Earth to see the day when its rotation matches that of the Moon's orbit, as by then the Sun will have begun to die and expand to a red giant engulfing both the Earth and Moon.

Now, this last—this talk of the Sun beginning to die—is not just some extra, pedantically added solely for completeness' sake: it is actually very germane when considering the co-evolution of life and this planet and the co-evolution of possible other life and their respective worlds. We will return to the lifetime of Earth's viability for life in Chapter 7. Meanwhile, let us bring this discussion back to the issue of the Earth's biosphere flipping states.

As we noted earlier, our large Moon is important to the nature of the Earth's biosphere because it confers stability to the Earth's axis, or angle of tilt, and it also provides tides. As we shall come to later, this could well have contributed to the first of the Earth's great flips if not slightly helped the second. Yet, as we have seen, the Earth-Moon system itself is evolving. This also is true of the Sun, which itself is changing through time as we shall see in the next section. That the Earth's biosphere flips between states does not mean that all else is standing still: to the contrary, though other things may change, the Earth still today remains conducive to life—even if it goes through change and flip states—this is a testimony to the co-evolution of life and planet relationship's being self-reinforcing and so keeping things conducive for complex life.

## The Earth and early biosphere

Today there are not many remnants of the early Earth's surface: as previously noted, much of the planet has been 'recycled' through subduction due to the plate tectonic movement that causes continental drift. Yet there are a few significant areas of ancient rocks that are as old as 3.8 billion years to be found in a handful of places, such as Greenland and Western Australia and even some even rarer 3.95 billion year old strata, with only a couple older still.

From our 4.6 billion year, arm's-length perspective of our planet's lifetime these 3.8 billion-year-old rocks are barely off the shoulder and not quite halfway down the upper arm to the elbow. Nonetheless, the chemical composition of these rocks does tell us a bit about the environment in which they formed.

Today our planet's atmosphere is an oxidising one, and there is free oxygen (molecules of just oxygen and not oxygen bound with any other element) in the atmosphere: you breathe it and it sustains you. Leave a lump of iron outdoors in the air and it rusts: it oxidises. Today the Earth's atmosphere contains about 20% free oxygen. This was not so with the Earth's primordial atmosphere, which was almost devoid of free oxygen; there was oxygen, but largely it was bound up mainly with carbon in the form of carbon dioxide as well as with hydrogen in the form of water vapour (and likely only just traces of free oxygen especially in the high atmosphere due to the action of sunlight on carbon dioxide).

Though there is almost no strata from much before 4 billion years old, we have found very tough and tiny crystals dating from that time. These crystals are so robust that they have survived reworking in the rock in which they were formed; its original surrounding rock having long since weathered away. These are zircon crystals, and we know how old they are because they were formed with uranium impurities, which—as discussed earlier—radioactively decays to lead in a precise way over time. Among the oldest of these zircon crystals come from Western Australia and date from around 4.4 billion years (approximately 100 million years or so since the formation of the Moon) and so we know that the Earth must have cooled enough for there to have been a crust at that early time. Furthermore, there is evidence that these zircon crystals formed in water and so it looks like there were oceans on this very early Earth (*17*). Having said that, it is thought that these early oceans were warm, if not hot, and they might well have for a while been over 200°C but kept liquid by the huge amount of carbon dioxide in the atmosphere, which exerted a far larger total atmospheric pressure at sea level than it does today. (The temperature at which water boils is determined by pressure, which is why a cup of tea in the low-pressure environment on the top of Mount Everest is a lukewarm affair.) The surface of the Earth in its very early days was a bit like a pressure cooker. Likely after some (possibly considerable) hydrogen loss during the Hadean (before 4 billion years ago [bya]), the Archaean atmosphere some 3.5 bya exerted possibly up to around twice the pressure of today's atmosphere with a far greater contribution from carbon dioxide and more methane than today with higher carbon dioxide persisting through to 2.5 bya (*18, 19*).

There is considerable debate as to how quickly the Earth's oceans cooled, and we will return to this shortly, as other events were to happen.

Before we continue, let us have a scale-check on this very approximate 100 million years between the formation of the Moon and the formation of these zircon crystals hence early Earth crust and oceans. On our arm's-length scale working back from the present, 100 million years takes us just halfway from the fingertip of the forefinger (representing the present time) towards the finger's first joint; alternatively, working the other way (representing the formation of the Earth) we are still barely off the shoulder. On the scale of the life of the

Earth, 100 million years is no time at all, but working back from the present 100 million years would take us to the middle of the Cretaceous era, which was the last era of the dinosaurs and the era following the Jurassic (as in the novel and film *Jurassic Park*). Put like that, 100 million years seems a longer time. Of course the dinosaurs could not have existed 100 million years after the formation of the Earth because, as said, there was no oxygen for them to breathe, let alone a biosphere like ours today with the necessary ecosystems to sustain them, and also because the Earth was about to come under extraterrestrial attack!

There is a bit of mystery arising from the age of the oldest rocks on Earth: there is a 700 million-year gap between most of the oldest rocks so far found on Earth and the age of the of the Solar System, yet it is thought that it took also just 100 million years for the Earth to form out of the Solar System's primordial swirl of dust and gas. Even if the formation of the Earth took twice this long—200 million years, as some estimate—there is still a substantive gap of a few hundred million years to the oldest rocks so far found. How do we reconcile this? In the 1970s the Americans helped.

The 15, 16, and 17 Apollo missions brought back rocks from large impact craters on the Moon, and these all have ages that cluster between 4.1 and 3.8 bya. What is thought to have happened (there are other theories) is that the very young Earth-Moon system experienced a substantial number of asteroid impacts during this time; for what affected the Moon surely affected the Earth. This storm of impacts is known as the Late Heavy Bombardment (LHB) that took place around this 4.1 to 3.8 bya time. It is thought to have been caused by the realignment of the early Solar System's planetary positions. The planets formed from the swirl of dust and gas but not necessarily in stable orbits. The planets gravitationally interacted with each other, moved, and continued to do so until gravitational resonances were reduced and their orbits became stable. Some think that the gas giants likely formed closer to the Sun and then moved further out to their present orbits: though some astrophysicists have other thoughts as to the exact pattern of migration. (If you feel inclined, you can search engine the words 'astronomical grand tack'.) Either way, in the course of this migration, they would almost certainly have disturbed the orbits of asteroids, if not some small bodies further out, and some of these would have fallen into the Earth-Moon system.

One advantage—if one can say that a heavy bombardment of asteroids has an advantage—is that some of these asteroids would have brought light elements such as hydrogen, oxygen (including in the form of more water), and carbon to the Earth. The early Earth therefore benefited from a 'late veneer' of these elements. There is a fair bit of evidence, in addition to that from the Moon, for this late veneer from the LHB. One recent piece of research looks at the heavy elements that would have accompanied these lighter ones. We can deduce from meteorites the proportions of various classes of asteroids' elements. Yet heavy elements, and lighter elements that become attached to heavy elements, would have largely disappeared from the Earth's surface in its first days when it was a completely molten ball: these heavy elements would have sunk to the Earth's core (a depletion of heavy elements from the crust phenomenon that—as we will see later in Chapter 6—helped explain what happened to the dinosaurs). Yet the Earth's surface does contain some elements that are either heavy or attracted to heavy elements like iron. Elements that have an affinity to iron are called siderophile and these include sulphur (S), selenium (Se), and tellurium (Te). While the heaviest of these should have sunk with iron towards the Earth's core when it was molten, yet—as research in 2013 from Germany has shown—they can be found in places in the Earth's crust today as well as in concentrations in a class of meteors known as carbonaceous chondrites. This suggests that meteor impacts originally brought these elements to the Earth after its surface had solidified and so is further evidence of the late veneer (*20*).

Having said all of the above, it must be borne in mind that there are many unknowns and there are rival theories that, while broadly telling the same story, do have a distinctive alternate nuance. Yet, there is evidence that is suggestive of water on Earth before the LHB. A question that remains is, how much water was there on Earth before the LHB and how much came to the Earth with the late veneer?

Among the very oldest, surviving crust of the Earth discovered is the Nuvvuagittuq greenstone, with an age of 4.28 billion years, by the shore of Hudson Bay, Quebec, in Canada. It is a type of rock we associate today with deep sea hydrothermal vents. Now, this is not to say that these were formed in early Earth 'deep sea' hydrothermal vents: the Earth back then was a very different place. The early Earth had not seen the billions of years of tectonic re-working of the planet's surface we have today. The early Earth had only just formed as a molten ball and so the early Earth's newly formed crust would have been thin. And so it is likely that its surface would have seen far more volcanic and volcanic-related activity (such as hot springs) in shallow places and not just the deep sea abysses. Also, the early atmosphere, with all the extra carbon dioxide, exerted a far greater pressure and so it could be that such normally deep sea geology formed nearer the surface than the deep sea vents we see today. And there are other peculiarities associated with the early Earth geology too.

Included with the Nuvvuagittuq geology are rock forms called banded iron formations (BIFs) that date from just after the LHB at around 3.8 billion years ago. BIFs here are thought by some to be a sign of life and so are important to our story. (Though it needs to be remembered that ultraviolet light can oxidise ferrous iron.) Currently, this is the earliest that we can possibly consider life having existed on the planet. It also is the sign of a flip of state of the planet from one that sees non-biological chemistry rule to one whereby life's chemistry affects its state.

A reminder, as previously mentioned, on the arm's-length scale of our planet's timeline 3.8 billion years ago is not quite halfway from your shoulder to your elbow.

Another reminder, the early Earth had no free oxygen in its atmosphere (possible aforementioned high atmosphere trace levels excepted) and so the atmosphere was not oxidising but, in chemistry terms, the exact opposite: 'reducing'. Ferrous iron (as opposed to the highly oxidised and insoluble ferric iron) would have been in solution in the early ocean. What is thought possibly may have happened to cause bands of iron to form was that early life—using a different metabolism to that dependent on oxygen, so employing a non-oxygen or anaerobic metabolism—had a similar effect to oxygen precipitating ferric iron. A process capable of producing BIFs so extensively (they have been found elsewhere) is photosynthesis (which is what bacterial algae and modern plants do, but back then, as we shall see in the next chapter, this was done differently). For example, photosynthetic, iron-oxidising bacteria can do this. Photosynthesis is the process by which energy from light (or infrared heat) is captured to drive a chemical reaction to create carbohydrates for a living organism: photosynthesis is done by forms of life. So, to sum up, some consider that extensively found BIFs are a sign of life that used some form of photosynthesis involving an oxidising agent that was not itself oxygen (*21*).

(A school chemistry reminder: some elements or compounds that tend to 'grab' electrons from atoms are called 'oxidising' elements or compounds, and those that tend to give their electrons away are called 'reducing'. Oxygen itself may not actually be involved.)

There is other evidence for this life. Just as there are isotopes of radioactive elements, so there are of non-radioactive isotopes of elements including the elements that make up life. For example, a heavier isotope of hydrogen is called deuterium, which is the form of hydrogen used to make heavy water used in early nuclear reactors. Then there is carbon, the fundamental element of life which is found in genetic material (such as DNA) and amino acids (more of which shortly). Carbon has three isotopes: carbon-12 ($^{12}C$) is by far the most the common form; $^{14}C$ is radioactive and used in carbon dating; and $^{13}C$. (The 12, 13, and 14 here relate to the atomic weight of the carbon, with $^{13}C$ having an extra neutron in its atom's nucleus compared to $^{12}C$.) Anything older than tens of thousands of years has lost nearly all its $^{14}C$ through radioactive decay, so that all older carbon is a mix of $^{12}C$ (nearly 99%) and $^{13}C$. The thing is that this mix you might think would be uniform across the Earth's surface, but it is not.

One thing in particular perturbs this carbon isotope mix, and that is life. The process of photosynthesis prefers $^{12}C$, so that carbon in life has a higher proportion of $^{12}C$ to $^{13}C$: $^{13}C$, being heavier than $^{12}C$, vibrates slower at a given temperature and enzymes and other reaction facilitators can pick up on this vibration difference. Photosynthetic mechanisms have evolved to use the far more common $^{12}C$: hence the discrimination against $^{13}C$ (see Figure 1.4).

Here, a clue for ancient life is that some carbon-bearing geological strata dating from 3.8 billion years ago are so depleted in $^{13}C$ that it suggests that the carbon had been processed by life. Again, though not absolutely conclusive (well-preserved fossil bacteria would be nice evidence), this does tie in with the notion given earlier that this time shortly after the LHB is when life got going. Also, the carbon in these carbon-bearing strata comes in the form of specks of graphite a few microns across, which happens to be the size of small prokaryote cells (such as today's bacteria) (*22*).

Among the oldest $^{13}C$-depleted carbon in grains within geological strata is that from Akilia island off West Greenland. It is thought that these grains could be old as 3.85 billion years (*23*, *24*). Further $^{13}C$-depleted carbon has been also been found in 3.7 billion-year-old rocks from Usua, again West Greenland (*25*).

Yet, carbon isotopes and BIFs are not direct evidence of life: direct evidence would be if we could see it. What we need are some really ancient fossils. In 2016, a small group of Australian researchers led by Allen Nutman and Vickie

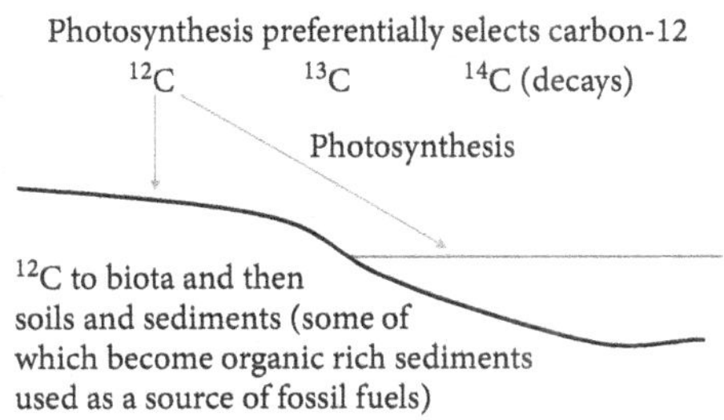


**Figure 1.4** Photosynthesis preferentially selects for the common $^{12}C$ as opposed to the rarer $^{13}C$. $^{12}C$ from carbon dioxide in the air ends up in plants and algae on the land and in water. This means that sediments made up of the remains of formerly living species (such as coal and oil) are even more rich in $^{12}C$. This leaves behind more of the rarer $^{13}C$ as carbon dioxide in the atmosphere. Alternatively, you can burn fossil fuels and release $^{12}C$-rich carbon dioxide into the atmosphere. Either way, some of this carbon ultimately forms carbonate (like the carbonate furring your kettle). The ratio of $^{13}C$ to $^{12}C$ therefore indicates the strength of flow of carbon between various biosphere reserves of carbon. Such flows can be driven by photosynthesis (which takes carbon from the atmosphere into plants) or burning fossil fuels (which takes carbon from former plant material and returns it to the atmosphere). We will return to this in subsequent chapters.

Bennett thought they had found just such evidence in at least 3.7 billion-year-old rocks from southwest Greenland. What they found were structures within the strata that looked like colonies of bacteria called stromatolites. Now, it has to be said that there is some debate as to attributing these structures as being bacterial/algal mats, but the researchers make a fair case, pointing out that the chemical composition of these structures is markedly different to the geology in which they are encased. Further, it does chime with other evidence for life arising early on the primordial Earth (*26, 27*).

Separate corroborating evidence came the following year in 2017 with a paper published by Matthew Dodd and Dominic Papineau of University College London and their small international team. They had found structures in part of the geology from the previously mentioned Nuvvuagittuq belt in Quebec, Canada, that was at least 3.77 billion years old and possibly even 4.28 billion years old. It has to be said that these structures were small but they were similar to those of filamentous microorganisms from modern hydrothermal vent precipitates and analogous microfossils in younger rocks (*28*). Again, though not conclusive evidence, as with the Allen Nutman and Vickie Bennett work as well as other evidence, it all points in the same direction, which makes it hard to dismiss out of hand.

It therefore seems that life became established—at the very least—quickly after the Late Heavy Bombardment. Yet if life was so good at getting going, could it have begun earlier still, before the LHB? What we do know is that many of the chemical precursors of life could have formed from the simplest chemicals present very quickly, and indeed easily. In fact, you could make them yourself.

Back in 1926, the Russian scientist Alexander Oparin suggested that the warm oceans under the carbon-dioxide-rich atmosphere, along with methane and sulphurous gases known to be emitted by volcanoes of the early Earth, would have been conducive to the establishment of life—an idea that was also championed by others including, notably, the British biologist J. B. S. Haldane. But it was two US scientists, Stanley Miller and Harold Urey at the University of Chicago, who decided to see what would happen if conditions on the early Earth were simulated. They created a gaseous mixture of one part hydrogen, two parts methane, and two parts ammonia over a flask of boiling water. (Though it has to be said that this mix of gases would not properly reflect the early Earth's atmosphere, there have subsequently been variations of this experiment.) Through this gaseous mix they passed electric sparks to simulate lightening and allowed the gas to circulate over another water trap before it was recycled to the flask of boiling water. They then let the experiment run . . . By the end of the week the water in the cool water trap had turned orange red. An analysis of this liquid revealed at least four amino acids (*29, 30*). Others repeated the experiment and managed to obtain even more amino acids, as many as a score; indeed, after

Miller died in 2007 the contents of sealed vials from the original experiments revealed the presence of 20 amino acids! As said, since then there have been many such experiments using different chemicals.

Amino acids are in essence a cross between a carboxylic acid (of which the simple two-carbon carboxylic acid is acetic acid, or vinegar, which you put on your chips) and a nitrogen-hydrogen amide group. (These last are related to ammonia, which itself is nitrogen combined with three hydrogen atoms and which is commonly found in some household cleaning products). The neat thing about amino acids is that the carboxylic acid part of one amino acid can react with the amide group of another so joining the two together. As a result, amino acids can form chains (proteins) that have a structure and, if you can make chemical structures, you can do things.

Now you may not think creating amino acids out of a mix of simple chemicals is a big deal but, be assured, amino acids are important: they are the fundamental chemical units of which proteins are comprised. And proteins in turn are used as enzymes to catalyse all life's metabolic reactions, as well as being involved in many structures in the tissues of animals, including muscle: amino acids are fundamental and sometimes referred to as the building blocks of life. They really are important. Just taking enzyme use alone, I remember from my pupil days one teacher telling us that 'without enzymes all the chemical reactions going on in your body would slowly grind to a halt, instantaneously!'

More recently, other evidence has been found as to the ease with which key metabolites form in an environment similar to the early Earth's oceans.

One key metabolic cycle is the tricarboxylic acid (TCA) cycle that is also known as the citric acid cycle as well as the Krebs cycle in honour of H. A. Krebs, who first proposed it in 1937. The TCA cycle is a key component of a biological respiration process that releases energy from the breakdown of sugars. It is a series of chemical reactions mediated by enzymes that essentially completely oxidises the biochemical acetate (formed from the breakdown of carbohydrates such as sugars) with the last reaction restoring the first chemical in the cycle which starts all over again. It is not necessary here to go into the detail of the cycle, what is important is that most of its components automatically arise in conditions similar to those in the early Earth's oceans but also possibly some salty terrestrial springs.

This discovery was made in 2019 by Kamila Muchowska, Sreejith Varma, and Joseph Moran of the University of Strasbourg. They found that two products of non-biological (purely chemical) carbon dioxide reduction (such as in the Miller-Urey experiment) in turn build up 9 of the 11 intermediates of the biological Krebs (or tricarboxylic acid) cycle, including all 5 of its universal metabolic precursors made of carbon, hydrogen, and oxygen: acetate, pyruvate, oxaloacetate, succinate, and α-ketoglutarate. (There is no need to worry about the actual biochemistry of these compounds but you can search engine them if you want.) These form when two simple biochemicals—glyoxylate ($HCOCOO^-$) and pyruvate ($CH3COCOO^-$)—are combined in water with ferrous iron ($Fe^{2+}$). The researchers simply mixed a solution of glyoxylate and pyruvate with some ferrous iron and warmed it (70°C) for a couple of days. At the end of that time, there were the other metabolites (*31*, *32*).

You will recall that ferrous iron was present, dissolved in the early Earth's oceans back when there was no free oxygen in the planet's atmosphere. The researchers proposed that their suite of chemical reactions might have formed a non-biological framework for metabolic pathways when life emerged on early Earth.

That nearly all of the amino acids (building blocks of life), and most of the molecular compounds involved in a fundamental energy-releasing process used by life as it metabolises, automatically arise out of simple chemistry of the early Earth, testifies as to the ease of life's beginning.

Amino acids and the tricarboxylic acid cycle biochemicals by themselves are not living things even though they are essential for life as we know it, and almost certainly (at least quite likely) are essential for carbon-based life elsewhere (on other worlds) even if it is not exactly as we know it.

Another recent discovery also corroborates the notion that the rise of life was an 'easy step' (comparatively speaking as opposed to a 'hard step' which meant that life arose later). In 2020 a largely British based team of seven together with two Polish-based researchers led by Jianfeng Xu, Václave Chmela, Nicholas Green, and David Russell, of a molecular biology lab in Cambridge, elucidated a series of chemical reactions under plausible prebiotic conditions. While not definite proof that this is what happened, their results are very intriguing: they managed to synthesise the precursors to both RNA and DNA at the same time!

Now, it had been thought that RNA was the first nucleic acid and that for a while there was an RNA world before DNA evolved. Instead, this work suggests that RNA and DNA came about at the same time. And, if that were not enough, their mechanism produces the 'handed' biomolecules life uses to build these nucleic acids.

'Handedness' comes about because carbon can have four atoms or molecular groups attached to it. These can be arranged in to two different, mirror-image ways, just as your hands are. Consider: the centre of your hand being a carbon atom; the top of (above) your hand being connected to another atom; the wrist being a second atom; your fingers (collectively) a third atom; and your thumb the fourth atom. No matter how you try, you cannot map your left hand onto your right hand: the thumbs do not match, being on opposite sides. True, you can turn one of your hands over and then map each hand's fingers and thumb on top of each other, but then the orientation of your hands is different: one hand is palm up and the other not.

A number of the biochemicals life uses, including the building blocks of nucleic acids (the precursors to RNA and DNA), have such handedness and use only one form and not (never) the other. Why? The Xu, Chmela, Green, Russell, and colleagues team's series of complex but possible—they did them in the lab—reactions explains this.

Interestingly, it only works if there is ultraviolet (UV) light, as this destroys unwanted chemical syntheses. There was ultraviolet light on the early Earth but only bathing its surface or penetrating shallow waters (*33, 34*). If this UV-mediated series of reactions was the one involved in life's establishment, then it means a marine cradle for life arising is unlikely. Indeed, previously other researchers have noted that a number of chemical elements in the contents of cells closely relate to the composition of surface ponds next to geothermal vents (not the environment around deep sea vents). The argument being that the earliest cells (from which true cells evolved) would not have developed cell membranes and so their contents would likely reflect that of their environment (*35*). Furthermore, that some of the Xu, Chmela, Green, Russell laboratory reactions take place at temperatures around the boiling point of water also speaks to a thermal vent environment in the real world. Finally, there is a problem with water. While water is needed, periods of drying are required for the formation of some key biomolecules. This too, suggests that life arose in and around a land-based thermal spring that varied in size through gaining and losing water and so saw some of its constituents exposed to periods of drying (*36*).

Having said this, while this series of reactions resulting in the precursors of DNA and RNA works in the lab, it is difficult to imagine the varied yet connected natural environment in which this could work: though the afore terrestrial thermal spring varying in size seems to be the best fit. Nonetheless, that it works in the lab, provides the precursors to both RNA and DNA with the correct handedness we find in nature, is compelling. Indeed, the least difficult thing to imagine is the necessity for UV light or the high temperatures. (We will very briefly return to this nucleic acid formation and the ultraviolet light mediation in Chapters 2 and 7.)

A third recent piece of research further illuminates the purely chemical rise of the molecules used by life. This work—led by Agnieszka Wołos, Rafał Roszak, and Anna Żądło-Dobrowolska, from Warsaw, Poland—used a chemistry computer model fact-checked with laboratory experiments.

Their computer programme contained the rules of chemistry (energetics, valency, hydrogen bonding, and so forth). They then began the programme with six, small, simple molecules (water, hydrogen sulphide, hydrogen cyanide, methane, ammonia, and nitrogen) to see how they would react. (This is a bit like the real-life Urey-Miller experiment discussed earlier but done in theory.) They found that allowing these chemicals to interact produced 12 new products. Then in turn, allowing these 12 to interact with each other as well as the original 6 molecules resulted in 20 more molecules. Included in these 20 is the first amino acid to be generated (as is in the Urey-Miller experiment), glycine.

Repeating this process 5 times resulted in around a thousand molecules, 28 of which are fundamental to life. As to why these molecules were used by life and not the remaining nine hundred plus or so is indicated by their properties. The life-using ones are all at least partially water soluble, synthesised in fewer steps, and many had the propensity for cyclic reactions spawning multiple copies.

Yet, a computer program is a computer program, but is it any good? Well, some of the chemical products the program predicted followed previously unknown pathways: these unidentified routes, were simply hypothetical. So the researchers checked these out and explored several of these in the laboratory (*e.g.*, prebiotic syntheses of acetaldehyde and diglycine, as well as malic, fumaric, citric, and uric acids), and they found by experiment that the theoretical chemical routes identified by the program worked in reality (*37*).

There has been a distinct gap in our understanding of the origins of life between the formation of amino acids—or indeed the formation of the precursors of nucleic acids (nucleic acids such as RNA)—and the ancestors of simple bacteria-like life. Nonetheless, these three recent discoveries—the possible route to the precursors of nucleic acids as well as handed amino acids, and the biochemicals associated with the citric acid cycle—provide intriguing possibilities. What we can do, by looking at how nucleic acid sequences of simple organisms have changed, is get an idea as to what the earliest, clearly living species might be like.

Simple bacteria clearly are a form of life. Biologists classify them as being Monera in the kingdom 'prokaryotae' or cousins to the Archaea, which are also prokaryotes (see Appendix 2 for use of biological nomenclature). It is thought that the earliest true form of life would be some simple form of prokaryote.

The simplest, and earliest, of prokaryote likely used chemosynthesis (perhaps using sulphur or ammonia compounds) to derive energy for their metabolism. These prokaryotes were chemolithoautotrophs: the 'chem' relates to chemicals (such as the aforementioned sulphur and ammonia compounds), the 'litho' relates to stone (as these prokaryotes also obtained chemicals from the geology around them), and 'autotrophs' is the term given to organisms that produce their own complex organic compounds (in this case from the aforementioned simple chemicals from the ocean and geology). They could have obtained their energy a number of ways, including using a reaction with hydrogen in thermal vents or at lower temperatures given off by the action of water on some minerals including olivine (*38*). Today the subsequently evolved methanogenic archaea are the only known microorganisms that are capable of performing biological methane production in the absence of oxygen, but they can do so over a wide range of temperatures and pressures: the early Earth was a far more extreme place than today. (We will return to methane-generating prokaryote analogues in Chapter 7.)

This amino-acid-to-simple-prokaryote gap in our understanding as to how life got going is a whole topic in itself that is fraught with far more speculation and uncertainty than the Earth flip narrative at hand. However, if you are interested then you can, using academic and scholarly search engines, seek out work on 'evolution' and/or 'rise' of 'early life'. Meanwhile, back at the plot . . .

While amino acids themselves do not constitute life, as said, they are fundamental and essential to it. That they can be produced so easily under conditions similar to those found on the early Earth is remarkable and, some might say, perhaps telling. Also, that a form of photosynthesis *might* have existed so soon (geologically speaking 'quickly') after the theoretical (remember, some have other theories) Late Heavy Bombardment (LHB) strongly suggests (even if it is not proof) that life arose quite easily: if the start of life were difficult then we would expect it to take a long time (*21*). Even if life arose much later—evidence of structures that are clearly fossils of algal mats (stromatolites) are only found much later—it could have been because the Earth was still too hot. Indeed, given that life rose *possibly* so soon after the Earth's formation and the LHB, considering Earthlike planets generally, could it be that life arises quite easily everywhere in the Galaxy given the right—early Earthlike—conditions?

This brings us back to the question that, *if* life arose so soon after the LHB, might it have arisen earlier and possibly even have survived the LHB? This last is not purely whimsical speculation: there are three lines of evidence that suggests that maybe life could have started earlier.

First, the nucleic acid (the genetic code) of some prokaryotes that live in hot springs suggests that its ancestry is older than the end of the LHB. A number of prokaryote nucleic acid sequences are common to a few species; the more similar the closer they are on the evolutionary tree. So, for example, more of your nucleic acid can be found in a chimpanzee than, say, compared to a banana. Identifying similarities and differences, it is possible to very approximately date when evolutionary branches took place (*39*, *40*, *41*).

While this is suggestive evidence, it is not firm evidence: the use of nucleic acid sequencing, or molecular clocks, relies in part on assumptions. So, while interesting, at best it is corroborative and so other lines of evidence and arguments are required.

Second, some computer models of what might have happened on the Earth during the LHB indicate that it was highly unlikely (some say there being 'no plausible situation') that the whole planet was at any one time completely biologically sterilised during that asteroid bombardment: the bombardment was not intense but leisurely paced albeit lasting millions of years with there being some time between the really big impacts. In short, life might have survived the bombardment and if so this would explain the genomic evidence that life's ancestry is older than the LHB. This is the contention of geoscientists Oleg Abramov and Stephen Mojzsis of the University of Colorado (*42*). Their modelling suggests that while the LHB would have been detrimental to some bacterial-like, prokaryote populations, it might have even been beneficial to the ones that live in hot springs. Habitats at the immediate surface of the early Earth were almost certainly repeatedly destroyed by the LHB, but not everywhere was destroyed at the same time: some areas would remain untouched for a while, even if ultimately they were hit. At the same time, however, new subsurface habitats would have formed from impact-induced hydrothermal systems, which perhaps, Oleg Abramov and Stephen Mojzsis further speculate, 'provided sanctuary to existing life or may even have been the crucible of its origin'. Meanwhile, in time during the bombardment, other areas would have recovered enough for life to colonise them. After a time the whole surface of the planet would have been reworked, but life could have survived throughout the heavy bombardment, a bit like WWI soldiers might shelter in a bombshell crater: some soldiers got killed by bombshells, but others might survive moving across no-man's land sheltering in shell craters where they could.

Third, and finally, there is evidence of periods of bombardment after the supposed end of the LHB, but life seems to have survived these later bombardments and so possibly could have during the LHB itself. At least seven spherule beds have been found that formed between 3.23 and 3.47 billion years ago, four between 2.49 and 2.63 billion years ago, and one between 1.7 and 2.1 billion years ago. It seems that most late impactors could have come from the E-belt, which is an extended, and now largely extinct, portion of the asteroid belt between 1.7 and 2.1 astronomical units from Earth (an astronomical unit being the mean distance from the Earth to the Sun) (*43*). Given that life seems to have got going well before these late impactors, it does suggest that it may have survived some of the earlier ones too.

This notion is an intriguing one and is supported by a suggestive piece of work by Japanese researchers led by Takayuki Tashiro in 2017. They found graphite in strata from northeast Canada (Labrador) that dates from 3.95 bya: this formed in the Late Heavy Bombardment. The thing about this graphite was that it was very depleted in $^{13}C$. Graphite is pure carbon and can be formed from iron carbonate at temperatures above 450°C. The idea is that the carbon must have come from life to have such a low $\delta^{13}C$. It was then geologically reworked over time, including a period of high temperature, to end up as graphite. It is a tantalising thought, but not as definitive as one might wish. So we must be cautious about forming any firm conclusions, just as we must about the similar specks of graphite described earlier in the strata from nearly 3.8 bya (*44*).

So, could life have got going during or even before the LHB? And, if so, was it wiped out and then new life arose? Or did it manage to cling on through the bombardments? Or does the very low $\delta^{13}C$ in these samples have another explanation?

Alas we do not have a firm answer for these questions, but the evidence we do have—though possibly considered as 'circumstantial' in a court of law—is suggestive, albeit not conclusive, that the initial rise of life on Earth was in terms of the planet's lifetime a 'likely' event and perhaps even an 'easy' evolutionary step, one of the first on the road where life and planet saw flips of states. This in turn speaks to the question of what might happen elsewhere than on our own world.

Finally, before we leave the early Earth and its proto-biosphere, let us consider a little more about what our planet would have actually been like.

You will recall the Earth's atmosphere was effectively devoid of free oxygen. The world was far warmer given the temperature of the seas, and very early on at a surface temperature we would consider today as boiling. (Remember high pressure elevates boiling points and it is possible that the early Earth's atmosphere was thicker.) The seas of the early Earth were shallower: today's larger continents had not accrued (the crust was not thick enough to sustain substantive mountains) and the volume of water not as great as today's until the end of the late veneer's final delivery by bombardment. So, as for the land there were no large continents as we see them today. It is almost certain that the depths of the seas varied (it is unlikely that the thin crust was smooth) and there may have been some land but no huge mountain ranges as we see today, in addition to the thin crust, plate tectonics (continental drift) had not got going, or had barely started and possibly in a different manner to crustal movement today. Conversely, it is likely that there was a greater extent of shallow seas compared to the present. Yet already even in the earliest of times there were some areas where more heat was coming to the surface from the Earth's mantle than others. The underlying geological circulation behind this would have brought new material to the surface and the beginnings of the first continents.

What would such early continents (more accurately proto-continents) have been like?

They would certainly have been smaller and, as said, without tall mountains: plate tectonic drift had yet to collide these early land masses to produce Himalaya-like mountain ranges with deep roots. Some think that a modern analogue of these truly ancient land forms, or at least the closest we are likely to get, is Iceland. Alas, no ancient rocks have been found with the same chemistry as Icelandic ones, that is, until somewhat recently. In 2014 Jesse Reimink and colleagues from Alberta University in Edmonton, Canada, discovered 4.02 billion-year-old rocks from western Canada that share the same geochemical signatures as those from Iceland, but which are distinct from other 3.9 billion- to 2.5 billion-year-old rocks (*45*).

This early, proto-continental land was different to that of present-day true continents. The continents of today consist of a far thicker body of solid matter resting on top of the hot mantle. Aside from places above mantle plumes sites (such as Hawai'i), only along the fractures of spreading tectonic plates, or near plates that subducted under their neighbours, do we tend to find volcanic activity. While the crust beneath the oceans is thin, the crust of the continents is thick, but early in the Earth's history it took a while for proper continents to evolve to the point when land masses—carried by plates—collide. We do have an idea when this happened, due to, of all things, diamonds. Diamonds are a form of carbon created under great pressure, hence depth, and so they only form beneath thick continental crust. In 2011, Steven Shirey from the Carnegie Institution of Washington university in the United States and Stephen Richardson from the University of Cape Town in South Africa, compiled isotopic analyses of silicate and sulphide inclusions within diamonds. (Inclusions are flaws and so you may not have seen them in jewellery, but they are common in diamonds found in rocks and they can be very informative to geoscientists.) Steven Shirey and Stephen Richardson found a change in inclusion compositions that took place around 3.0 billion years ago; before 3.2 billion years only diamonds with one type of composition (peridotitic) formed, whereas after 3.0 billion years other (eclogitic) diamonds became prevalent (see Appendix 2 for a technical definition of the afore specialist terms, but do not worry about them too much). They suggest that this transition marked the onset of the Wilson cycle of plate tectonics by which continents migrate over the Earth coming together to form a single supercontinent (like Pangaea 250 million years ago) and then splitting (as the continents are today, scattered across the globe) (*46*). In 2012, other isotopic evidence was published that lent support to the idea of a transition and the start of the cycle of continental growth around 3.2 billion years ago (*47*). Prior to this time, it is thought that there was a different primordial form of plate movement of the Earth's then-thinner crust.

Before continuing, we need to briefly return to the question of how warm were the oceans. We have already noted that they most likely were hot shortly after the Late Heavy Bombardment, but how soon did they cool? Here there is much debate as to how much and how fast it cooled from 3.8 bya (*48*). It is also likely that the Earth's surface temperature decreased further over geological time from around 75°C in the Archaean (around 3.0 bya) (*49*).

Returning to the onset of plate tectonics, it is perhaps worth noting that the above works' conclusions—that plate tectonics did not properly get going until somewhere around 3.2–3.0 bya—were based on observational evidence: an actual isotopic date gleaned from real geological strata. They also chime with earlier theoretical work. In 2008 Patrice Rey and Nicolas Coltice theoretically calculated that it was unlikely that the early Earth's thin crust would have become thick enough without more cooling to support continents with major mountains above 2.5 km until at least 2.5–2.8 bya. (Here there is further evidence for the evolution of continental sizes to which we will return in Chapters 3 and 5.) To put this in perspective, 1 km is roughly around the height of the tallest of mountains in Wales's Black Mountains, and 2–3 km the height of the Sierra Madres in Mexico, the Pindus Mountains in Greece, and many mountains on all continents today. Meanwhile, the tallest of Himalayan mountains are well over twice as high! Indeed, all the major continents today sport mountain ranges over 4 km with some peaks even higher. Tall mountains are important, as the taller they are the more weathering to which they are subjected. Weathering in turn relates to the transport and exposure of chemical elements (including those that are nutrients for life) that interact with water and the atmosphere. This means that tall mountains are important to the geochemical cycling of many elements including carbon, which is fundamental to life (*50*).

Robert Holder and colleagues in 2019 came to a broadly similar conclusion that plate tectonic cycling of thicker continental crust began 2.5 bya at the latest. In essence, they looked at metamorphic rocks, as different metamorphic rocks form at different pressures and temperatures. With thicker continental crust being recycled at different pressures and temperatures than thinner, it is possible to elucidate the pressure and temperature at which they were formed. Further, as the crust cooled and thickened, so there was a greater diversity of pressure and temperature environments in which such rocks formed: in the very ancient, thinner-crusted Earth there was a smaller range of such formation environments. Dating such strata, it was possible for the researchers to see how the formation environments and their range varied with time (*51*).

Taking all this together, the picture that emerges is that a little before 3 bya (somewhere almost as far as your elbow on our arm's-length scale of the history of the Earth starting from the shoulder): there was little plate movement, continents lacked mountains as tall as those we find today, and the capability for the scale of geochemical cycling that we see today did not begin until somewhere before 2.5 bya (even if there was geochemical cycling on a different scale prior to then).

The important thing is that, irrespective of when life started (possibly quite early), roughly somewhere around 3.2 bya or so, we had the all the makings of the first proper, sustainable biosphere (atmosphere, land, oceans, continental geochemical cycling, and early life globally interacting) in place. All we need to do now is to consider a bit more as to life's relationship to, and within, the biosphere, and then we can investigate the Earth's biosphere's forthcoming big flip and next change of state.

## References

1. López-Corredoira, M., Allende Prieto, C., Garzón, F., *et al.*, Disk stars in the Milky Way detected beyond 25 kpc from its center. *Astronomy & Astrophysics*, **612**, L8 (2018).
2. Xu, Y., Newberg, H. J., Carlin, J. L., *et al.*, Rings and radial waves in the disk of the Milky Way. *The Astrophysical Journal*, **801**, 105 (2015).
3. Hu, S., *Hawking Says Humans Must Go into Space*. Associated Press 13th June (2006). http://apnews.myway.com/article/20060613/D8I7ADB81.html.
4. Lissauer, J. J., Barnes, J. W., & Chambers, J. E., Obliquity variations of a moonless Earth. *Icarus*, **217**, 77–87 (2012).
5. Kruijer, T. S., Touboul, M., Fischer-Gödde, M., *et al.*, Protracted core formation and rapid accretion of protoplanets. *Science*, **344**, 1150–1154 (2014).
6. Lunine, J., Physical conditions on the early Earth. *Philosophical Transactions of the Royal Society B*, **361**, 1721–1731(2006).
7. Melis, C., Zuckerman, B., Rhee, J. H., *et al.*, Rapid disappearance of a warm, dusty circumstellar disk. *Nature*, **487**, 74–76 (2012).
8. Cameron, A. G. W., & Ward, W. R., The origin of the Moon. *Lunar Science*, **7**, 120–122 (1976).
9. Canup, R. M., & Asphaug, E., Origin of the Moon in a giant impact near the end of the Earth's formation. *Nature*, **412**, 708–712 (2012).
10. Herwartz, D., Pack, A., Friedrichs, B., & Bischoff, A., Identification of the giant impactor Theia in lunar rocks. *Science*, **344**, 1146–1150 (2014).
11. Fell, B., *Life, Space and Time: A Course in Environmental Biology*. Harper & Row: New York, USA & London, Great Britain (1974).
12. Wells, J. W., Coral growth and geochronometry. *Nature*, **197**, 948–950 (1963).
13. Scrutton, C. T., Periodicity in coral growth. *Palaeontology*, **7** (4), 552–558 (1964).
14. Kant, I., *Whether the Earth Has Undergone an Alteration of Its Axial Rotation*, Wochentliche Frag- und Anzeigungs-Nachricten (Königsberg), No. 23–24 (1754); English translation in W. Hastie, *Kant's Cosmogony*, James Maclehose: Glasgow (1900)
15. Darwin, G. H., *The Tides and Kindred Phenomena in the Solar System: The Substance of Lectures Delivered in 1897 at the Lowell Institute, Boston, Massachusetts*. John Murray: London, Great Britain (1898).
16. Darwin, G. H., On the secular changes in the elements of the orbit of a satellite revolving about a tidally distorted planet. *Philosophical Transactions of the Royal Society London*. **171**, 713–891 (1880).
17. Wilde, S. A., Valley, J. W., Peck, W. H., & Graham, C. M., Evidence from detrital zircons for the existence of continental crust and oceans on the Earth 4.4 Gyr ago. *Nature*, **409**, 175–178 (2001).
18. Catling, D. C., & Zahnle, K. J., The Archean atmosphere, *Science Advances*, **6**, eaax1420 (2020).
19. Payne, R. C., Brownlee, D., & Kasting, J. F., Oxidized micrometeorites suggest either high $pCO_2$ or low $pN_2$ during the Neoarchean. *Proceedings of the National Academies of Sciences USA*, **117** (3), 1360–1366 (2020).

20. Wang, Z., & Becker, H., Ratios of S, Se and Te in the silicate Earth require a volatile-rich late veneer. *Nature*, **499**, 328–332 (2013).
21. Lenton, T., & Watson, A., *Revolutions That Made the Earth.* Oxford University Press: Oxford, Great Britain (2011).
22. Rosing, M. T., $^{13}$C-depeleted microparticles in >3700-Ma sea-floor rocks from west Greenland. *Science*, **283**, 674–676 (1999).
23. Mojzsis, S. J., Arrhenius, G., Mc Keegan, K. D., *et al.*, Evidence for life on Earth before 3,800 million years ago. *Nature*, **384**, 55–59 (1996).
24. McKeegan, K. D., Kudryavtsev, A. B., & Schopf, J. W., Raman Ion microscopic imagery of graphitic inclusions in apatite from older than 3,830 Ma Akilia supracrustal rocks, west Greenland. *Geology*, **35**, 591–594 (2007).
25. Hassenkam, T., Andersson, M. P., Dalby, K. N., Mackenzie, D. M. A., & Rosing, M. T., Elements of Eoarchean life trapped in mineral inclusions. *Nature*, **548**, 78–81 (2017).
26. Nutman, A. P., Bennett, V. C., Friend, C. R. L., van Krandendonk, M. J., & Chivas, A. R., Rapid emergence of life shown by discoveries of 3,700 million-year-old microbial structures. *Nature*, **537**, 535–538 (2016).
27. Allwood, A. C. Evidence of life in Earth's oldest rocks. *Nature*, **537**, 500–501 (2016).
28. Dodd, M. S., Papineau, D., Grenne, T., Slack, J. F., Rittner, M., Pirajno, F., O'Neil, J., & Little, C. T. S., Evidence for early life in Earth's oldest hydrothermal vent precipitates. *Nature*, **543**, 60–64 (2017).
29. Miller, S. L., Production of amino acids under possible primitive Earth conditions. *Science*, **117** (3046), 528–529 (1953).
30. Miller, S. L., & Urey, H. C., Organic compound synthesis on the primitive Earth. *Science*, **130**, 245–251 (1959).
31. Muchowska, K. B., Varma, S. J., & Moran, J., Synthesis and breakdown of universal metabolic precursors promoted by iron. *Nature*, **569**, 104–107 (2019).
32. Pascal, R., A possible prebiotic basis for metabolism. *Nature*, **569**, 47–48 (2019).
33. Xu, J., Chmela, V., Green. N. J., *et al.*, Selective prebiotic formation of RNA pyrimidine and DNA purine synthesis. *Nature*, **582**, 60–66 (2020).
34. Le Vay, K., & Mutschler, H., A plausible route to the first genetic alphabet. *Nature*, **582**, 33–34 (2020).
35. Mulkidjaniana, A. Y., Bychkovc, A. Y., Dibrovaa, D. V., *et al.*, (2012) Origin of first cells at terrestrial, anoxic geothermal fields. *Proceedings of the National Academies of Sciences USA*, **109** (14), 5156–5157. https://doi.org/10.1073/pnas.1117774109.
36. Marshall, M., The water paradox and the origins of life. *Nature*, **588**, 210–213 (2020).
37. Wolos, A., Woznak, R., Żądło-Dobrowolska, A., *et al.*, Synthetic connectivity, emergence, and self-regeneration in the network of prebiotic chemistry. *Science*, **369**, eaaw1955 (2020). https://doi.org/10.1126/science.aaw1955.
38. Mayhew, L. E., Ellison, E. T., McCollom, T. M., Trainor, T. P., & Templeton, A. S., Hydrogen generation from low-temperature water-rock reactions. *Nature Geoscience*, **6**, 478–484 (2013).
39. Julie Marin, J., Battistuzzi, F. U., Brown, A. C., & S. Hedges, B., The timetree of Prokaryotes: New insights into their evolution and speciation. *Molecular Biology and Evolution*, **34** (2), 437–446 (2016).
40. Betts, H. C., Puttick, M. N., Clark, J. W., Williams, T. A., Donoghue, P. C. J., & Pisani, D., Integrated genomic and fossil evidence illuminates life's early evolution and eukaryote origin. *Nature Ecology & Evolution*, **2**, 1556–1562 (2018).
41. Moody, E. R. R., Sandra Álvarez-Carretero, S., Mahendrarajah, T. A., *et al.*, The nature of the last universal common ancestor and its impact on the early Earth system. *Nature Ecology & Evolution*, **8**, 1654–1666 (2024).
42. Abramov, O., & Mojzsis, S. J., Microbial habitability of the Hadean Earth during the Late Heavy Bombardment. *Nature*, **459**, 419–422 (2009).

43. Bottke, W. F., Vokrouhlicky, D., Minton, D., *et al.*, An Archaean heavy bombardment from a destabilized extension of the asteroid belt. *Nature*, **485**, 78–81 (2012).
44. Tashiro, T., Ishida, A., Hori, M., Igisu, M., Koike, M., Méjean, P., Takahata, N., Sano, Y., & Komiya, T., Early trace of life from 3.95 Ga sedimentary rocks in Labrador, Canada. *Nature*, **549**, 516–518 (2017).
45. Reimink, J. R., Chacko, T., Stern, R. A., & Heaman, L. M., Earth's earliest evolved crust generated in an Iceland-like setting. *Nature Geoscience*, **7**, 529–533 (2014). https://doi.org/10.1038/ngeo2170.
46. Shirey, S. B., & Richardson, S. H., Start of the Wilson Cycle at 3 Ga shown by diamonds from subcontinental mantle. *Science*, **333**, 434–436 (2011).
47. Næraa, T., Scherstén, A., Rosing, M. T., *et al.*, Hafnium isotope evidence for a transition in the dynamics of continental growth 3.2 Gyr ago. *Nature*, **485**, 627–630 (2012).
48. Sleep, N. H., The Hadean-Archaean environment. *Cold Spring Harbor Perspectives in Biology*, **2**, a002527 (2010).
49. Garciaa, A. K., Schopfa, W., Yokoborie, S-i., Akanumaf, S., & Yamagishie, A., Reconstructed ancestral enzymes suggest long-term cooling of Earth's photic zone since the Archean. *Proc. Natl. Acad. Sci.*, **114** (18), 4619–4624 (2017).
50. Rey, P. F., & Coltice, N., Neoarchean lithospheric strengthening and the coupling of Earth's geochemical reservoirs. *Geology*, **36** (8), 635–638 (2008).
51. Holder, R. M., Viete, D. R., Brown, M., & Johnson, T. E., Metamorphism and the evolution of plate tectonics. *Nature*, **572**, 378–381 (2019).

# 2
# Towards a Flip

As we continue our journey down the arm's-length of the Earth's history, we start to find in its seas the spread of earliest prokaryotic life (one of the ancestors of the subsequent bacterial Monera and their cousins the Archaea). It is certain that these earliest forms of life—we will skate over the question of whether viruses (that possibly arose beforehand) are alive—were able to elaborate all their structural and metabolic chemical constituents from simple inorganic compounds such as carbon dioxide. Such species—as mentioned in the preceding chapter—are autotrophs, and the earliest cells were prokaryotes, simple bacteria-like cells without a nucleus and without internal membrane-bound specialist structures (organelles). All well and good, but what was the biosphere like?

## Carbon and the early biosphere

As noted in the preceding chapter, the very early oceans were warm (if not hot); the atmosphere was devoid of oxygen, so it was a chemically reducing environment, as were the oceans. There was next to no life on what little land there was, though it was possible for some bacteria and other early prokaryotes to live in freshwater had they chemical nutrients and an energy source to draw upon: hot springs are one such candidate environment. But, by and large, once life reached the seas (should life have arisen in terrestrial thermal springs) nearly all the early Earth's biomass was in the oceans.

Though there was this early life on Earth, there was not much of it; not nearly as much as there is today. In 2005 the Royal Society held a discussion meeting on the topic 'Major steps in cell evolution'. In the proceedings of this meeting, James Kasting of Penn State University and Shuhei Ono of the Carnegie Institute of Washington noted that their model of the Archaean (roughly around 3.2 billion years ago [bya]) agrees with the view put forward by others that the Earth's biological net primary productivity (see Appendix 2) was then at least an order of magnitude (a power of ten) less than modern values today (*1*). So, it is reasonable to assume that before this time, during the earliest time of life on our planet, the Earth's total biomass was smaller still.

You will recall from the previous chapter, and Figure 1.4, that the strength of carbon flow from the atmosphere due to photosynthesis affects the ratio of

*Earth's Flips of State*. Jonathan Cowie, Oxford University Press. © Oxford University Press (2026).
DOI: 10.1093/9780197825402.003.0002

carbon-13 to carbon-12 ($^{13}C$ to $^{12}C$) which in turn is seen in sediments such as carbonates. (As previously mentioned, this is because photosynthesis has a preference for capturing $^{12}C$.)

You might, therefore, think that a large biomass on the planet Earth—such as we see today—compared to the smaller biomass the Earth sported over 3 billion years or so ago, would see more of the $^{12}C$ locked up as living matter than it did back then, and indeed it does. This in turn would perhaps suggest that the ratio of $^{13}C$ to $^{12}C$ would change with time as the planet's total biomass grew. But is this so? How has the ratio of $^{13}C$ to $^{12}C$ actually changed over nearly 4 billion years?

To answer this, we can look at the ratio of $^{13}C$ to $^{12}C$ in carbonates from fossils of the past 500 million years and, from before then, carbonate sediments and forams (an order of single-celled species with carbonate shells) as well as carbonates of billions of years ago. Over the past few decades scores of researchers have looked at many hundreds of sediments and fossil strata and analysed many thousands of samples. Over the years, these were reported in a number of academic papers in scientific journals. Fortunately for us, we do not have to read all these, as two Canadian researchers, Graham Shields and Ján Veizer, and colleagues have compiled the data from many of these papers, from a decade or so up to 2002 (together with just a few from as early as the late 1960s), and have put this on their university's website, so that now anyone can use this data to compile their own graph showing how the $^{13}C$ to $^{12}C$ ratio has changed over nearly 4 billion years. Even though they have undertaken the herculean task of bringing together all this data, transcribing it to an excel sheet in a form that can be turned into a graph is itself not a trivial task as around some twelve thousand data points are involved. However, if you have a few days to spare, it can be done, and if you do it then you will end up with something like Figure 2.1.

If we look at Figure 2.1 and the broad sweep of the $^{13}C$ to $^{12}C$ ratio in rocks across billions of years, the ratio is remarkably constant even though there may be short-term—that is the geologically sense of short-term of some hundreds of thousands to millions of years—variations (called carbon isotope excursions). We will return to some of these variations much later, but for now note that in terms of the big sweep of the Earth's lifetime the $^{13}C$ to $^{12}C$ ratio is remarkably stable with the principal exceptions of just two periods of very large variation.

Now, given you might think that if the Earth's living biomass had been growing over the planet's lifetime, you might conclude that more and more $^{12}C$ would be trapped photosynthetically, leaving behind $^{13}C$ and so changing the ratio of the two isotopes in untrapped carbon in carbonate and taken up (in a non-photosynthetic way) by the carbonate-rich shells of some creatures. Yet the $^{13}C$ to $^{12}C$ ratio is for much of the time roughly constant! This at first might suggest that the total biomass of the early Earth was most likely similar to that of today.

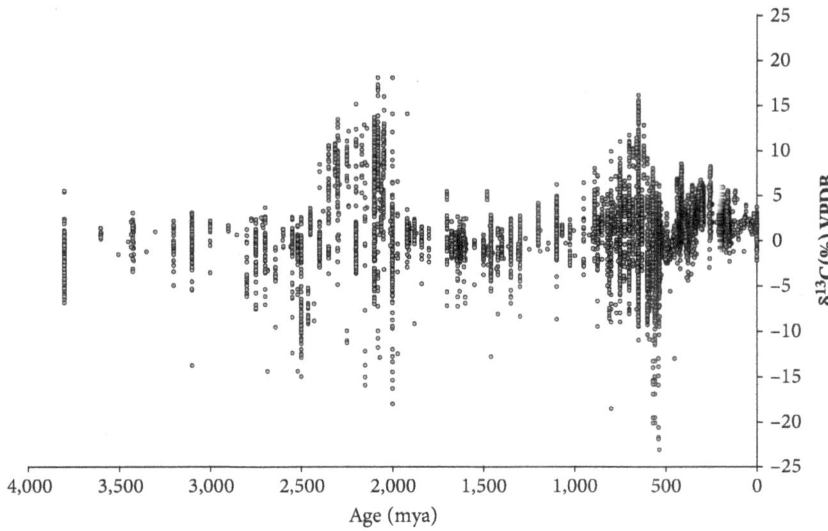


**Figure 2.1** The ratio of $^{13}C$ to $^{12}C$ in carbonate fossil and sediment strata over nearly 4 billion years. Compiled by the author from Graham Shields and Ján Veizer data at mysite.science.uottawa.ca/jveizer/isotope_data/ (accessed November 2014). The graph depicts carbon isotope changes compared to the VPDB geological standard. (VPDB stands for Vienna Pee-Dee Belemnite.) There are some 12,000 data points, which, spending a few hours each day, took a couple of weeks to process. I cannot claim to have captured them all but have the majority The reason why you can see only a few hundred data points is because many studies gave multiple similar values and so the above graph shows many (most) data points stacked one on top of another. See also Veizer, J., *et al.*, 87Sr/86Sr, d13C and d18O evolution of Phanerozoic seawater, *Chemical Geology*, **161**, 59–88 (1999), and Shields, G. A. & Veizer, J., The Precambrian marine carbonate isotope database: Version 1.1., *Geochemistry, Geophysics, Geosystems*, **3**(6) (2002), https://doi.org/10.1029/2001GC000266.

Further, if there were no trees and extensive terrestrial ecosystems, then much of the Earth's early biomass was aquatic and marine-based. According to this line of thought, you might reasonably conclude that the early Earth's seas would therefore sport a large biomass of bacteria, equal to the Earth's biomass today. This is wrong.

Yes, over 3 billion years ago the early Earth's amount of *living* biomass was smaller compared to the Earth's today. The modern Earth has plants and animals, including considerable terrestrial life with vast, biodiverse rich tropical rain forests; conversely the very early Earth just had simple cells and the very earliest of life in a comparatively few specialised habitats. So, the total amount of life (by weight) must surely have been small even once it had colonised the sea surface. Indeed, as previously noted, its total biomass has been estimated to be very roughly less than a tenth of that today. This is the prevailing view which James Kasting and others hold and is considered the most likely.

This view seems to make sense. Yet, how is this reconcilable? How do we explain this against the more-or-less constant $^{13}C$ to $^{12}C$ ratio throughout the geological record? And even if we can explain this, why is there this long-term stability in the $^{13}C$ to $^{12}C$ ratio? Fortunately, there are reasons that explain the long-term constancy of the carbon isotope ratios.

First, not all photosynthetically processed carbon exists solely in living plants. For example, all the carbon in your body is photosynthetically processed carbon, as you eat plants as well as animals that in turn eat plants. Fair enough, yet there were no animals 3 billion years ago, let alone one as complex a creature as yourself. What there was though, even after the first billion years of life, was plenty of dead organic matter. Today, there is still plenty of dead organic matter, formerly living biomass. For instance, natural gas, coal and oil, which is formed of the remains of living creatures. This also has a higher ratio of $^{12}C$ to $^{13}C$ ($^{12}C$ becomes even more common) than your average carbonate. And then, additionally today there is the organic carbon stored in soils. The total amount of carbon stored in soils (and excluding deeper fossil fuel strata) today is somewhere in excess of around 1,500 gigatonnes of carbon (GtC) or petagrams of carbon (PtC), and this alone is over twice that stored as living biomass, and incidentally is also more than that residing in the atmosphere as carbon dioxide. Some estimates of the total amount of carbon stored in soils globally (from permafrosts to tropical soils) range between 3,500–4,800 petagrams. This compares with 450 to 650 PgC in living plant biomass and about 829 PgC in the atmosphere, to which is annually added nearly 8 PgC from fossil fuel combustion and cement production (estimated as of the middle of the 21st century's first decade) (2).

Importantly, this soil carbon, formed from the remains of vegetative and microbial matter, also has a higher ratio of $^{12}C$ to $^{13}C$ compared to the atmosphere. Even so, as terrestrial life (other than microorganisms) was not prevalent prior to before 600 million years ago (mya), the $^{12}C$ rich organic carbon—if it existed—must have been elsewhere, and one obvious answer here is that much non-living $^{12}C$ rich organic carbon might have been buried, and then trapped, in marine sediments. (This is especially likely as the early Earth did not have the developed biological decay [detritovore] systems we have today). Another is that, as well as in the atmosphere, much carbon existed in the early Earth's crust in a deeper geological form including the mantle, and had yet to be released to the atmosphere and then drawn to the surface and processed by life. Indeed, once plate tectonics had got going, carbon could be returned to the Earth's mantle through plate subduction.

In short, the amount of $^{12}C$ rich organic carbon does not solely exist as living biomass, but living biomass and dead biomass incorporated into marine sediments and, much more limited back then, terrestrial soils. And then there was carbon in inorganic form deeper in the Earth that still had to make its way to the surface following the Earth's formation. But this is only part of the answer and is a far-from-complete explanation for the constancy of the $^{13}C$ to $^{12}C$ record.

Far more important is a second reason: there are Earth-system processes that continually reset the atmospheric $^{13}C$ to $^{12}C$ ratio (which ends up as the marine geological record). Carbon from organic strata does leak to the atmosphere, as does carbon from the living biomass itself: you, for instance, are right now breathing out carbon as carbon dioxide. Indeed, the bigger the amount of living biomass, the bigger the carbon leakage, as some prokaryotes (heterotrophs) feed off the remains of other prokaryotes (autotrophs). There is a saying, 'what goes up must come down'. But, in global biosphere carbon terms, you could easily say, 'what comes down (from the atmosphere) eventually goes up'.

Here, remember another important atmospheric-return process on a geological timescale is that of volcanoes that continually spew out carbon from deeper in the Earth's mantle. Further, the mantle itself sees carbon continually returned to it through subducting plate tectonics. You see, in addition to the fast carbon cycle—that can be summarised as photosynthesis trapping carbon on one hand and respiration and the breakdown of dead biomass releasing carbon on the other—there is also a 'slow' or 'deep' carbon cycle, in which dead organic carbon in sediments is slowly recycled into the mantle due to plate tectonic movement. This transports marine sediments to subduction zones and then carries them down into the Earth's mantle. There they are warmed and eventually returned to the surface in a mantle mix in volcanoes and volcanic vents. Given, broadly speaking, that supercontinents form, break up, and re-form every 600 million years or so, the bottom of the oceans must be mostly recycled over a similar timescale. And also, it is to the oceans that sediments from the continents are carried by rivers with, over the ages, ancient continental mountain ranges being eroded and new mountain ranges rising in their place. Of course, this renewal process is continuous and even though there are the two periods of major variation in Figure 2.1, as well as (as we shall eventually discuss) minor ones, there is a continual resetting trend of the $^{13}C$ to $^{12}C$ ratio as the deep carbon cycle slowly keeps churning. In short, the $^{13}C$ to $^{12}C$ ratio over many millions of years tends to hover close to a zero standard norm.

And so, what the $^{13}C$ to $^{12}C$ graph more properly represents is organic, formerly living, carbon $^{12}C$ *burial* and also *return* to the atmosphere: in short, carbon flux. Points in the top of the graph represent times when there is greater organic burial, and points at the bottom where there is return of $^{12}C$ to the atmosphere.

Better still, we might view Figure 2.1 as depicting short-term deviations from the long-term average of the balance between carbon burial and return. (Again, we need to be mindful that the use of 'long-term' and 'short-term' is in the geological sense, with 'long-term' relating to many millions of years and longer and 'short-term' relating more to tens or even hundreds of thousands of years.)

If you are still uncertain about the question of size of the increasing total biomass on the planet as our Earth evolved trapping more and more $^{12}C$, we might consider the $^{13}C$ and $^{12}C$ ratio graph as that of a see-saw in balance: though here the balance is not of equal amounts but about a ratio of relative amounts (in which $^{13}C$ is always the minority component) that has remained, very loosely speaking, more or less broadly constant (with a few exceptions) on a 'long-term' basis for much of the Earth's lifetime.

A see-saw can be in balance a number of ways. The early Earth billions of years ago had a far smaller total biomass than today, but there was, broadly speaking, still balance between biological carbon drawdown and release. If you wish, you might consider this as two small children balancing the see-saw. Similarly, much of the past couple of hundred million years with the Earth sporting a far greater biomass (tropical rain forests, seas teeming with fish, extensive plains of grassland, *etc.*), though broadly seeing much balance—there have been exceptions which we will return to later—might be viewed as also having a balanced see-saw, but, instead of children on it, imagine two giant sumo wrestlers. (Incidentally, this see-saw analogy can also be used to help understand the significance today of the comparatively small additions of carbon dioxide from burning fossil fuels compared to the large, annual photosynthetic drawdown of carbon dioxide that is largely balanced by a similarly large respiratory and anaerobic breakdown carbon release.) Here the two children (or the two sumo wrestlers) represent the size of biological $^{12}C$ drawdown through photosynthesis on one hand and the release of $^{12}C$ through living respiration and combustion (among other ultimately oxidative processes) on the other. (Relating this to our present time, it just takes someone to press lightly on one arm of the two sumo-wrestler balanced see-saw with their finger—or release a small amount of greenhouse gases into the atmosphere's carbon see-saw—to tip it away from being balanced.)

Early in the Earth's history there was little life and so little drawdown and so little biological, $^{12}C$ rich, carbon to return to the atmosphere. More recently (the past 500 million years or so) there was much more life and so greater drawdown that was balanced by greater return to the atmosphere, where ultimately it was rained out to form the carbonates, or rained out into the seas, where it was picked up by calcareous organisms (species), both of which we see in the geological record.

With this understanding we can begin to build up a picture of the evolution of carbon-based life on Earth from the $^{13}C$ to $^{12}C$ record. True, we have no detailed data (few data points) very early in the Earth's history (say prior to 3 bya) and so do not have an accurate view of $^{13}C$ to $^{12}C$ changes even if we knew exactly when life began (which we do not). But starting a little way from the left of Figure 2.1, we have a small living component of the biosphere rich in $^{12}C$, together with another larger reservoir of $^{12}C$ in dead (formerly living) biomass and then the large reservoirs of inorganic carbon. Taking into account both these biomass components, we can explain the early part of Figure 2.1 as best we can. What seems likely to have happened (but is not clear on the graph with few data points before 2,750 mya—the ones after that time consist of many data points stacked on top of one another, so you don't see them) is that as life got going it rapidly drew down more $^{12}C$ (than $^{13}C$) from the atmosphere. But the living component of this early biosphere would be small compared to today. This last is the view to which James Kasting and Shuhei Ono refer (*1*). However, as the dead remains of successive generations built up within marine sediments then eventually (geologically speaking within tens or hundreds of million years) both the *dead* remains and smaller *living* biomass together broadly equated to the size we find today (albeit that today we have a larger living component) and then after the first few hundred million years or so of life on Earth the long-term, deep carbon cycle returns started to kick in and these added to the fast carbon cycle returns. All the very early carbon dynamics happened at the far left end of the graph, a region for which we have all too few data points and so, even if there was a very early difference (a smaller global living biomass and hardly any dead biomass) from the broader sweep of Earth history, we cannot see it. Subsequently the Earth reasonably quickly (again in geological terms) established a $^{13}C$ to $^{12}C$ balance that has for the most part largely persisted for over 3 billion years to today.

Today the amount of dead biomass carbon over the surface of the Earth (as opposed to buried deep, such as fossil fuel) is so great that there is considerable 'leakage' from this biosphere store. For example, carbon (that was formerly incorporated in life) leaves swamps and woodland as methane, as it does leaving melting, organic-rich permafrosts. Of course, even deeper buried dead organic matter also releases carbon through oil seepage into oceans, where it eventually decays, and so forth.

Now, the above is a very simplistic rationalisation of what might have been going on, and note the use of the weasel words 'for the most part' at the end of the paragraph before last and 'simplistic' in this sentence. The carbon cycle is actually more complicated than just carbon in the atmosphere (be it as methane or carbon dioxide), carbon in the living biosphere, and inorganic carbon, and there have been times of deviation from the norm: times when either there was more $^{12}C$ return to the atmosphere and times when there was more net-$^{12}C$ drawdown and burial so relatively boosting $^{13}C$ in the atmosphere and oceans which then became new carbonate sediments. But we do not need to worry about the full complexities to learn something from this broad-sweep graph; though we will look at some of the major but very brief (and not in this graph) deviations in due course.

Despite hovering close to a zero VPD standard (see Appendix 2 and also the caption to Figure 2.1), on the grandest of geological perspectives of the Earth's history there are two (geologically speaking at this grand scale) 'brief' periods of large $^{13}C$ to $^{12}C$ scattering. Here, for now, just note these two brief episodes, around just before 2 bya and again before 600 mya, of broad scattering, as these are very much related to the two biggest 'flips' of state the Earth underwent. Not surprisingly—and logically, given that Figure 2.1 relates to carbon—both these regions of scattering coincide with changes in the way the biosphere processes carbon. And, to take another unsurprising step of logic, both these regions coincide with major developments in biological evolution; it is unsurprising because life on Earth is carbon-based and so evolutionary developments will, *ipso facto*, be bound to affect the way life processes carbon in some way or another.

Ultimately, what enables life to process carbon is energy. So how life is able to harness energy is key to how life processes carbon. Consequently, a reasonable assumption would be that major evolutionary developments in the way energy is harnessed by life will affect life's drawing down of carbon, and with more carbon drawn down come greater opportunities for carbon return. This great drawdown and great return explains the deviations above and below the long-term norm.

In short, these two major flips of the Earth's state are revealed by two major periods of disruption in the $^{13}C$ to $^{12}C$ ratio that in turn relate to changes in the way life processes carbon, that yet again in turn relates to the way life harnesses energy through evolutionary innovation: in other words, the disruption periods relate to major evolutionary steps!

Given all of this, we need to return to our examination of the evolution of life on Earth, and especially the way life harnesses energy, so as to appreciate what were the evolutionary developments at the time of these two 'flips' or changes of state. And then also to take a close look at the geological record to deduce the Earth system science of these 'flips' and changes of planetary state. In fact, for comparison, we also need to look at systems that stay in the same (non-flipping) state—stability is as important a matter as is system instability—if we are to understand the science of 'flips' or 'change of system state'. As we shall see, it looks very much as if the major 'flips', or 'changes of system state' the Earth has undergone since its formation, might be down to life and its evolution. Let's examine this early evolution of life before we look at the two major flips of the Earth's state.

## Energising early life

That life needs energy is such an obvious statement that it has almost become an iconic notion if not a cliché: who can forget the numerous cinematic portrayals of Mary Shelly's Frankenstein imbuing his creation with life using the energy of lightning.

As noted, so far in this story of Earth flips the focus has been more on life being carbon-based and only briefly touched upon the energy it uses to process this carbon. Let's look at this now.

To return to the beginning, the first—the earliest—of the precursors to bacterial life probably drew on hydrogen (an electron donor) from undersea thermal vents, or hot springs on land, and even at low but useful-to-life concentrations in the early Earth's atmosphere, as a source of energy to power metabolism. Indeed chemolithoautotrophs use hydrogen released at low temperatures from some minerals (see Chapter 1). The equation that releases 135 kilojoules would be:

$$4H_{2(g)} + CO_{2(g)} \rightarrow CH_{4(g)} + 2H_2O_{(aq)} \quad \Delta G^0 = -135 \text{ kJ mol}^{-1}$$

Here: H = hydrogen; C = carbon, O = oxygen, g = gas, aq = liquid or aqueous with $\Delta G^0$ being the change in (Gibbs free) energy expressed in kilojoules *per* mole.

As noted in the last chapter, there was more hydrogen in the very early Earth's atmosphere and possibly enough for life to involve its use as a source of energy.

As previously mentioned, these were not necessarily around the 'deep-sea' vents often talked about today. Though remember that the early Earth's oceans were shallower then (there were no substantial continents compared with today's) and, with the crust thinner, there would have been plenty of thermal vents nearer the sea surface and not just confined to the abyssal deep seas, as many are today. Indeed, for the earliest of all life it is likely that these thermal vents were near the surface and, if not more likely, even on land-based thermal springs given the laboratory-proven—hence in nature theoretically possible—chemical pathways to precursor RNA and DNA requiring ultraviolet light. (See Chapter 1 [3, 4].) As high-energy (higher-frequency) UV light cannot penetrate far into water, these reactions must have taken place near the surface. Although, once life got going it could colonise the environments around deeper vents.

Hydrogen, in its pure ($H_2$) state, is today not a widely available element on our Earth, though there may very well have been more of it in the early Earth being released from volcanoes than today after billions of years of crustal and upper mantle reworking and venting of volatiles. Nonetheless, even today some hydrogen is released from vents: it comes from the chemical breakdown of water as it passes through hot geological strata. Yet it is soon lost. Here hydrogen, being light, has a tendency to slowly escape into space and so in its pure elemental form today it is not plentiful near the surface. (In its bound form with other elements—such as with oxygen as water—it is plentiful.) But there was more of it in the primordial Earth's atmosphere, as more was being geologically vented by the younger Earth.

Given that the supply of hydrogen in its pure state was limited even if there was more of it escaping the early Earth's mantle to the surface (its primordial reservoir had yet to escape into space), other sources of energy would be needed for life to affect the oceans to such a degree as to change sediments as seen. Hydrogen sulphide and elemental sulphur are two other energy source candidates that were likely used providing the electron transfer mechanism, such as from converting (soluble) ferrous iron to (insoluble) ferric iron. Remember from Chapter 1 we have our BIFs (banded iron formations) that first appeared around 3.75 billion years ago. This is a potential, if not likely, signature of early life—early life that was able to affect the environment, if not at the level of oceans, certainly at the level of large seas. BIFs are possibly a sign that life was altering the planet's state from that where pure-chemistry—here meaning chemistry unrelated to life—determines matters.

One key reason that the amount of life was small compared to today was due to the ways it was deriving its energy. Even though early life employed a number of energy-trapping and energy-generating techniques, none of these techniques garnered or generated much energy compared to mechanisms generating oxygen today. But how much energy does using sulphur, or iron, or neutral hydrogen, garner? And how does this compare to how today plants get their energy through slightly positively charged ($\delta^+$) hydrogen bound with oxygen—or splitting 'water' ($H_2O$), as you and I commonly know it?

It all boils down to the energy—either consumed or released—in chemical reactions. Such reactions are all to do with the change in electrons surrounding the nucleus of atoms: the change that is the essence of a net chemical reaction. In summary, life's energy comes from the transference of an electron from one atom or molecule (the electron donor) to another (the electron acceptor).

Assuming that much of life (through metabolising carbohydrates) was getting its carbon from carbon dioxide (the electron acceptor), then straightforward chemistry can inform us of the energy involved using a variety of electron donors. Table 2.1 depicts the energy involved with a number of electron donors using the carbon dioxide as the electron acceptor reaction with a redox potential of −0.434 eV to create carbohydrates that life can then use. (Here you only really need to look at the numbers, but, if you recall your school science the energy can be measured in electron volts [eV] with 0 eV being the standard electrode potential. One electron volt itself is equivalent to the energy needed to move an electron across a potential difference of one volt: this equals $1.6 \times 10^{-19}$ joules. Those of you who knew that, knew that. And those of you didn't need not worry. As said, just look at the numbers' size.)

Another bit of science jargon, that might help should you decide to read around the subject, relates to the terms 'oxidising' and 'reducing'. In chemistry something (an atom or a charged molecular group) is said to be oxidised when it loses one or more electrons, and reduced when it gains them (see Chapter 1). You might think that oxidising is to do with becoming chemically bound with oxygen, and indeed sometimes it is, though sometimes it is not: oxygen is an element that has a considerable tendency (the appropriate electron orbital energetics) to grab electrons, but other elements do so too; 'oxygen' is a good 'oxidiser', but it is not the only oxidiser and when binding with other elements it goes through an energy change.

Indeed, you may recall from your early school science, oxidising hydrogen by igniting hydrogen in the presence of oxygen releases a tremendous amount of energy, and the right mix of oxygen and hydrogen is decidedly explosive! Actually, it is the very reverse of this that today's plants do—splitting water (or reducing water's slightly positive ($\delta^+$) hydrogen) to liberate oxygen. The energy today's plants need to split water (and the amount gained by the plant—or the animal that eats the plant when the reverse, respiration, reaction takes place) is obtained from sunlight through the process of photosynthesis, in this case oxygenic photosynthesis. Sunlight is a good source of energy to split water as well as other chemicals.

Table 2.1 clearly shows this. Other than in the top row that depicts reducing oxidised hydrogen (water, $H_2O$) by removing the oxygen (the process used by today's plants), the other reactions (the early-life reactions) involve far less energy.

Using the figures in Table 2.1, splitting water as today's plants do to make carbohydrates ($CH_2O$) requires a net energy of $+0.82 - (-0.434) = 1.254$ eV. Conversely, using hydrogen requires $-0.414 - (-0.434) = 0.02$ eV, it is easier (requires less energy) to grow biomass using hydrogen than by splitting water: early prokaryotes needed only infrared from thermal vents, whereas today's plants need higher energy visible light from sunlight. So, early life likely soon started using neutral hydrogen, as the energy involved is a lot less: roughly an order of magnitude less. While on one hand this meant that, molecule-for-molecule, not much energy could be harvested by early forms of life, on the other it meant that life's macromolecules (big molecules) did not have to contend with much energy in the processes. It is a bit like the difference between using a

**Table 2.1** The energy change involved from various electron donor reactions compared to the carbon dioxide/carbohydrate electron acceptor reaction and measured in electron volts. Note, that these are the stripped down, bare essential reactions. For example, the first one is used by modern plants. Table based on Lenton, T., & Watson, A., *Revolutions That Made the Earth.* Oxford University Press: Oxford, Great Britain (2011).

| | | |
|---|---|---|
| *Electron donor* | | |
| Hydrogen | $H_2O \leftrightarrow ½O_2$ | +0.820eV |
| Ferrous iron | $Fe^{2+} \leftrightarrow Fe(OH)_3$ | −0.236 eV |
| Sulphur | $HS^- \leftrightarrow S^0$ | −0.278eV |
| Neutral hydrogen | $½H \leftrightarrow H^+$ | −0.414 eV |
| *Electron acceptor* | | |
| Carbon dioxide | $CO_2 \leftrightarrow (CH_2O)$ | −0.434 eV |

hand-held torch powered by batteries and wielding a stadium floodlight plugged into the mains: you may not get nearly as much light with a battery torch, but neither will you suffer the risk of an injurious electric shock.

So, all the early-life core energy reactions—using sulphur, iron, and neutral hydrogen—did not involve much energy. Another advantage of early life harvesting so little energy *per* molecule was that if photosynthesis was used (rather than purely deriving energy chemically) the light used to drive this could itself have a lower energy content. Using 'light', or to use a more precise term 'electromagnetic radiation', with a lower energy content meant that sources other than sunlight could be used. This is because the amount of energy carried by a photon of light is directly related to the frequency (the inverse of wavelength), hence colour, of the light.

Meanwhile, heat energy (infrared, which is electromagnetic radiation with a lower frequency than light) was available in hot vents, and those hot vents—that also could give hydrogen, iron, or sulphur compounds and other minerals with carbon dioxide in solution—provided everything that early photosynthetic life needed. Conversely, using photosynthesis to split water, as done by today's plants, involves so much energy that higher-energy, or higher-frequency, visible light is required, and there is not a lot of that around submarine vents deeper than a few tens of metres from the sea's surface and none at all around abyssal deep-sea vents, compared to those close to the sea surface.

Now, you might be wondering why very early life was unlikely to use sunlight. Well, there were several reasons. One good reason, as alluded to above, was that only the chemical-giving vents in the shallowest waters would have received sunlight, and even then that sunlight, unless in the shallowest of water, would not have much infrared. Water with infrared shining through it quickly becomes hot, but infrared cannot penetrate more than a metre or two of water; visible light can penetrate further. (On the other hand, water-based thermal solar panels work so well because not only does the water travel through black coloured tubes—with the black absorbing heat—but also the width of the tubes is such that not much water is needed to absorb the infrared.) Conversely, the temperatures around thermal vents are often much higher: swim by many of those and you would quickly be stewed to death: so there is plenty of infrared that some organisms can use. Though early life, the ancestors of bacteria-like organisms—types of prokaryote—in these hot environments would today be considered extremophiles, 'liking extremes', for them extremes were normal and they could thrive in such environments. Yet thermal vents only occupy a tiny proportion of the Earth's surface. Again, it was this rarity of environments that could provide energy for very early life that was another good reason why the very early Earth's total living biomass was far smaller than today's.

Theoretically, it presumably would not have taken long for some precursor to bacteria, or early prokaryotes, to get carried to near the sea surface and the potential to photosynthesise using sunlight. But remember that early on the Earth this was anaerobic photosynthesis, capturing carbon from carbon dioxide in a process that did not generate oxygen. There was no (or next to no) free oxygen in the atmosphere (hence sea) to poison these anaerobes. Also remember that even if some early prokaryotes did get carried to the sea surface away from the vents, though they would gain access to sunlight, they would lose access to the high concentrations of sulphur and/or iron coming from the vent, though there would still be ferrous iron in solution at lower levels as well as sulphur in the sea. Photosynthesising prokaryotes would need to do a number of things very differently before they were able to take advantage of the higher-energy (higher-frequency) sunlight in a controlled way to split water and generate oxygen.

And so matters continued for around a billion years or so during which time evolution was taking place with species acquiring new tricks (abilities), and the genes for such tricks were being swapped around with mutations providing both successes and failures. Here, in true Matthew-Wallace-Darwin 'natural selection' evolutionary style, species' successes survived (with the species' failures dying out) until the time came when a major development took place and the Earth went through a major transformation and an associated flip of state: the first major life-induced flip (not counting the change from a purely non-biological chemistry-determined world likely indicated by the aforementioned BIFs).

Before continuing this narrative, and the discussion of Earth's first big flip of state, it should be noted that there are other explanations as to how and why the Earth underwent a major flip of state over 2 billion years ago. Here, the thinking up to now has largely had it that this flip took place for reasons having nothing to do with life, but that life was spurred to evolve from the flip to a new, purely geologically instigated, change of state (we will visit the other ideas in

Chapter 5). However, an alternative view—that is beginning to gain some traction and which is the one being presented here—is that life itself caused this flip and that life and geological processes combined with regards to the timing of this event.

Yet before we look at this flip we do need to appreciate some aspects of 'stability' as well as 'instability' of a system's state. And we need to note that while matters on the Earth roughly between 3.5 and 2.5 billion years ago appeared comparatively stable, there were still changes taking place.

What follows in the next section may seem a digression, but it is necessary so as to understand the nature of 'flips'.

## Biosphere stability

Most intuitively seem to have an idea as to what is 'stability', hence, what is 'instability'? Stability, we might reasonably suppose, is when nothing changes and instability is when change happens all the time. However, such intuition is simplistic. It is possible to have not just stable and instable but also semi-stable sates. It is also possible to have small changes taking place even within a comparatively stable set of circumstances.

Consider this. A ping-pong ball perched on a small eggcup that has been placed on a plate that is at rest on the table. The ping-ball may seem perfectly stable just sitting there but, in actuality, much of this supposed stability is illusionary. Give the table a thump with your hand and the ping-pong ball may very likely fall out of the eggcup and onto the plate; perhaps that ping-pong ball was not in such a stable state after all?

So now we have this ping-pong ball on the plate sitting on the table. This too may seem a stable relationship, but a door may open and—due to the resulting draft—the ping-pong ball may move about the plate. This also might be considered a semi-stable state, as the ping-pong ball is moving about albeit confined by the plate's rim. But do remember that even when the ping-pong ball was on the eggcup it was not perfectly stationary. Examine it carefully and you would have discovered that it was vibrating, buffeted both due to the sounds in the room and due to its atoms' own thermal energy.

So, even in what we might think of as a semi-stable state on the plate, the ping-pong ball can easily move around due to the vagaries of small drafts, and may even do so with some speed. Indeed, with a sudden gust the ping-pong ball might even get blown off the plate and on to the table. Now the ball is in a very unstable state, as it will take little for it to roll further and even fall off the table to the floor. There it may bounce up and down a while before it properly settles.

On the floor, the ping-pong ball is in a state of greater stability than it was before, but it still may not remain stationery with the small drafts. Indeed, now it has the entire floor of the room about which to move (and not just the small confines of the plate); it may have greater stability being at a lower energy state, but has a greater freedom to move around given the input of just a small amount of energy.

The ping-pong ball's state has changed from being on an eggcup to that of a plate on the table, and finally from that to the state of being on the floor.

Then consider this, it is conceivable that someone passing could give the floor-bound ping-pong ball a kick so that it bounces and ends up back on the table. It might even, if luck would have it and the kick was precise enough, end up back on the plate. Of course it would be very unlikely but—and do note this—not impossible for it even to end up back on the eggcup! True, this would be an extremely chancy event, which colloquially you might call a 'fluke', but it would not be impossible. Indeed, kick the ping-pong ball around for long enough then the chances do increase that at some stage (possibly after a very long time) it will end up back on the plate, if not on the eggcup. Such rare-but-not-inevitable chances are the basis of lotteries in which the chance of an individual win is extremely low, but the chance of a win among very many people is very much a likelihood.

What we might deduce with these scenarios are four notions.

First, stability is comparative, with some states being more stable than others: the ball on the table is less stable than the ball on the floor, and it would be a highly unlikely circumstance for the ball on the floor to end up back on the plate on the table.

Second, there can still be some change or movement when within a particular state: so the ball can move about the floor or about the plate with little energy.

Third, different states have different energy levels: the ball on the table has a higher energy level than the ball on the floor, and energy has to be added (or, to be more exact in physics terms, work done) to raise the ball from the floor to the table (the unlikely circumstance needs to involve energy, such as giving the ball a kick).

This leads us directly to the fourth point: it is possible to move from one state that has a certain stability and energy level, to another with a different degree of stability and energy, and you can even move back again to the previous state, but this return may be less likely.

And then there is this business of the process of actually changing state: if you like, the process of flipping from one state to another. Often these flips happen very fast: one second the ping-pong ball is on the plate and the next on the table; one second on the table and the next the floor. One minute it is on the floor, but give it a kick and the next second it is sent flying across the room.

That states can 'flip', or change dramatically, is something we come across daily, even if we do not always fully appreciate it. Put water into a freezer to make ice cubes and at first the water will cool from room temperature's several degrees centigrade without (other than temperature) much change. However, when the water reaches 0°C it suddenly changes state from liquid to ice: it has flipped state from being a liquid to a solid. Oddly, even though overall it is cooling, at the point of freezing the water emits a little heat: the latent heat of cooling. This takes place at the point of transition between liquid and ice. Such 'complications' close to the point of transition between states are, as we shall shortly see, common in systems that flip between states.

In the context of this book it is the 'Earth system' that is the ping-pong ball and 'life' that is the 'force' giving the kicks, as well as life itself being a component of the Earth system. Additionally, other things can give the Earth system kicks too, be it other-than-life components of the Earth system (*e.g.* eruptions of supervolcanoes) or even external factors (*e.g.* an asteroid strike). While these other 'kickers' do impart real kicks (you would not want to be by Yellowstone when it blows) it is the major kicks that life itself has imparted during the Earth's history that are interesting because—as the argument presented by this book goes—such life-induced kicks are due to the way life has evolved and these kicks affect the whole planet in a lasting way. In essence, these kicks are both natural and quite possibly an inevitable part of evolution on a planetary scale of an Earthlike planet! That is to say that they may apply to other planets, in other star systems, not too dissimilar to Earth.

Given this line of thought, and should the notion stand up to scrutiny, it would be reasonable to expect the Earth (or indeed any Earthlike exoplanet harbouring life) to experience a number of changes of state, or flips during its natural evolution.

A further matter to consider is that the corollary of something to 'flip' means that there has been both something to flip 'from' and something to flip 'to' as well as the flip itself; the latter being a period of intense instability in between two periods of comparative stability. Consider our water on the point of freezing. It will freeze if it loses just a small amount of thermal energy; conversely, it will remain a liquid with just the addition of a small amount of energy. And if a flip itself is a period of instability, then what causes this instability, and why is there a preceding period of stability (in the instance of early life on Earth, as discussed in the last chapter, around a billion years of stability, or at least comparative stability in $^{13}C$ to $^{12}C$ terms.)

Such stability changes are even of importance to us modern humans today because we have flourished—at least in terms of population—in the roughly 10,000-year Goldilocks period of current climate stability afforded by the Earth system since the end of the last ice age glacial. (An epoch of time known as the Holocene.) It matters because if our species has flourished due to climate stability then surely we would want to avoid climate instability? Of course ice ages themselves—within which we do see considerable temperature variation—are comparatively stable compared to the far more dramatic flips associated with the key events of the evolution of life on our planet, but such smaller glacial-to-interglacial flips within ice ages are still important to us, such is the comparative fragility of our existence on Earth (we will return to glacials and interglacials shortly). Indeed our *Homo* spp. (which includes Neanderthals and other hominins earlier as well as anatomically modern humans *H. sapiens*, and the extant species of humans, *H. sapiens sapiens*) evolved considerably across a number of glacials over the past 2 million years of our current Quaternary ice age (part of Chapter 8 will look at this in greater depth). If you will, humanity has had to jump through a number of evolutionary hoops thrown up by climate-change-induced environmental change: it has had to adapt each time, with those that could not adapt dying out, and this spurred our species' evolution. In this light, the concerns over post-20th century global warming accrue an even greater urgency: there will be losers.

This stability of environment also occurs in spatial (distance) terms as well as time. The fiction author Terry Pratchett provided another excellent illustration of the Goldilocks fragility of our human existence on the Earth's surface.

> Two miles *down* there you'd fry and two miles *up there* you'd gasp for breath, and there's a small but, given the consequences to us, significant chance that in the next thousand years a large comet or asteroid will smack into the planet. I'm not making it up. I don't lose sleep worrying about it. But finding this out when you're 13 or so is a bit of an eye-opener. It puts acne in its place, for a start.
>
> Terry Pratchett (1994)
> www.concatenation.org/articles/pratchett.html

But still, we have this question as to why the Earth's thin planetary skin of the biosphere, in an overall sense, has comparative stability.

The answer may be surprising in that planetary stability—such as the period between 3.5 and 2.5 bya (see the end of the previous subsection)—is not due to any lack of changing circumstances of the Earth; circumstances have changed throughout the Earth's history, but there are still long tracts of comparative stability between really big flips in the planet's state.

For example, in the grand scheme of things, across the roughly 4 billion year arm's length of the Earth's life since cooling to form the first surface rocks, the Sun has increased its output (grown warmer): it has not remained the same! Also, the Earth's days have lengthened (as noted in the previous chapter). These alone both affect the amount and the way energy reaches the Earth's surface. (4.5 bya, a day was only about 6 hours long.) (5)

Of course changing day-length did, in terms of daily growth rings as we saw in the last chapter, affect species, but broadly speaking life, in the BIG flip sense, has continued across the past few—indeed many—millions of years very much the same in key aspects. True, as we shall see, there have been hiccoughs and even minor flips. And, true, the nature of species have changed as they have evolved. Yet there are these protracted periods of comparative stability in-between the biggest of these planetary flips. (These we will shortly explore.)

This stability is due to the many systems, and their subsystems, that make up our biosphere, themselves interacting in a complex way to cause islands of stability within a sea of instability. These systems are both abiotic (not involving life) and biotic (of life), and there truly are many of them, which in turn makes the whole, overall Earth-system very complex indeed. Here, an illustrative example or two will help.

Consider just one aspect of climate. You may have noticed that in the sunshine it is nice and warm. Yet a passing cloud's shade sees a noticeable cooling. (It should be noted for any meteorologist and climatologist readers that that this is a simplified picture involving comparatively low-level clouds: some clouds, such as high thin cirrus clouds, warm the planets as their greenhouse impact is greater than their albedo effect.) This shade is quite simply because the cloud stops the sunlight from reaching the Earth's surface and, being white, reflects energy back out into space. Nothing remarkable here, and you will have seen pictures of Earth from space, hence seen the vast swirls of white clouds obscuring a fair proportion of the Earth's surface. It is a picture with which you are undoubtedly familiar, but just a generation or two ago, before the 1960s, this perspective from space of our world was unknown. Indeed, if you ever see a science fiction film from before the 1960s that shows the Earth from space then invariably these depicted an Earth with few if any clouds and the continents were clearly portrayed. Yet clouds exist, and we patently knew that before we ventured into space. The Earth's clouds are a small part of what confers our living planet's climate stability. But how does this work?

Imagine our world getting just a little bit warmer. This sees more evaporation from the ocean. This in turn leads to increased cloud formation. And this ends up with more sunlight reflected back out into space and so cooling of the Earth's surface back down again. Conversely, cool the Earth sees less ocean evaporation, less cloud formation, less energy reflected, and so there is some additional warming. It is a bit like a thermostat on your home's boiler keeping the temperature constant. Thermal cloud formation in the Earth system is one of stabilisation processes, and there are many such stabilising feedbacks: as indicated, the Earth's stabilising thermostat is complicated.

Life too has some Earth system stabilising properties. Both these many biotic and abiotic feedbacks lead to states of comparative stability within other less stable (or even unstable) states. This was the basis for the idea British engineer James Lovelock had and which he presented in 1979: that the Earth's biosphere consisted of numerous feedback systems which together conferred stability. This complex, of subsystems within an overarching system, he called 'Gaia', a term with which he titled his 1979 book (*6*). That he called this stabilising, or homeostatic mechanism, Gaia, after the Greek goddess, was in part a little unfortunate, as some people assumed that he was saying that these 'controlling' systems meant that there was some sort of godlike planetary consciousness, or even spirit, controller at work. Actually, he meant nothing of the sort; being an engineer, he was aware of cybernetic systems (such as the thermostat on your home's boiler) and recognised that there were analogous systems in nature such as the cloud example above; some sort of controlling intelligence, or spirit, was certainly not required. Yet many were still enchanted with the notion of Gaia as some exotic kind of semi-sentient control, and those scientists in the 1970s and 1980s critical of Lovelock's notion used the popular, folklore-type interpretation to dismiss the idea of Gaian cybernetic feedback.

Yet Lovelock knew that any stabilising cybernetic system can only be partial: exceed the system's parameters and stabilisation will fail! Following failure, what happens next is that the system changes state, and here it might be—if there is a sufficient balance of new positive and negative feedbacks—that a new degree of system stability becomes manifest with the Earth in a new, different state.

Let's take you as an example: after all, you are a biological system with built-in stabilising feedbacks. If it gets warmer, you will begin to sweat and this will help cool you down; if it cools, you will get goosebumps and your hairs stand and this will in part trap an insulating layer of air and so help keep you warm. You might even use your naked ape intelligence to put on extra clothing, or take it off to keep you cool. However, exceeding your system's parameters in extreme cold (and without the necessary survival gear) you run the risk of hypothermia, or in extreme heat symptoms akin to sunstroke. Yet, within the system parameters that allow functional cybernetic feedback, there will be greater stability than would otherwise exist being subject to external environmental change without such feedbacks.

What Lovelock needed to explain was that Gaia was a simple example of biological stabilising feedback, and so he came up with the now classic thought-experiment of 'Daisyworld'. He first presented Daisyworld in 1983 underpinned with some maths provided by Andrew Watson (who went on to co-author *Revolutions That Made the Earth*, which itself is a key text for students exploring the science of Earth flips) (*7*). In 1988 Lovelock presented Daisyworld to the public in his follow-up book to *Gaia* called *The Ages of Gaia: A Biography of Our Living Earth* (8).

Daisyworld works like this. Imagine a world at a certain temperature populated by two types of daisy: black daisies and white daisies. If this world is in a steady state then clearly there must be a thermal energy (heat) balance, yet the two types of daisies interact with thermal energy quite differently. On one hand, the colour black absorbs high-energy heat from the world's sun, so black daisies warm themselves and re-radiating the heat as lower-frequency (lower-energy) heat, warming the surrounding environment. On the other, white daisies reflect heat back up into space and so keep themselves and their surrounding environment cooler. All well and good for the daisies, but what happens if Daisyworld planet warms or cools due to changes in its sun?

In the event that Daisyworld's sun becomes brighter, and Daisyworld warms, so the heat-absorbing black daisies eventually get too warm and die. Indeed, well before they get too warm to die from actual heat exhaustion, they begin to get stressed, metabolising suboptimally; meanwhile, the heat-reflecting, white daisies fare better and can out-compete the black daisies. This leads to the proportion of white daisies to black daisies increasing, and Daisyworld overall becomes whiter: its albedo has changed. This in turn leads to more heat energy being reflected away from Daisyworld as a whole, keeping Daisyworld's temperature cooler than it would otherwise be under the brighter sun.

Similarly, if Daisyworld's sun dims and the world cools, so the proportion of black daisies increases and the world as a whole absorbs more solar energy and becomes warmer than it would otherwise be. So the changing mix of black and white daises helps stabilise Daisyworld's temperature.

Daisyworld is a cartoon-like (very simple) illustration of biologically mediated thermal control, but it does *not* require some kind of overarching intelligence for its function; though intelligence is required to understand what is going on, this does not make the system itself intelligent. Similarly, when considering the Earth system, Lovelock's Gaia is not intelligent or even a barely conscious entity even if some of its feedback mechanisms have living components. However, this did not stop some of the public in the 1970s and 1980s becoming entranced with the notion of the Earth having some sort of overall consciousness and they, contrary to what Lovelock was actually saying, considered Gaia in this erroneous way.

Of course, the real world is more complex than the simple two-types-of-daisy model. But then you can make Daisyworld more complex for instance by adding grey daisies to the mix of black and white daisies; add even more tints and colours and what happens is that even more stability is conferred to the Daisyworld system.

In short, non-biological (abiotic phenomena such as clouds) as well as biological feedback factors (such as our Daisyworld theoretical example) can confer biosphere stability. Consider a number of these factors together (as happens in the real world), and stability can become very significant.

However, other feedbacks work the other way and bring instability to the biosphere system: these are positive feedbacks and they can lead to runaway change.

An example of a positive feedback is that of sea ice melt. (See the central pair of feedbacks in Figure 2.2) A sea without ice is a dark colour and so absorbs more heat than a sea covered in white, heat-reflecting ice. Take the North Pole as an example. In the winter it is completely covered in ice. Part of this ice melts in the summer. But, in an ever so slightly warmer world, a lot more than you might think of this summer-melt takes place; disproportionally so, due to a positive feedback effect. What is happening is this: as some of the reflective ice melts it uncovers dark, heat-absorbing water and the temperature rises further, which in turn causes more melting. The reverse happens in the autumn, when the ice sheet begins to grow, reflecting more heat that in turn causes further cooling, hence even more ice formation, more heat reflection, and even further cooling.

When looking at changes in Earth's biosphere, consideration needs to be given to all these stabilising and destabilising factors both biotic and abiotic. What emerges is a system that has, as one variable is changed (and energy input is the key one commonly explored), areas of stability as well as other areas of instability.

What this means is that as one gently tries to warm our world—be it through the way the planet receives solar energy or greenhouse traps it—then the actual warming we see varies. There are certain temperature ranges within which these various feedbacks serve to counteract each other. Here, we can slightly change energy flows, yet the biosphere seems comparatively resistant to temperature change: these are temperature ranges that seem to have some stability. Conversely, within other ranges of energy flow it is comparatively easy to warm or cool the planet: these are temperature ranges that confer instability. Our planet seems to switch between comparatively stable (semi-stable) modes once a key factor becomes too extreme. At this point the Earth tends to move rapidly through the range of instability to a new stable (or semi-stable) state.

Take the past 2 million years. As mentioned earlier, this Quaternary period has seen the Earth in the depth of an ice age. We are in it now, though you may not realise it, as the Quaternary ice age has lengthy (roughly a hundred-thousand-year) cold episodes, called glacials, and much shorter warm periods—we are in one now—called interglacials. Both the interglacials and glacials seem to be more stable states than in-between them. So what we see over time are glacials and shorter interglacials with the rapid transition through a zone of less stability

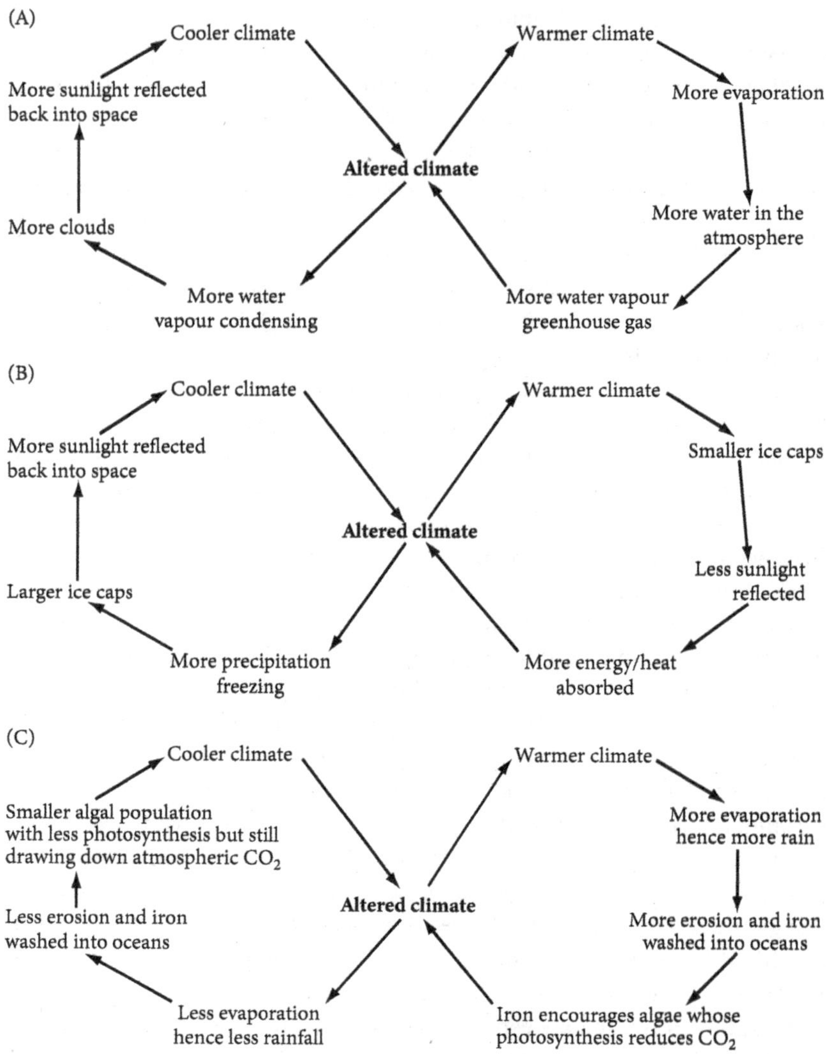


**Figure 2.2** Examples of how interacting positive and negative feedback mechanisms affect climate. There are many such feedbacks.

(instability) between these two states. The feedback systems operate slightly differently depending on whether the Earth is in a full glacial or interglacial mode. (For example, in glacial mode more carbon dioxide can be stored in the oceans, keeping it out of the atmosphere and so making the Earth cooler. The opposite is also true, as warming impedes ocean storage and carbon dioxide vents into the atmosphere, furthering the warming.)

(A)

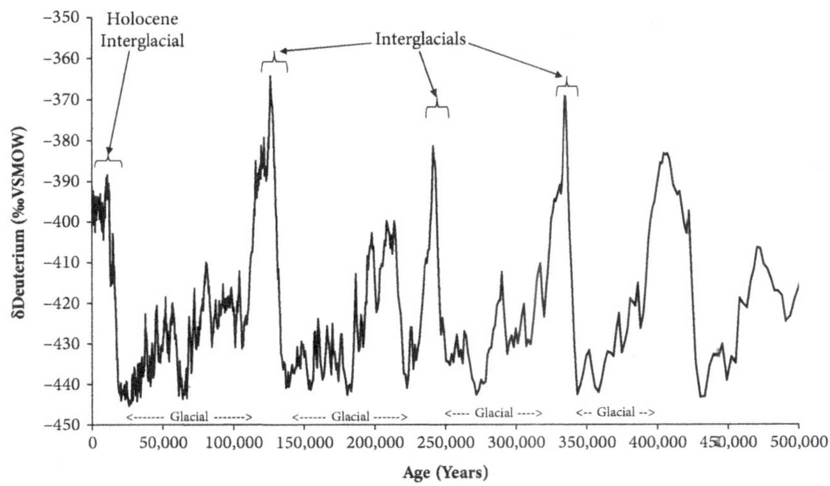


**Figure 2.3A** The way temperature change has taken place past half a million years of the current Ice Age. Deuterium in ice cores from Antarctica serve as an analogue for temperature change. See Appendix 1 for units. The current Ice Age's rapid flipping between two semi-stable states of cold glacials and shorter warmer interglacials is represented by the deuterium concentration in the ice. Data source: Jouzel, J., *et al.*, Ice cores deuterium data, *Data Center for Paleoclimatology Series #2004-038 Program.* NOAA: Boulder, CO (2004).

Figure 2.3A shows how the Earth has flipped between glacials and interglacials the past 500,000 years of the current ice age. Our current ice age has been more pronounced the past 2 million years in what is called the Quaternary period. We are currently in an interglacial called the Holocene epoch. Figure 2.3A depicts deuterium (heavy hydrogen) in Antarctic ice's $H_2O$. (Deuterium was mentioned briefly in Chapter 1, and we will encounter it again in Chapter 7.) A small proportion of deuterium exists in all water. Because deuterium is twice the atomic weight of hydrogen, more energy is needed to evaporate a water molecule containing deuterium than a molecule of normal hydrogen. When snow falls over Antarctica, further snow compresses it as ice. As the years pass, more layers of snow are added, so building up the inland Antarctic ice sheets. Ice cores drilled out of these sheets have layers of different deuterium concentration, and this reflects the temperature difference between that of the distant ocean from which the water evaporated and the temperature at which the water condensed out as snow. The deuterium concentration in the ice therefore reflects global hemispheric temperature. Northern hemisphere ice cores from Greenland also show a similar pattern of glacials and interglacials. This is firm evidence that glacials and interglacials are global climatic events.

The flipping—the geologically comparatively fast transition—of states between the cold glacial mode and warmer interglacial one, is driven by Earth orbital variations: they are, if you will, the 'pacemaker of the ice ages' (*9*, *10*). Currently, the continents are positioned so that the majority of the land is in the northern hemisphere, while the darker, more heat-absorbing ocean dominates the southern. Orbital variations due to the cycle of small but regular, waxing and waning of angle of tilt, the regular variation of the orbit being slightly elliptical to being more circular, and what is known as the precession of equinoxes, all contrive to vary the amount of the Sun's energy received in the northern hemisphere summer (see Figure 2.3B). Summer is important, as it is the amount of ice that survives from winter through the summer into next year that counts. If it grows then more sunlight is reflected and the Earth cools, so enabling more ice to survive the next summer, so enhancing ice build-up the next annual cycle. If ice surviving the summer decreases then less ice is available for the next annual cycle, there is less heat-reflecting ice, and so the Earth warms. The pacemaker of the flipping between glacial and interglacial states is the orbital variation affecting summer solar energy arriving in the northern hemisphere (*9*, *10*).

To return to our earlier ping-pong ball on a plate analogy, it is as if the changing way the Earth receives sunlight due to these orbital variations, are the rhythmic gusts of draft moving the ping-pong ball about the plate.

That there are these governing feedback (Gaian) processes regulating the Earth is therefore most certainly *not* supernatural or mystical. That the life-planet system—the Earth system—exhibits aspects of self-regulation equally does not require some kind of overseeing intelligence. This should not come as a surprise to anyone with a school-level, or even a popular science level of understanding of biology, and so does make it rather worrying that the idea that Lovelock's Gaia (*6*) was somehow sentient was accepted by some of the general public: even a few scientists thought that this was Lovelock's hypothesis. Having said that, the same perceptual difficulty of the need to invoke sentience is found in other areas of science. Similar to Earth system self-regulation processes not requiring any external guidance, there is the whole issue of Darwinian evolution: that too requires no overseeing intelligence or, as Richard Dawkins puts it, no blind watchmaker creating and maintaining the mechanism (*12*). Yet a number still fail to grasp this idea. Nonetheless, even though there is no overseeing intelligence behind evolution, it does not mean that evolution cannot lead to a species sufficiently intelligent to generate technology. (We will return to this much later.)

Yet, even if Gaia itself is not sentient, this does not mean that a small part of the Earth system cannot be sentient, or even feature sentient generating technology.

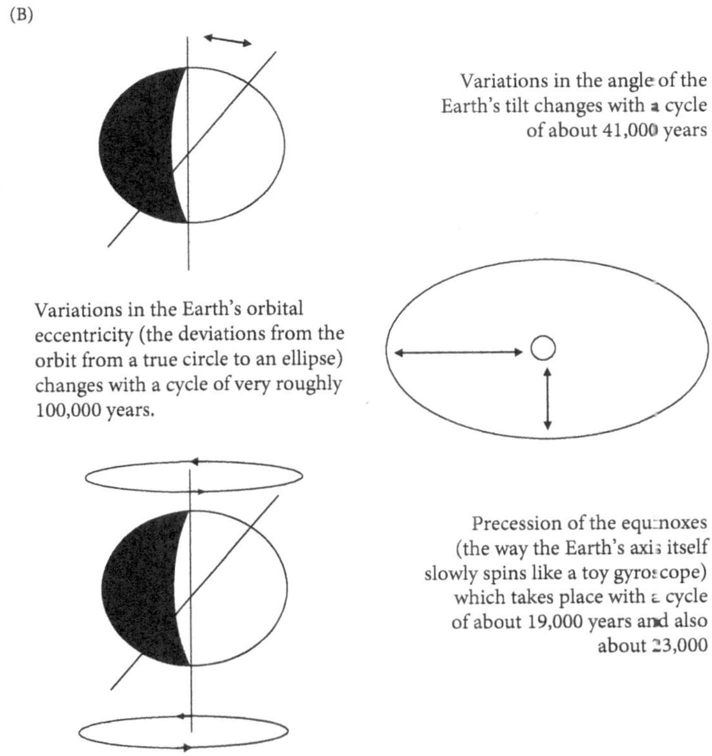


**Figure 2.3B** The three Earth orbital parameters that affect the amount solar energy either the northern or southern hemisphere receives. This would not matter if the Earth's land were equally distributed between the northern and southern hemisphere. However, the present continental positions see a land-dominated northern and ocean-dominated southern hemisphere. Land and ocean absorb solar energy differently. Palaeoclimatologists commonly use midsummer (July), northern hemisphere insolation (solar energy) calculations based on these three parameters. This is a simplified portrayal, as there are other factors going on: for example, precession has a longer period but it is modulated by the precession of perihelion in the opposite direction. The values have over the years been refined a number of times since the 1976 'pacemaker of the ice ages' paper (*10*) and are summarised here (from refs. *7* & *11*). There are other longer resonances.

James Lovelock himself said that if humanity managed to deflect an incoming asteroid that would otherwise lead to an extinction event, then this could be considered as Gaia defending itself (*13*).

The bottom line is that Gaia itself—as a whole—does not need to be intelligent or sentient to confer stability. This is so even if biological intelligence can naturally evolve.

Later on, further examples of stability and instability in the Earth system will be cited and a number of other feedback mechanisms given. Meanwhile, having considered islands of stability within a sea of instability within the Earth system, arguably the next logical issue to consider is how the system shifts from one stable, or semi-stable, state to another? In other words, how systems—including the Earth system—flip, or transition, from one state to another.

## Flips—thresholds—critical transitions (and chaos)

As flips in the Earth system are central to this book, there is no escaping having to come to grips with the topic of critical transitions (the sharp change/flipping from one state to another). This is a little more complex than it at first seems. In scientific terms, it can be defined mathematically, and a maths explanation is needed for a more detailed understanding. However, because not all scientists are expert mathematicians, and because a range of scientists are required to gather the data as well as to interpret what is happening in real-world terms, a number of explanatory simple (cartoon) critical transition models have been developed that help explain the way flips between different states take place. What follows largely draws on these, but if you would like to find out about the underpinning mathematics some technical terms will be provided that you can enter into a search engine, as well as look up this text's references, to find out more.

The very term 'flip' itself suggests a rapid change of state: one 'flips a burger' or 'flips a coin', and both are quick actions. Yet, in the context of the broad sweep of the Earth's history, we again need to remember that in terms of the geological record 'quick' or 'rapid' itself might refer to change over a few (several) thousand years (such as the transition from an ice age glacial to a warmer interglacial state) or even shorter periods of time, but not at all as fast as flipping a burger. Equally, other flips, such as those that greatly change the planetary state, can take a few millions of years, though still at most only 1% or 2% of the Earth's history. Further, as we shall see, each flip may have distinct episodes within it.

From our early 21st-century perspective, sudden jumps in the global climate, or regional conditions and processes, at first might appear surprising given that the UN Intergovernmental Panel on Climate Change (IPCC) forecasts depict seemingly gradual change of a few degrees centigrade over a century. Yet the IPCC, especially starting in its 1995 assessment, does warn of 'surprises' and sudden jumps that can take place as the biosphere system, or components thereof, goes through a critical transition (*14*). A good example of a climate threshold at the global scale, and the associated critical transition, is that between the aforementioned current (Quaternary) ice age's glacial and interglacial states of the past 2 million years. The last of these glacials ended before 11,700 years ago, when, within a few thousand years (which in geological terms is rapid), the Earth system flipped from its glacial mode, crossing the climate threshold, and into the Holocene, the interglacial mode which we are in today. (We will briefly return to this in Chapter 8.)

The phenomena of critical transitions may seem new to you but you have undoubtedly come across them before many times, even if you did not realise it. There is the earlier example of water freezing. There is also the example from when you go to make a cup of tea. Here you supply heat at a steady rate to water and the temperature rises in a near-linear way. But continue to supply heat and it stops warming near-linearly and suddenly the system flips to a new state: the water boils. Indeed, at the point near boiling it takes extra energy (latent heat) to get the water to actually vaporise: the water has reached a critical threshold and is transitioning to a new state, 'steam': the water molecules have flipped state from moving in the loosely bound way that molecules of liquids do, to a mode in which molecules are more isolated as they are when they are part of a gas.

In the 1980s the idea of critical transitions in complex systems came more to the fore in popular science with 'catastrophe theory' and 'bifurcation theory'. Then in the 1990s there was 'chaos theory', that described the dramatic movement of a system's state between what are more stable states in a 'Lorenz attractor'.

Broadly speaking, there are two types of transition: non-critical (sometimes called 'non-catastrophic') and critical (or 'catastrophic') transitions. An illustration of a near-linear and a non-critical transition is given in Figure 2.4A. (The example of a purely linear transition is not given, as you should be able to surmise that this would simply be a straight line.)

Conversely, Figure 2.4B shows a critical transition (or catastrophic transition in the parlance of catastrophe theory). This shows two states that are not directly connected. Consider two variables 'V' and 'T'. If it helps you might consider 'T' as temperature and to take a purely hypothetical example in the form of a component of the Earth system, 'V' might be the annual end-summer volume of ice on some high-latitude island. To continue this purely invented analogy, as the temperature warms, the ice volume on the island decreases. However, as regional warming continues we may get more ocean evaporation, hence precipitation as snow, and so the rate of ice depletion on the island slows. But a point will come with continued warming, when the extra snowfall cannot counteract the effect of the warming, and suddenly all the ice on the low-lying area of the island melts. Indeed, that this can happen quite fast may not come as a surprise given local albedo (light and heat reflection) effects: less ice and there is less sunlight reflected by the landscape and so even more warming (something similar to what we mentioned a little earlier with polar sea ice). Finally, when all the white snow

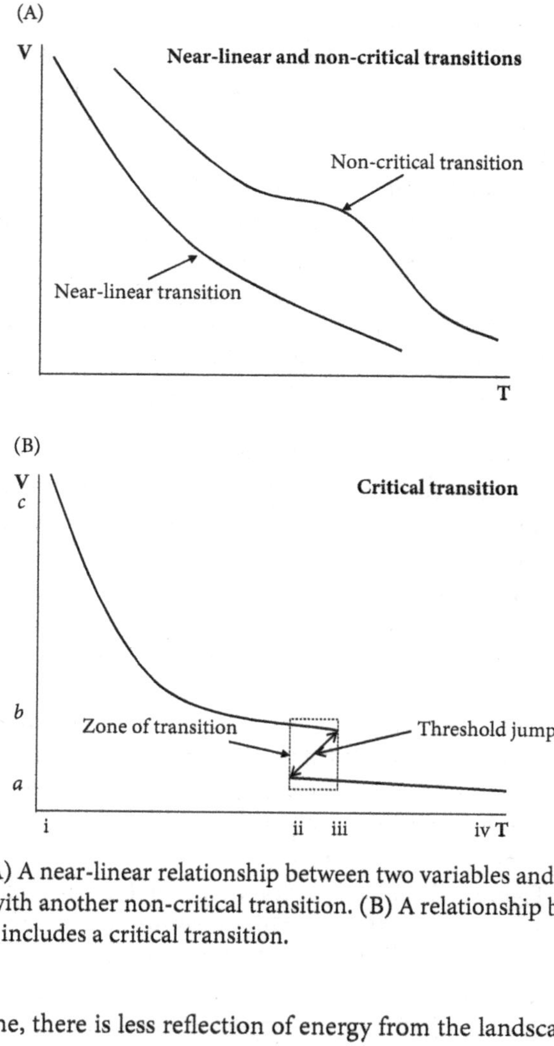


**Figure 2.4** (A) A near-linear relationship between two variables and also a relationship with another non-critical transition. (B) A relationship between two variables that includes a critical transition.

and ice is gone, there is less reflection of energy from the landscape and so the system is less prone to the retention of snow.

There are two important points about critical transitions. First, the transition between two states does not take place at a specific point but in a zone. (This is why the term 'tipping point' strictly speaking is not appropriate, as the system does not move about a single point: instead, there is a zone of transition.) Second, change is sudden in the zone of transition. (This is why some do use the term 'tipping point', as it conveys the notion of a large change in one variable for a very small change in another.) It is as if the system has two overlapping states: between the area encompassed by i–iii/b-c on Figure 2.4B and the area encompassed by ii–iv/a and the x-axis. These two states of the system within

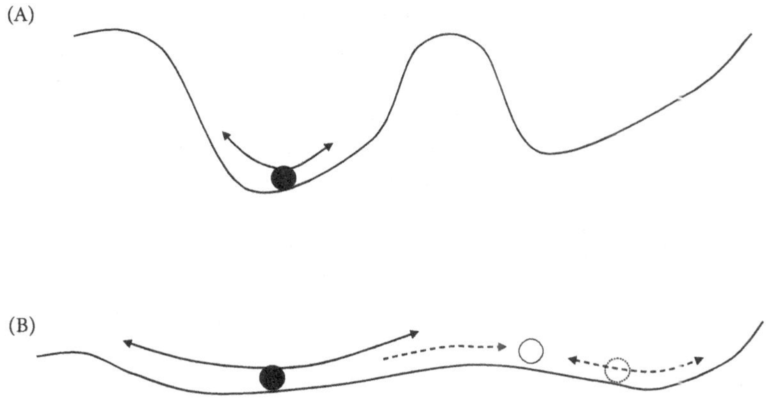


**Figure 2.5** (A) System far from crossing a critical threshold. Far from a critical transition the system only has small fluctuations and returns quickly to its central state. (B) A system nearer to crossing a critical threshold. Closer to a critical transition the fluctuations become greater and the system takes longer (gets slower) to return to the central position. Given enough impetus, the system can flip to an alternate (the neighbouring) state, crossing a critical transition, which is what the white ball above has done. The system might then stay in this new state (which is what the dotted ball has done). Alternatively, if given a sufficient impetus, it might 'flip' back to its original state. Much depends on the stability (the depth of the new valley) in which it finds itself.

these two areas might be considered analogous to areas of comparative stability respectively around an 'Lorenz attractor', in the parlance of 'chaos theory', outside of the transition zone. (And in Figure 2.5 the Lorenz attractor would be the areas of stability the ball tends to at the bottom of either of the two 'valleys' across the intervening 'hill'.) Leaving the mathematics aside, the key thing to note is that a system close to a critical transition can suddenly change state in a very marked way: the kettle boils dry; the drought dries the soil and it blows away in a dust storm; too many cars crowding into a city centre result in gridlock; a plane slows until it suddenly stalls; a person's heart becomes increasingly arrhythmic, finally resulting in a heart attack.

The earlier example of water in a fridge freezing to become an ice cube is an example of a critical transition. And you will recall that close to the transition there is a complication in the cooling as the water loses extra energy (latent heat). This demonstrates some of the complications that arise around the zone of transition in critical transitions.

An example of components of the Earth system thought from time to time to potentially exhibit critical transitions is the down-welling of water in the N.

Atlantic that helps draw warm water from further south north (this is the Meridional Atlantic Turnover). Other possibly forthcoming critical transitions that may occur if global warming continues to elevate the Earth's temperature for a few thousand years include Greenland and Antarctic icecap melt and also the SE Amazon ecosystem with its drought prone regions. The former may see Greenland lose its ice sheet (which would include an albedo-driven warming force on the Earth system) while the latter may see parts of the Amazon lose parts of its tropical rain forest (which would release a volume of carbon that would also serve to further warm the Earth system). We will return to the question of forthcoming critical transitions in Chapter 9 (see also Figure 9.2).

The thing about the Earth system is that many of its components, in a mathematical sense, are 'chaotic', and this led in the 1990s, when climate models were becoming more complex, to the development of chaos theory. To put it simply, the precise outcome of the system can vary considerably even though the system's initial conditions were similar. Try as hard as you can, when you start a game of billiards you cannot replicate exactly, or even approximately, where all the balls will end up from a previous attempt: hitting hard to break into all the other balls grouped together, the balls will end up in very different places when you try to repeat the process. This is because even though you may try to hit the standard billiard arrangement in the same way at the start of different games, each of your starting shots will be ever so slightly different and the subsequent collisions between balls become even more changed. And so the standard opening billiard ball arrangement will, after the first hard opening shot, end up in an unpredictable (chaotic) pattern. However, such systems do exhibit some clear mathematical patterns.

Given this, the question you may well have is whether it is possible to predict which changes in a system with chaotic properties are driving it towards a critical transition? In other words, can we somehow forecast when a potential forthcoming critical transition will take place?

Because systems capable of a critical transition do exhibit some mathematical patterns, the short answer is 'yes'. Figure 2.5 provides a pictorial illustration that will be explained.

In 2008 a team of German and Dutch researchers, led by Vasilis Dakos and Marten Scheffer (*15*) from the Netherlands and Germany, examined data from eight ancient and (geologically) abrupt climate change events. They showed that these abrupt events were all preceded by a characteristic slowing down of individual climate fluctuations' return to their mid-fluctuation state (the 'valley' midpoints in Figure 2.5), and that this slowing down started well before the actual shift across the threshold (into the second 'valley' in Figure 2.5B). By 'slowing down' they mean that approaching the transition zone and abnormal event (be it a drought or whatever) the system only slowly returns to the mid-point of the fluctuation or deviation (such as moderately wetter conditions or whatever): whether or not the intensity of individual fluctuations themselves decreases, it is the time within a fluctuation that counts.

If we can observe a trend in how long the system takes to return to its midpoint position (the 'valley' bottom in Figure 2.5A) then we might be able to predict the system approaching a possible critical transition (to leave the first 'valley' and move into the second, Figure 2.5B).

Another way to look at it pictorially is to consider something like a river valley. First, consider water flowing down the valley and think about where it goes. Up in the mountains where the river is more a large stream, the valley's walls are high and the valley floor is not wide. Lower down, the mountains are not so high and eventually become hills. High in the mountains the stream wriggles along the centre of the valley floor. Lower down among the hills, with valleys with more spacious, flatter bottoms, the river begins to meander. Finally, when it is on the wider plains it meanders considerably. Indeed, in the mountains a river in a neighbouring catchment is quite separate: there would literally be a mountain between the river and its neighbour. However, going downstream, as it too approaches the sea it might come close to the adjacent valley with its own river and eventually, on the coastal plain, might even merge with its neighbour. There are many examples of this in real life—such as the Ganges and Brahmaputra—that come together near the sea and actually share a common delta. And so it is possible for a boat going down one river to the delta to make their way to the other river without having to go to sea and come up a different estuary. This is graphically analogous, in a very loose sense, to crossing a critical threshold. (Though in chaos theory terms a Lorenz attractor diagram with the attractor representing the two states is a far more accurate graphic representation. See Appendix 2 if you are interested for starter detail/terms before exploring further on the internet.)

Let's look at this quasi-river analogy more closely. The watercourse, when a stream, high up in the mountains, as noted above, tends to stay in a tight region on the valley floor with steep sides to either side. Conversely, down on the broad flood plains the river in time meanders all over the place, even leaving oxbow lakes behind. And then it comes to the delta. Here the river can carve a channel almost anywhere in the delta, and after a time the channels it first carves silt up and a new channels elsewhere are created. Here in the delta, the river does not necessarily have its principal channel in the centre: its distance from the central line fluctuates. This distance is slower to return to the centre in the delta than it is higher up the river catchment. This propensity to return to the centre is defined, and can be quantified, by what mathematicians call 'autocorrelation'. Autocorrelation change prior to the crossing of a critical transition can be a mathematical diagnostic predictor of a forthcoming transition.

A third way of considering critical transitions is, instead of a river, to consider a ball in our 'valley'. High up in the mountains when our system is firmly in one state (one valley) a ball can only move tightly like the river. Lower down when the valley is broader and its bottom flatter, the ball is more free to move horizontally (as the river does with its meandering). Finally, near the river delta it shares with its neighbour, there are only the smallest of—if any—hills and the ball is almost free to move across from one river valley (state) to another (state), and indeed can, if given enough energy (a push), flip from one valley to the other. Here, the ball can be said to have crossed a critical transition. Just prior to this, the ball moving to either side of the valley moved slowly (more correctly, the ball moved further across the valley so taking more time): this is the slowing down to which Vasilis Dakos and Marten Scheffer refer (*15*). Higher up the catchment in the mountains the ball moves more quickly and not far from the stream. See Figure 2.5.

As said, such slowing down, measured as what mathematicians call increased 'autocorrelation', the researchers showed this to be a hallmark of climate critical transitions. The following year after Vasilis Dakos and Marten Scheffer's work, this question was reviewed in the journal *Nature* with the inclusion of other possible mathematical diagnostic uses (*16*). These range from ecosystems to financial markets as well as climate change to biomedical predictions: these are real-life potential applications and not some obscure mathematical curiosity.

The problem is that to make such forecasts of whether a system is about to undergo a change we need to see how the system behaves normally (when it is not near a transition or flip). For that, we need a long-term database so that it can be analysed for these autocorrelation and other diagnostic signals. But long-term research projects needed to obtain data over a considerable time are expensive. Though it is easier to predict a system about to undergo a change—as close to a flip some changes will already be taking place—to start monitoring a long time in advance when things will seem normal with only typical vagaries (as opposed to catastrophic ones), is difficult: not only is a long-term investment in monitoring required, but the decision has to be made as to what variables to monitor. Yet, the real value in prediction comes when we are farther in advance of the event: predicting who wins a general election a year in advance is much more useful and impressive than predicting it halfway through polling night, when a good number of the constituency results are already known.

There is a further complication, in that chaotic systems, such as the Earth system, are themselves extremely complex with many interacting subsystems and feedbacks. As previously noted, sometimes these individual interactions serve to make things more unstable and sometimes less so. Hence, again as discussed earlier, together these interactions within such complex systems tend to have islands (or ranges) of comparative stability within a sea (or other ranges) of instability. What this means is that such systems can see periods of calm (or stability) before flipping (or transitioning) quickly to a new, albeit possibly semi-stable, state when subjected to pressure.

As mentioned above, with Earth flips (our planet flipping from one state to another), one class of flip is of that between cold glacial times and warm interglacials (such as the one we are now in) within an ice age. As we shall see (Chapter 6) there are also flips, lesser known to the public, that go the other way with the Earth suddenly entering a very warm time (a hyperthermal), as opposed to a very cold (glacial), state. However even these flips between states are nothing compared to the changes that the Earth went through from what it was originally like, billions of years ago, when life first evolved, to the present, or other major changes in the Earth system due to biological innovations arising in the course of evolution.

As we shall also see, leading up to the rise of our species, there were two major flips in our planet's lifetime, both of which saw the Earth go through a major change of state as the planetary system crossed a critical threshold. (Again, discounting the change of state of the planet from one where it was determined by non-biological chemistry to one affected by life's chemistry as indicated by banded iron formations, or BIFs.) Both these major flips were non-reversible: the Earth never returned to the state it was in before each flip—it had evolved. Indeed, even the initial, primordial flip of the Earth's state from one where non-biological chemistry ruled, has never seen the planet return to this 'pristine' state.

There were other unique characteristics to these two major flips of state. Both were far greater than the Quaternary glacial-to-interglacial changes of the past 2 million years; far greater than the climatic upheaval following the asteroid impact that killed the dinosaurs; and far bigger that the hyperthermal episodes of planetary warming seen in the past few hundred million years and at some other times. (We will look at some of these comparatively minor flips causally unrelated to life's evolution in Chapter 6 and identify some characteristics that differentiate them from life-evolution-driven flips.) Both the big flips were associated with the biosphere developing in new ways, and, as we shall eventually see, they were also associated with changes in information processing . . . . But let's not get ahead of ourselves.

Following the shift of the planet's state from being one in purely chemical equilibrium to one with the equilibrium perturbed by the rise of life, the first of these big flips took place over a billion years ago or so after life began on Earth (depending on exactly when life first arose). And this flip was associated with the way life processed carbon and energy.

To understand this we therefore need to know how life processes carbon to provide energy.

## References

1. Kasting, J., & Ono, S., Palaeoclimates: The first two billion years. *Philosophical Transactions of the Royal Society B*, **361**, 917–929 (2006).
2. Intergovernmental Panel on Climate Change, *Climate Change 2013: The Physical Science Basis. Contribution of Working Group I to the Fifth Assessment Report of the Intergovernmental Panel on Climate Change.* Cambridge University Press: Cambridge, Great Britain (2013).
3. Xu. J., Chmela, V., Green. N. J., *et al.*, Selective prebiotic formation of RNA pyrimidine and DNA purine synthesis. *Nature*, **582**, 60–66 (2020).
4. Le Vay, K., & Mutschler, H., A plausible route to the first genetic alphabet. *Nature*, **582**, 33–34 (2020).
5. Lambeck, K., *The Earth's Variable Rotation: Geophysical Causes and Consequences.* Cambridge University Press: Cambridge, Great Britain (1980).
6. Lovelock, J. E., *Gaia: A New Look at Life on Earth.* Oxford University Press: Oxford, Great Britain (1979).
7. Lenton, T., & Watson, A., *Revolutions That Made the Earth.* Oxford University Press: Oxford, Great Britain (2011).
8. Lovelock, J., *The Ages of Gaia: A Biography of Our Living Earth.* Oxford University Press: Oxford, Great Britain (1988).
9. Milankovitch, M., *Théorie Mathématique des Phénomènes Thermiques Produits par la Radiation Solaire.* Gauthier-Villars: Paris, France (1920).
10. Hays, J. D., Imbrie, J., & Shackleton, N. J., Variations in the Earth's orbit: Pacemaker of the ice ages. *Science*, **194**, 1121–1132 (1976).
11. Burroughs, W. J., *Climate Change: A Multidisciplinary Approach.* Cambridge University Press: Cambridge, Great Britain (2001).
12. Dawkins, R., *The Blind Watchmaker.* Penguin: London, Great Britain (1986).
13. Lovelock, J., *Novocene: The Coming Age of Hyperintelligence.* Allen Lane: London, Great Britain (2019).
14. Intergovernmental Panel on Climate Change, *Climate Change 1995: The Science of Climate Change.* Cambridge: Cambridge University Press, Great Britain. (1995).
15. Dakos, V., Scheffer, M., van Nes, E. H., *et al.*, Slowing down as an early warning signal for abrupt climate change. *Proceedings of the National Academies of Sciences USA*, **105 (38)**, 14308–14312 (2008).
16. Scheffer, M., Bascompte, J., Brock, W. A., *et al.*, Early-warning signals for critical transitions. *Nature*, **461**, 53–59 (2009).

# 3
# Towards a Snowball

The two really big flips of the Earth's state saw times when our planet became largely (some say completely) covered in ice: the Earth became a 'snowball' (some say 'slushball' as there may have been open water and ice-free land near the equator). The idea explored in this book is that life's evolution effectively drove this process (though geology did greatly facilitate matters). So, to understand these flips of planetary state, we need to look at life.

## Prokaryote (the first cell type) evolution and early photosynthesis

Let us remind ourselves that this story of our planet flipping states is intrinsically bound up with our being carbon-based life forms; more accurately speaking, it is because *all* life on our planet is carbon-based (and not just us humans).

To see why, if not how, the Earth flipped between states, we need to appreciate how carbon-based life evolved beyond its origins. This is important because carbon-based life utilises energy to *process* carbon. As such, biological forms of carbon—even non-living biological forms of carbon—can themselves be energy sources. So it is likely that planets teeming with carbon-based life—such as the Earth—have reservoirs of energy in the form of biologically made carbon compounds.

Yes, this seems—and is—blindingly obvious, but it is perhaps so much so that we tend to overlook it. Yet, if you follow this line of reasoning to its logical conclusion then we arrive at 'us' (humans), carbon-based intelligence, seeking an energy source for our emerging technology eventually, if not inevitably, coming up with the idea of harnessing such bio-energy; first by burning wood and then fossil fuels.

Looked at this way, the industrial processing of carbon for energy could well be an inevitable consequence for any world that sees the rise of a technology-wielding sentient life form: on Earth it happens to be us humans. Therefore, it could very well be that the global warming problems we now have (caused by our evolved intelligence) itself has evolutionary roots that go back through the entire history of carbon-based life on our world.

*Earth's Flips of State*. Jonathan Cowie, Oxford University Press. © Oxford University Press (2026)
DOI: 10.1093/9780197825402.003.0003

Yet another, a completely different, way to look at it would be to say that life uses energy to generate information: a bacterium, plant or animal, indeed anything, can be defined as, and through, information.

A physicist might define this information as that coding for the spatial coordinates and types of atoms that make up the object (even a biological one). A biologist might define this information as the instructions on genetic code (RNA or DNA) that direct the manipulations of chemicals enabling a species to live, grow, and reproduce.

Using this perspective, as evolution has taken place, with organisms becoming more complex, so more information is processed and generated; a bacterium contains less, or requires less, information to define it than a human being. Even allowing for the mass difference, a huge colony of bacteria weighing the same as a human contains less information: it needs less information to define it even if that information is copied many times (with there being many bacteria in the colony).

We will return to this idea of processing carbon and information in Chapters 4 and 8, but for now will continue with the matter of how life obtains its energy and how this has changed over our planet's history. The contention here being that with each of Earth's flips of state an evolutionary leap took place involving a substantive increase in information as well as energy processing.

To appreciate this journey we must look at some of the Earth's flips in turn: certainly the two major ones were, according to the new narrative, driven by leaps in evolution, but also some were not (these last are explored in Chapter 6). Examining both types, we might then be in position to see if there are any characteristics enabling us to differentiate between fundamental (that is, driven by innovation in biological evolution) flips in the Earth state and those driven by wild-card, joker events. If so, we could even go further and see how our current (today's human-driven) extinction event and global warming (itself signalling a change in the Earth system) fits into the grand scheme of things.

But this is perhaps getting ahead of ourselves. Let's return to the beginning.

Before life's first cells, there were intermediate steps between what clearly were non-living chemicals and possibly living viruses. (Viruses simply being nucleic acid molecules surrounded by a coat of proteins, but as I mentioned earlier I am not going to get into the question as to whether or not we might consider them to be truly alive or not. This is more to do with our not having a clear definition as to what life is than any lack of understanding as to what is a virus. Indeed, one of the more humorous—but still with some validity—definitions of life is that one cannot define it but one knows it when one sees it. Whatever the debate, biologists are clear that cells are living and that it was with cellular life that the Earth first saw carbon [$^{12}C$] processing on a significant scale.)

Likely, somewhere in this primordial mix, there were self-replicating protein-like molecules vaguely akin to prions, and while these may have played a significant role very early on in life's rise they are unimportant to our narrative.

As discussed in Chapter 1, the earliest, and simplest of cells were the prokaryotic cells that were the ancestors of the prokaryotic bacterial Monera and the prokaryotic Archaea (archaebacteria as they were formerly known). The term 'prokaryote' comes from the Greek, with 'pro' meaning 'before' and 'karyote' meaning 'nut' or 'kernel'. Here the 'nut' or 'kernel' is used as a stand-in word for 'nucleus' (specifically a membrane-bound nucleus) which more evolved (eukaryotic) cells have (see Figure 4.1b cartoon sketch in the next chapter). The nucleus is the structure that contains eukaryotic cells' nucleic acids that carry its genetic information. Prokaryotes lack a nucleus. Figure 4.1a is a very simple depiction of an idealised prokaryote. Primitive archaebacteria such as *Methanopyrus* that live in hot water—as high as 100°c environments—are about 8–10 μm (micrometres) long and 0.5 μm across and are typical in size of most prokaryotes: they are small.

As a brief aside, unlike some prokaryotes, no eukaryotic species can thrive at such high temperatures. There is an argument that one reason for the delay in the rise of eukaryotes might be due to a hot-ocean early Earth, as noted in Chapter 1 (and to which we will return shortly).

As was previously noted when discussing the early biosphere, it is likely that early life drew on hydrogen and then hydrogen sulphide to secure energy for metabolism through chemosynthesis. But a more abundant and more intense energy source was needed, and light energy (though not necessarily *visible* light) was the key. Using the energy from light's photons it is possible to boost an electron's energy and so drive a chemical reaction to synthesise energy-giving organic compounds in a more effective way than chemosynthesis. This light-driven process is called 'photosynthesis'.

Key to the capture of energy from light in plants is a pigment called chlorophyll, and it plays a central role in photosynthesis. Photosynthetic bacteria have similar light-sensitive pigments called 'bacteriochlorophylls'.

We do not know exactly how photosynthesis became established as it all happened so long ago in Earth's primordial days and there are no surviving biological samples for us to examine. However, we can look at present-day photosynthesis in plants and present-day photosynthesis in a number of species of photosynthesising bacteria. Indeed, we have a good understanding of the chemical and energetic mechanisms involved. We can also look at the genes involved, and so we can deduce a likely narrative as to how matters might well have evolved.

Whatever photosynthetic mechanism one chooses to examine, they all in essence do the same thing: they use light to increase the energy of an electron

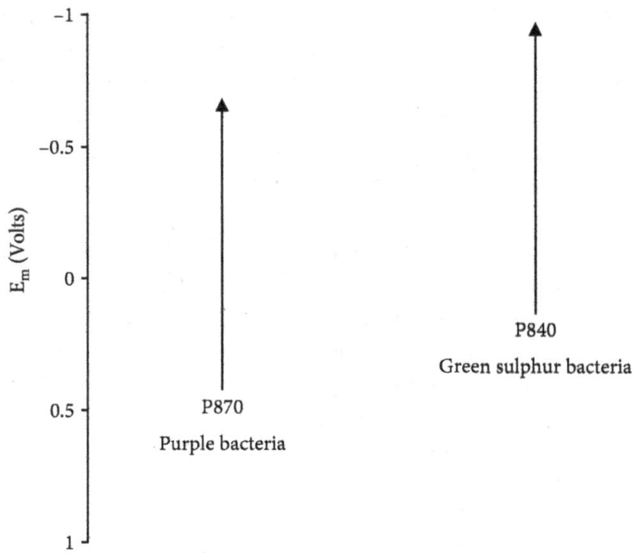


**Figure 3.1** Comparison of photosynthetic electron transfer in purple and green bacteria. The photosystem number is equal to the wavelength of light in nanometres (nm) that is absorbed by the bacteria's photosynthetic pigment (bacteriochlorophyll) in question: For example, here the purple bacteria's chlorophyll captures photons of light with a wavelength of 870 nanometres, which is in the infrared. (And in case some of you wondered, the energy gradient in biochemistry is a redox potential, with zero here being defined as the potential between two solutions of positive ions on one hand and negative ions on the other, at the same concentration. But don't worry about this too much: it is just a scale.)

so that the now energetic electron can then drive a chemical reaction. There are mechanisms in bacteria that use different photosynthetic pigments that absorb light of different wavelengths but the process is essentially the same. (See Figure 3.1.) Several photons of light are needed to energise four electrons across a voltage gradient to reduce one molecule of carbon dioxide.

Given that variations in global carbon cycling, as revealed by $^{13}C$ analysis (see Chapter 2), evolved early in Earth's history, it does suggest that the development of photosynthesis was an 'easy' evolutionary step: if it had been a difficult step then arguably it would have taken much more time for the necessary mechanisms to evolve.

Looking back at Figure 2.1 and how $^{13}C$ changed across time, there is a short vertical line of points at 3,800 million years ago (mya; 3.8 billion years ago [bya]) which some think as being indicative of photosynthesis: no other mechanism it is thought could processes that much carbon to give such isotopic variation. Conversely, there are only a few data points from this time and so the evidence is not as compelling as we would like. (Many of the graph's points, especially the middle and towards the right side, cannot be seen as they are stacked on top of one another, but to the left of the graph—way back in the Earth's history—there are fewer data points hidden beneath others.) Nonetheless, there is additional (some say circumstantial) evidence for early carbon processing by photosynthesis that comes from the 'banded iron formations' (often denoted by the abbreviation BIFs) of 3.8 bya mentioned earlier (Chapters 1 and 2). If photosynthesis had got going 3.8 bya then it would have arisen soon after the Late Heavy Bombardment (LHB) and that means that not only did it evolve quickly in the Earth's existence, it evolved *very* quickly (geologically speaking) after the LHB. (Incidentally, if this is so, it also has implications for the possibilities and likelihood of types of life on other worlds, including Mars.)

An alternative hypothesis is that photosynthesis evolved later around 3.0–3.1 bya. This was the view put forward by Michael Tribe, Andrew Morgan, and Peter Whittaker of Sussex University (Great Britain) and St. Patrick's College (Ireland) in 1981 in an Institute of Biology 'Studies in Biology' short-book series; a series very commonly used by British bioscience undergraduates between the mid-1960s and early 2000s (*1*). Then, as this book is now, the authors were cautious about being firmly wedded to a specific evolutionary timeline and so expressed caution using adverbs such as 'likely' and phrases such as 'postulated evolution'. However, then, like now, it was still recognised that photosynthesis evolved early (geologically speaking), and within several hundred million years from the end of the LHB.

If one does not accept that photosynthesis was behind the carbon processing revealed by isotopic evidence prior to 3.0–3.1 bya, then one has to assume that the magnitude of the very early, non-photosynthetic, carbon processing mechanisms (such as chemosynthesis) previously mentioned was sufficiently great, and isotopically discriminating, as to leave this marked isotopic record: this last is questionable. And so the debate among early-life specialists continues.

Returning to Figure 2.1, more varying $^{13}C/^{12}C$ data points are found 3.0–3.5 bya and it can far more firmly be assured that photosynthesis had become established then. Even if photosynthesis had become a global carbon processing factor later, in the great scheme of things it was still not that long after the Late Heavy Bombardment: half a billion years is not that long a time in grandest of early geological terms where the experimental error in dating a sample can sometimes be as great as millions of years.

Even if photosynthesis arose a little later, it was still not quite halfway down from the shoulder to elbow on our arm's-length scale of the Earth's lifetime, and can still be considered early in life's own history.

Given the probable ease of the fundamental photosynthetic evolutionary step, it is possible (if not likely) that (anaerobic) photosynthesis evolved a number of times. Certainly today we see a number of different types of bacterial photosynthesis. In addition to photosynthesis evolving comparatively early (quickly), that there are many species all told utilising a number of different photosynthetic systems, so photosynthesis likely arose a number of times, also suggests it was an 'easy' evolutionary step. (But note, this was anaerobic photosynthesis, and not oxygenic photosynthesis that we will come to later.)

If this multiple evolution of early photosynthesis was truly so, then there may have been an original ancestral pool of different species of photosynthesising bacteria with considerable gene transfer between them. These then speciated into the different phyla of photosynthesising bacteria whose descendants we see today. These include: the Chlorobiaceae (an anaerobic family of green bacteria); the Chromatiaceae (purple sulphur bacteria); Rhodospirillaceae (purple non-sulphur bacteria); and Heliobacteria. It is not necessary for you to remember these names, but do note that all are strictly anoxygenic bacteria: that is they photosynthesise without releasing oxygen, and none can use water as a substrate (unlike cyanobacteria and today's plants, more of which later).

So (anaerobic) photosynthesis came about comparatively early (even if we cannot pinpoint with surety exactly when). But with time, the Earth itself was changing: the oceans were cooling, the crust was thickening and, finally, plate tectonics became more established around 3.0–3.2 bya but with thicker continental plate crust not existing until at least 2.5–2.8 bya (see Chapter 1).

But at this early time, what was the Earth like for early life?

Work by Eric Gaucher, Sridhar Govindarajan, and Omjoy Ganesh in 2008 (2) elucidated the theoretical temperature-sensitive proteins that early bacteria may have used. These suggest an early Earth living environment in the seas where these bacteria lived of around 65°C–73°C and that this bacterial environment cooled over the next billion years or so to 51°C–53°C. Later that year a French and French Canadian team led by Bastien Boussau (3) had a paper published that looked at ribosomal RNA and protein sequences (similar to Gaucher and colleagues) from evolutionarily ancient but extant bacteria and, seeing how these differed between species, they created a model taking evolution backwards and deducing likely early proteins. This work suggests that the earliest forms of living bacterial cells may have thrived in an 'optimal growth temperature' of up to 50°C. As the bacterial species adapted, a second 'thermophilic' (warmth liking) phase of evolution then may have taken place in which the precursors to today's bacteria (and then later still Archaea and Eukaryota) thrived in a thermophilic environment of 50°C–80°C. This speculative, tentative theory has the benefit of reconciling the then previous, seemingly contradictory, perspectives from biology and geology, though it has other disadvantages.

That these early bacteria lived in a warm environment does not necessarily mean that all the oceans were that warm. It might have been (if not more likely) that these bacteria lived in water near to thermal vents and other large geothermal areas. Remember the very early Earth's crust was thinner, and plate tectonics had not been properly established (hence large continents had not developed) and so it is probable that there were extensive shallow seas that may have been geothermically warmer and certainly warmer in some places than others. Today thermophilic bacteria are common in hot springs and, popularly often associated with, deep-sea vents, but back then—just as today—deep-sea vents were not the only place where geothermal activity was common.

Nonetheless, as time progressed the early Earth changed and cooled. These changes were of the non-critical transition kind and the planet was not yet going through its first major critical transition or flip. (Actually, though major, this was the planet's second transition with the first being the Earth system gently shifting from one purely being in chemical equilibrium to one in which the equilibrium is perturbed by life.)

In 2006 French researchers François Robert and Marc Chaussidon (*4*), using $^{18}O/^{16}O$ and $^{30}Si/^{28}Si$ isotope analysis of cherts (a type of geological strata), deduced that the ambient temperature of sea water around 3.4 bya was, they suggest, roughly 70°C having possibly cooled roughly 10°C from nearly 80°C the previous 500 million years, and which subsequently further cooled to around 20°C about 800 mya later. So, even with such simple (given the paucity of data) analyses, it does seem as though there was a time early in the Earth's history when seawater might have been around 70°C–80°C and this fits with the Boussau team's idea that there was an early-life phase, living in an environment of 50°C–80°C. However, what this early hot ocean notion (should it be true) does provide is a way to make it easier for early life to leave the neighbourhood of volcanic vents and if there were anaerobic photosynthesisers then, then they could near the sea surface use the red wavelengths of sunlight and solar infrared heat before evolving ways to use visible light wavelengths. Nonetheless, the evidence is being challenged—as in all good science—and work to develop a firmer picture continues.

For example, one challenge came in 2009 when the US team of Hren, Tice, and Chamberlain used a combination of deuterium (see Chapter 2) and oxygen isotopes to analyse South African Buck Reef Cherts (a geological formation) from 3.42 bya in an attempt to get around some of the assumptions made by many earlier workers (including that the early Earth was essentially ice-free). They concluded that the sea temperature was 40°C or less (*5*). I mention this because there is (rightly) debate in the scientific literature, although of course it is important not to make the 'It rained on Mongo' mistake of assuming that the sea temperature everywhere in the primordial Earth was the same.

(Here a word of explanation is in order. 'It rained on Mongo' is writer and space engineer Jerry Pournelle's way of explaining the oversimplification poor science fiction authors sometimes use—describing planets where local or regional conditions are applied to the whole world. This is also a valuable lesson to be kept in mind by scientists when considering climate and environmental proxies: some environmental proxies and palaeo-indicators represent more local conditions than general planetary ones; so, does a palaeo-indicator suggest it is cold because it was from a cold place, or because the whole Earth was entering an ice age?)

Nonetheless, there is a general consensus in the academic literature that over 3 bya sea temperatures were significantly warmer than today and that there was subsequent cooling over a period of around a billion of years. Here the eagle-eyed among you will recall (see Chapter 2) that the Sun has been slowly increasing its output over its lifetime. (Currently, as it proceeds through its main sequence phase, the Sun's energy output is increasing at roughly 1% every 100 million years.) This then raises the question as to why it was warmer back in the early-Earth's times when the Sun was cooler? This question is known as the 'faint Sun paradox'. The answer likely lies in the higher concentration of atmospheric greenhouse gases back then (especially carbon dioxide and methane) and possibly there being more nitrogen in the atmosphere (nitrogen has been slowly leaving the atmosphere to accrue in minerals across geological time). A thicker atmosphere would keep the Earth warmer.

Yet while the Earth was not (yet) going through a critical transition (just non-critical change), a lot was going on on the biological front. Perhaps the best way to explain this is to provide a *likely* narrative of what *might* have been happening between 2.5 and 3.8 bya. This narrative includes what we do know with fair certainty but includes some speculation, especially with regards to exact timing (reader beware).

If we suppose that that (non-oxygen-generating) photosynthesis arose around or after 3.8 bya, as discussed, then—irrespective of the average ocean temperature—in all likelihood it arose around areas of high geothermal activity. These areas would provide the pre-photosynthesisers, from which the photosynthesisers evolved, that used hydrogen (which was likely more available in the early Earth with hydrogen yet to be out-gassed) and hydrogen sulphide as energy sources. These areas, being warm, would also have infrared heat radiation that could be used.

Now, we know that today some bacterial chlorophylls do absorb in the infrared and so it is not at all unreasonable to suppose that the first photosynthesis used infrared rather than visible light, given infrared was universally available in all thermal vents but not all thermal vents (unless near the surface) had visible light.

As mentioned in Chapter 2, the problem for any early infrared photosynthesiser was that there was not that much of the kind (the intensity and the frequency) of infrared light these prokaryotes needed except that found close to these geothermal vents or possibly near the ocean surface (within the first metre or so) due to sunlight in the tropics and seasonally in the summer outside of the tropics. This is enough for life to exist, but it is rather restrictive: geothermal vents, though more common in an early, thinner-crusted Earth than today, were still not that common as a proportion of the Earth's surface. The ocean surface is also restrictive: infrared light only has significant penetration within a metre or so depth of water (it has much less penetrating power than visible red light). Conversely, *all* the Earth's surface during the day was, as is now, bathed in more energy rich sunlight (visible light) and visible light (red through to ultraviolet) penetrates to 7.5 metres and over, even orange-through-to-blue frequencies penetrate over 15 metres depth: green to blue can penetrate to over 22 metres and greeny-blue can just reach over 37 metres, but high-frequency UV can only penetrate a couple of metres.

While deep-sea vents never see sunlight, life by vents deeper than just a metre or two under water would be protected from the Sun's harmful ultraviolet (UV): the higher frequencies of UV. (Remember there was no protective ozone layer in the oxygen-free atmosphere, even though there likely was some other, at least partially, protective mechanism: a tholin/methane haze as seen today on Saturn's moon Titan is a possibility—we will return to this in Chapter 7.) Early-Earth bacterial colonies might even have survived brief partial surface exposure such as at times of exceptional low tide. Yet if such bacteria were in waters shallower than, say, around 12 metres, they would (once visible light photosynthesis had evolved) still potentially get the light they needed to photosynthesise (though remember, red light cannot penetrate greater depths). Depths of less than somewhere around half a dozen metres would get the red part of the solar spectrum which the earliest bacterial photosynthesisers might well use if there was an evolutionary drift from the infrared provided by thermal vents. (Blue light penetrates more deeply into the water column, even up to around 37 metres.)

This spectral gradient of infrared penetrating water less, and blue light further, is interesting because such gradients facilitate evolution. We see this today with the rise of antibiotic resistance. Go to a place where there are gradients of antibiotic concentrations and antibiotic resistance will evolve. In hospitals, there are places where ambient antibiotic concentrations are low, so enabling some bacteria to survive as a low-level antibiotic resistant strain. The hardiest of these can then go on to colonise a place in the hospital where the antibiotic concentration is a little stronger, and so the process repeats until full-blown antibiotic resistance results.

In a similar way, in vents near the surface, early photosynthesisers could have evolved from having to rely on the comparatively little infrared from a few vents to the plentiful and more energetic (in the frequency sense) red light bathing the tropical daytime planet where seas were shallow enough to permit them to grow. Those earliest mutants (the new photosynthesisers) that could use this more plentiful sunlight resource could thrive; those that could not remained confined to the thermal vents. Today the ancestors of this latter, vent-confined group can be found in deep-sea vents as the oxygen at the surface of our current Earth kills them: remember, back then there was no (or only the very faintest of traces of) free oxygen in the atmosphere. Therefore deep-sea vents are today a refuge for the ancestors of some of the early bacteria species.

As noted above, that there are many types of bacterial photosynthesis, employing different photosynthetic pigments and energising electrons across different parts of the redox potential gradient. (For now, consider this gradient as one in which moving between two points on it either requires, or gives out, energy.) This raises the question as to whether they arose from multiple ancestors—the primordial ancestors to all different types of photosynthesising life—or whether there was a single ancestor that then evolved different systems with different bacterial species?

Alas, we have no definitive answer to this question. On balance though, as alluded to earlier, given that life and photosynthesis arose so early in the Earth's history, there is still merit to the notion of (anaerobic) photosynthesis being 'easy' to evolve. Irrespective of whether it arose independently a number of times, prokaryotic gene transfer would nonetheless further facilitate variations.

And so matters, very broadly speaking, continued for hundreds of millions of years.

To a casual, fictional, alien astronaut, or human time traveller, to the Earth over these millions of years, billions of years ago, it might seem that there was little change. Of course the Earth and life on Earth was not static. For instance, as noted in Chapter 1, plate tectonic movement properly began somewhere around, if not before, 3 bya with thicker continental crust. And there were other key developments too, life was changing.

Due to the lack of space in this book, and because it requires much hypothetical theorising, it is not possible to detail all the changes that went on. But, one of them was the evolution of 'light harvesting complexes' in cyanobacteria. These not only control the way light photons reach the photosystems of photosynthesis but help capture other, higher-energy (higher-frequency) light. Life was beginning to be able to capture more light energy.

## The evolution of oxygenic photosynthesis

Somewhere around 2.7 bya (I contend much earlier, a little before 2.9 bya and possibly even earlier, before 3.1 bya—also see Figures 4.3 and 8.8) there was a major development in photosynthesis. Prior to then photosynthesis did not result in oxygen. Yet oxygen oxidising hydrocarbons and carbohydrates through respiration produces a lot of energy. For example, today you oxidise hydrocarbons using oxygen to power your car (petrol is a hydrocarbon mix you burn). You know yourself that when you ignite petrol in our modern, oxygen-rich atmosphere, the energy release is considerable. If life could oxidise (grab electrons—see Chapter 2) using oxygen, then it would have a considerable energy source.

The problem for early life (had they had a conscious thought about the matter, which they did not) was that there was no free oxygen to use. The problem here is that energy can neither be created nor destroyed, just transformed. To create oxygen out of the chemicals available back in the Archaean (see Appendix 2), oxygenic (oxygen generating) photosynthesis would need to energise an electron across far more of the redox potential gradient than bacterial photosynthesis had ever managed.

However, biology found a way.

The early bacterial anaerobic photosynthesis of the anaerobic Earth could not boost an electron's energy to the high levels needed to split water ($H_2O$) into its components of hydrogen and oxygen: the energy needed is equivalent to that of burning hydrogen (in our present-day, oxygenated atmosphere), which is the reverse chemical reaction to water splitting. Indeed, as mentioned in the previous chapter, you may have done this reaction at school, with the ignited hydrogen gas going pop, or seen old news reels of the hydrogen-filled Zeppelin conflagration tragedy: the energy released from burning hydrogen to form water is considerable.

Coming up with a brand new method of photosynthesis—evolving oxygenic (oxygen generating) photosynthesis (OP)—to span this greater energy gradient would be difficult for evolution. You might think that this might mean going back to the drawing board and—instead of evolving the comparatively easy anaerobic photosynthesis—creating from a standing start something far more powerful. However there is an easier evolutionary way: the adding, or if you will 'bolting', together pre-existing photosynthetic mechanisms to make a supercharged system. This easier route to oxygenic photosynthesis is what happened.

Once life on the Earth had had the best part of around a billion years of various anaerobic photosynthetic systems that boosted electrons a small way up the redox potential gradient. All it needed was to add two of these existing systems together to boost electrons further up the energy gradient, far enough so that it could split water to release oxygen.

Cyanobacteria (or the ancestors of cyanobacteria) were the first to do this, and all subsequent OP sprang from this development: unlike anaerobic photosynthesis, oxygenic photosynthesis can be traced back to a single origin.

You will recall a little earlier the mention that cyanobacteria evolved 'light harvesting complexes' and so they were already beginning to, literally, harvest more light energy than other photosynthesising prokaryotes.

Now, before we go any further (and in case you decide to read around the topic), what you need to know to avoid misunderstanding is that an alternative name biologists have for cyanobacteria is 'blue-green algae'. Calling cyanobacteria 'algae' can, and does, cause confusion. True algae—which are not cyanobacteria—are traditionally associated with the plant kingdom (Plantae), and this last is the sense the term 'algae' is used for the remainder of this book: to make the distinction clear, I will not call cyanobacteria 'blue-green algae'.

This bolting of two different types (Figure 3.1) of non-oxygenic photosynthetic systems together (Figure 3.2C) can boost electrons' energies to the necessary degree to split water and release oxygen. This bolting together provides a redox span of 1.25 volts (see Figure 3.2B).

What happens is this: through photosynthetic system II (PSII) light boosts electrons' energy levels. These electrons are then rapidly passed by a series of molecules called cytochromes to photosynthetic system I or PSI (see Figure 3.2C), which then further boosts the electrons' energy. If you like, it is a bit like preparing the perfect roast potato with a soft fluffy centre and a crisp coating. First there is the par boiling of a potato in a pan of water which softens the potato. Then to get the crisp coating the cook removes the hot potato from the pan and quickly passing it from one hand to another (so as to avoid burning their hands) places it in the oven for roasting. In this crude analogy, PSII is the boiling pan, the cook's hands are the series of cytochromes (which are special big molecules), and PSI is the oven.

Now, you might think that one photon of light stimulating photosynthesis results in the combination of one molecule of carbon dioxide ($CO_2$) with one of water ($H_2O$). Actually several (typically eight or more photons) are required. Not only is the process inefficient (few things are perfectly efficient due to the laws of thermodynamics) but also four electrons are required (see equations 'i' and 'ii' in Figure 3.2A). The overall reaction is given in Figure 3.2A equation 'iii'.

Having created this carbohydrate fuel, early cyanobacteria could then 'burn' (through a process called respiration) the carbohydrate using the oxygen to release far more energy than other metabolic processes that did not use oxygen as the oxidising agent. This is the reverse of reaction 'iii' in Figure 3.2A and it releases back the energy (allowing for inefficiencies) of the light used

(A)

$$i)\quad 2H_2O \longrightarrow O_2 + 4H^+ + 4e^-$$

$$ii)\quad CO_2 + 4H^+ + 4e^- \longrightarrow \underset{\text{carbohydrate}}{(CH_2O)} + H_2O$$

$$iii)\quad CO_2 + 2H_2O \longrightarrow O_2 + (CH_2O)$$

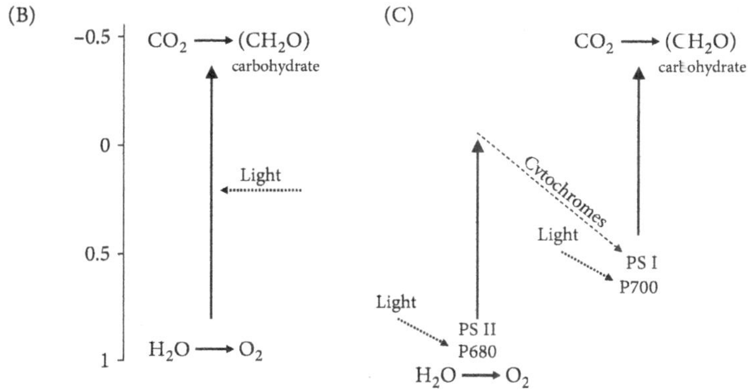


**Figure 3.2** (A) The summary chemical reactions for oxygenic photosynthesis: (*i*) splitting water, (*ii*) reducing carbon dioxide, and (*iii*) the overall reaction. (B) The total redox potential required. (C) The bolting on of two photosystems—PSI (P700) using light with a wavelength of 700 nanometres and PSII (P680) using light with a wavelength of 680 nanometres—to the sufficient redox potential for oxygenic photosynthesis.

in photosynthesis, which the organism can use to drive the biochemical reactions of life. Because more energy is harnessed and then used through oxygen metabolism (than non-oxygen or anaerobic metabolism), it gave these early cyanobacteria an advantage over other species that did not use OP. On the other hand, as OP requires light of a higher energy (frequency) than prokaryotic non-oxygenic photosynthesisers, more energy is required to split a molecule of water than the other molecules used in anaerobic metabolism. Consequently, in addition to bolting together two photosystems, oxygenic photosynthesisers need to use higher-frequency (hence higher-energy, as mentioned in Chapter 2) light in the visible part of the spectrum compared to the lower-frequency, and lower-energy infrared light (or heat radiation) non-oxygenic photosynthesisers used in places of geothermal activity. (Therefore, non-oxygenic photosynthesisers have bacteriochlorophylls that absorb in 800–950 nanometres wavelengths and 1,000–1,150 nanometres, whereas cyanobacteria and the much more subsequent plants use visible light with photosynthetic pigment absorption peaks in the lower-wavelength and higher-frequency visible wavelength range between 400 and 700 nanometres.)

One can speculate parts of the scenario as to how this use of visible light probably came about. The precursors and precursor cousins of cyanobacteria, employing non-oxygenic (or anaerobic) photosynthesis using heat (infrared light) and living near shallow areas of geothermal activity, over generations threw up variations of their chlorophylls (called bacteriochlorophylls) through natural mutation. Most of these mutant chlorophylls did not work (or did not work so well) and the bacteria using these died out. Nonetheless, some visible light-harvesting pigments worked; some very well. Some mutants continued to use their existing, working bacteriochlorophylls alongside the successful new chlorophylls. If the new mutant chlorophylls (and light-sensitive pigments in the light-harvesting complex) could use visible light (and thermal vents in shallow waters could do so) then the photosynthesiser had a competition advantage over its compatriots only accessing infrared radiation and these were confined solely to the locality of the geothermal activity.

Finally, one prokaryote acquired two systems using visible light. This could have happened a number of ways. One of the most likely may have been two bacteria using different photosystems engaging in horizontal gene transfer (a way bacteria can exchange genes) and one ended up with the genes for two different photosystems that, with further mutations, generations later, became bolted together.

Either way, this new photosynthetic mutant not only was freed from the constraint of having to live near the geothermal activity (it could now use sunlight), but it could also split water and so create, and then use, oxygen. This in turn could react with carbohydrates (through respiration) to release more energy than is available to its anaerobic bacteria precursors and evolutionary cousins.

Fossil evidence suggests the possible existence of organisms somewhat similar to modern-day cyanobacteria (that engage in oxygenic photosynthesis) possibly arose as early as 3.5 bya, but it is unlikely that oxygenic photosynthesis (OP) itself developed quite that early with the rise of true cyanobacteria. Yet, almost certainly oxygenic photosynthesis had already arisen by at least somewhere around 2.7 bya. The period—~300 my long—following 2.7 bya saw considerable carbon drawdown. (See Figure 2.1.) So oxygenic photosynthesis most likely arose somewhere around this time (even if it could have been a little later: there is some debate).

My personal take is that OP arose (well?) before 2.9 bya, but by how much I am uncertain. There certainly is isotopic evidence suggestive of (but not conclusively proving) OP at 2.9 bya (6). There was also a Pongola glaciation around this time. This leads to the speculation that OP evolved before 2.9 bya and that there was significant carbon dioxide drawdown to cause this glaciation. Tim Lenton and Andrew Watson in *Revolutions That Made the Earth* (2011) identify a possible scenario of OP arising by 2.9 bya that drew down carbon dioxide possibly so much that its concentration equalled that of methane so producing a stronger, upper-atmosphere organic (tholin) haze. Both the upper atmosphere haze and the carbon drawdown, and I would add together with oxygen converting the methane to carbon dioxide and water, would have both served to cool the planet. Subsequently, with so much organic carbon sediment accumulation, methanogens would have released methane, so re-warming the planet (7).

Irrespective of the exact timing of the rise of OP (which is still debatable), this bolting together of two photosystems is now the respected hypothesis, the exact finer details as to how oxygenic photosynthesis arose are not known, though there are some notions that researchers discuss (which are beyond the scope of this book). Nonetheless, what you might want to consider is that the evidence for oxygenic photosynthesis 3.5 bya (such as $^{13}C$ variations 3.5 bya that could have other explanations) is less convincing than later evidence, and that as we move forward in the Earth's history from there the body of evidence mounts. This later evidence includes that of microfossils, further and more extensive $^{13}C$ variations, the return of banded iron formations—BIFs—which admittedly could have been produced by anaerobic photosynthesis, and various biochemicals associated (some weakly, some strongly) with aerobic photosynthesis.

## Oxygen in the atmosphere

Oxygen need not be produced by life; it can be generated by purely abiotic, chemical means, such as the splitting of water molecules in the atmosphere by ultraviolet light. However, such production of oxygen is minimal, and 3 billion years ago in the Archaean any oxygen entering the reducing atmosphere would be quickly mopped up by the methane (resulting in carbon dioxide and water). So any free oxygen back then could only exist in very low concentrations, measured in parts *per* trillion: the faintest of whiffs. Today 21% of our atmosphere is oxygen; not so back then around 3 bya and earlier.

Oxygen has (as previously noted in both Chapters 1 and 2) a propensity to want to *grab* electrons when reacting with other elements (or, again as previously mentioned, oxidising them, as chemists like to say about elements and compounds that attract electrons from other elements and compounds), and so, also as mentioned earlier, iron rusts in the modern atmosphere to form iron oxide.

Equally, today's oxidising atmosphere sees just low concentrations of reducing compounds; compounds like methane that have a propensity to *give* electrons when they react with other chemicals. Today, compounds like methane have an atmospheric concentration of just a couple of parts *per* million. Today methane is a minor component of the atmosphere and oxygen a significant one: back in, and before, the Archaean matters were reversed. And today the reason methane can still be measured in parts *per* million, and not parts *per* trillion, is that life—such as methanogenic bacteria that were far more abundant in the Archaean and earlier—still today generates methane. However such 'methanogens' (methane generating bacteria) today are restricted to those habitats that are devoid of oxygen: organic rich sediments beneath stagnant water, in the guts of ruminants, within deep sea sediments, and so forth.

So prior to the evolution of oxygenic photosynthesis, when matters were reversed, atmospheric oxygen could have at best been measured in parts *per* trillion: initially, but with the initial evolution of OP it would have been measured in parts *per* billion. This was still but the faintest of faint whiffs.

Obtaining evidence for the faint whiff of oxygen in the Archaean atmosphere is understandably difficult; it was a long time ago and a faint whiff of oxygen does not leave much (if any) of a long-standing mark in the geological record. However, looking hard, there is some evidence to be found. For example, the remains of ancient soils (palaeosols) from 2.2 billion years ago show no iron loss: the iron had been fixed by oxidation. This indicates that there was enough oxygen at that time to oxidise the iron and so by somewhere around 2.2 bya oxygen in the atmosphere could possibly have been measured in parts *per* billion. Indeed, a little earlier, from 2.32 bya oxidised iron (rust) exists in the form of clearly visible 'red beds' (*7*).

Make no mistake, the jump from parts *per* trillion to parts *per* billion is a big one, it being of two or three orders of magnitude: that is a leap in concentration of hundreds to thousands of times greater than before. Such an increase in oxygen would begin to mop up methane. This jump, was driven by oxygenic photosynthesis being firmly established in the biosphere and we have evidence for this from what is called the 'mass independent fractionation' (MIF) of sulphur isotopes.

Queue another perspective . . .

Like carbon, which has a number of isotopes, so has sulphur. Just as carbon's most common isotope is $^{12}C$, with $^{13}C$ being one of its rare isotopes, so, in addition to sulphur's common $^{32}S$, there are minority isotopes $^{33}S$ and $^{34}S$.

Normally the various isotopes of sulphur are processed by the Earth's elemental cycles in an orderly way so that the ratio between sulphur isotopes is constant. However, this can be disrupted with novel reactions of sulphur compounds in the presence of intense ultraviolet light.

Today the Earth's surface is protected from the worst of ultraviolet (UV) light by ozone ($O_3$) in the atmosphere. The ozone itself is formed high up in the atmosphere from molecules of oxygen ($O_2$) intercepting UV light, so generating radicals of oxygen (O) that then react with oxygen molecules to form ozone ($O_3$). However, prior to around 2.6 bya there was virtually no free atmospheric oxygen and the surface of the Earth was shielded from UV by other mechanisms.

Back in most of the Archaean (2.6–4 billion years ago), and all the time since the Late Heavy Bombardment, it is thought that the surface of the Earth was partially protected from the Sun's UV light possibly by a brown haze of organic (in this case just hydrogen and carbon) compounds high in the then methane-rich atmosphere. This was possibly a bit like the orange-brown tholin haze seen high in the atmosphere of Titan (briefly mentioned earlier). The methane absorbing the UV, as well as the tholins themselves, shielded the Earth's surface from much of the harmful UV light. (We shall return to Titan in Chapter 7.)

Looking at geological strata across time, we can see that strata that are younger than 2.2 bya have no variation of their sulphur isotopes' proportions. Yet for a short while (in the geological time sense) up to this time there is variation and a transitory MIF of sulphur signal, but none before 2.6 bya (*7*).

This signal is clear and so suggests that for a while (somewhere around 2.6 and 2.4 bya) some high-energy UV was getting through the atmosphere and disrupting the orderly cycling of sulphur isotopes. In other words, some time before or around (the dating of events so long ago is not exact) 2.2 bya there was not sufficient oxygen for a protecting ozone layer as there was after 2.2 bya and up to the present.

Also, after 2.6 bya there was no protective brown haze of organic compounds formed from atmospheric methane and carbon dioxide.

Indeed, for a period between 2.6 and 2.4 bya there was a really strong MIF of sulphur signal. This suggests a brief (in the grandest of geological terms sense) period where there was no UV protection of either kind: no ozone layer and no tholin haze (*8*, *9*).

To sum up, what seems to have been happening was that prior to 2.6 bya there was some partial UV protection given by a high-altitude (tholin) haze in the methane atmosphere. Then from around 2.6 bya there was no protection from this high-altitude haze of organic compounds in the atmosphere (so giving a strong MIF of sulphur signal). After 2.2 bya there is no MIF signal and there are clear signs of low-level oxygen, so suggesting the reestablishment of UV protection but this time through the new ozone layer generated from the new excess of oxygen.

So the notion is that this flip of the Earth's state—the sudden jump in oxygen concentration from parts *per* trillion to parts *per* billion—took place around 2.6 and 2.2 bya. Prior to 2.6 bya there was not enough oxygen being generated to mop up the methane responsible for the organic compound haze conferring UV protection. Yet after 2.2 bya there was enough oxygen for a fully UV-protecting ozone layer. But between 2.4 and 2.6 bya the initial surge in oxygen was being mopped up by the organic compounds in the atmosphere which itself was destroyed by the said oxygen. During this time there was little protection from the haze and not yet enough oxygen to confer the protection from an ozone layer, and so high-energy UV light penetrated right through the atmosphere to the surface, disrupting sulphur's isotope fractionation and leaving a high MIF signal.

After 2.2 bya, oxygen generation—mopping up much of the organic methane generation—continued, allowing oxygen to build up to a couple of per cent of the atmosphere.

This alteration in atmospheric composition 2.2 bya is known as the Great Oxidation Event. It is also known in various other ways as, the Oxygen Catastrophe, Oxygen Crisis, Oxygen Holocaust, Oxygen Revolution, or just the Great Oxidation. So if you want to learn as to how thinking on this has changed with time, as well as other theories (including those that overlap with the narrative presented here), you will need to search the literature on all these terms.

## Summary of the evolution of bacterial photosynthesis

Bringing all this together, in brief, we have the following plausible scenario. Life got going not long (in geological terms) after the Late Heavy Bombardment (if not hypothetically earlier). Again, not long after this (again geologically speaking and so many millions of years later), early bacterial and anaerobic (non-oxygenic) photosynthesis arose. Rising so early in the Earth's history, anaerobic photosynthesis must have been a comparatively easy evolutionary step to make and so likely arose a number of times. Even if it did not arise a number of times, mutations led to variation that resulted in a number of different bacterial species employing different photosystems. Either way, this meant that there were multiple photosynthetic species with different photosystems over 3 bya.

Early bacterial photosynthesis could have existed by marine thermal vents using infrared light. Those vents near the surface would not be far from sunlight and so bacteria carried away from the vents to the nearby sunlight would have a potential alternative source of energy to draw on: visible light. The warm early oceans would have also facilitated these thermophilic bacteria's survival. Eventually, after several hundreds of millions of years, gene transfer enabled an early species of cyanobacteria (or precursor thereof) to bolt together two photosystems and so boost an electron sufficiently far up the redox energy gradient to split water and release oxygen: the process of oxygenic photosynthesis had evolved! These cyanobacteria, possibly as early as 3 bya, could harness more of the electromagnetic spectrum than the other non-oxygenic photosynthesisers.

Translating this into a more of a biological picture (as opposed to a more geochemical one involving BIFs and MIFS) is harder, but it is not impossible to come up with a narrative that is both biologically plausible and which fits in with the geological evidence as we currently know it.

Though arising in terrestrial thermal springs, later early prokaryotes probably also lived around marine thermal vents using chemosynthesis Some then became photosynthetic using infrared energy from the said thermal vents. In turn, some of those that then got swept away by currents towards the surface began using sunlight formed colonies in the shallows. Or—as we shall return to in due course—some thermal vents were in shallow enough water to receive sunlight, an alternative, richer energy source to the lower-energy infrared coming from the thermal vents. Today we see similar bacterial colony mats in the shallows along some coasts. Fossilised versions of these are called stromatolites. It is likely that the first stromatolites consisted of prokaryotes that included anaerobic (non-oxygenic) photosynthesisers and they did not release any oxygen directly to the atmosphere as it was used *in situ* or was mopped up by the reducing environment.

When oxygenic photosynthesis did evolve in the ancestors of cyanobacteria, other bacteria used the oxygen they generated as part of a new, high-energy metabolism within the microbial mat. Any oxygen that did escape these early oxygenic photosynthesisers would have been toxic to the other strictly anaerobic bacteria of the time that could not harness the oxygen. This, as we shall shortly see, would not matter prior to 2.6 bya. However, it would not take long for some prokaryotes to mutate and survive so as to take advantage of the high-energy metabolic prospects now available.

That bacteria can adapt to toxins is evident today with biomedical scientists' on-going battle with antibiotic resistance (which was touched on earlier). These oxygen respiring prokaryotes within stromatolites would have a symbiotic relationship with the oxygenic photosynthesising cyanobacteria, and so it is likely that at first at least little oxygen would escape from algal mats into the atmosphere. Furthermore, any oxygen that did escape would be mopped up by the atmosphere's methane that was back then a significant atmospheric component. (Oxygen would also have been mopped up oxidising the until-then chemically reduced parts of the Earth's surface.) This, reasonably effective containment of oxygen to within mats and any mopping up, explains the delay between the rise of oxygenic photosynthesising bacteria at the latest by 2.7 billion years ago (I suggest by around 3.1 bya or possibly even earlier) and the first significant (but still small) amount of oxygen in the atmosphere somewhere after 2.2 bya.

Once again, it is important to note the caveat terms in the preceding sentence: 'at the latest by' and 'somewhere'. Tim Lenton and Andrew Watson in their book *Revolutions That Made the Earth* (2011) suggest oxygenic photosynthesis arose somewhere around 2.9 bya (7), conversely, as I previously noted, I suspect it might have been just a little earlier, and other scientists have their own opinions. This does not mean that we are all doing bad science but that these issues are at the cutting edge of science and we would all like more evidence so as to devise improved scenarios. Nonetheless, most agree that there was sufficient oxygen being generated by 2.7 bya to start successful mopping up of much atmospheric methane. The tables were turning—atmospheric methane was losing and oxygen winning—and so we get the start of MIFs of sulphur. In short, oxygenic photosynthesis must have arisen *before* 2.7 bya.

After this, as time progressed more and more oxygen was produced and the Earth flipped state from a highly reducing atmosphere to one with somewhere around just a couple of per cent of oxygen from around 2 bya or a little earlier. This oxygen concentration was still thousands of times more than it had been earlier, prior to the earliest (I suspect somewhere around 3.1 bya if not before) spreading of oxygenic photosynthesisers. As previously noted, prior to this time the atmosphere contained but the faintest of traces of oxygen, yet after 2.2 it was a couple of per cent oxygen. (This oxygenation prior to 2.2 bya is well established.) By comparison, today the Earth's atmosphere is close to 21% oxygen, and so the Earth of 2 bya still needed to see a ten-or-more-fold increase in oxygen concentration for atmospheric oxygen from a couple of per cent to reach today's levels. That was to happen much later.

## The first big flip

So, the Earth's first dramatic change of state took place before 2.2 bya from a world whose atmosphere lacked oxygen to one where there was free oxygen present at the level of a couple of per cent. This was a critical transition (see Chapter 2), and the Earth system oscillated as the threshold was crossed: this was not an instantaneous change and the system wavered around the critical threshold. Clearly, because this flip was from a world (almost) without oxygen to one with some (albeit a little) in the atmosphere, the evolution of oxygenic photosynthesis must have been involved. And while we cannot be certain as to when exactly oxygenic photosynthesis did evolve, it obviously must have been before this transition began to take place: that is to say, it must have been before 2.9 bya and the Pongola precursor glaciation of Snowball I. Pongola was a major glaciation before the 2.4–2.2 bya Snowball I and it is likely carbon drawdown from oxygenic photosynthesis was involved. If so, OP must have evolved before then.

Because Earth sports carbon-based life, and oxygenic photosynthesis was a new, high-powered way of capturing carbon, this first flip can be clearly seen in the $^{13}C$ to $^{12}C$ record. Indeed, we can see a very marked spread of $^{13}C$ to $^{12}C$ data points in the record between 2.9 and 1.9 bya (see Figure 2.1): there were oscillations in the Earth's carbon cycling and this was due to the Earth system wavering as it crossed a critical threshold and shifted between states.

As noted, around this time conditions on the Earth changed markedly. Before this flip the anaerobes ruled; after the flip, with oxygen being toxic to many of them, they were forced to retreat to where there was little or no oxygen. Actually, there was probably little retreating: much anaerobic life must have died *in situ* with only those that happened to be separated from the new atmospheric oxygen surviving.

But there were other changes too and, according to this perspective of life and planet, it is no coincidence that this period saw considerable climate change and even Earth's Snowball I.

With oxygenic photosynthesis (OP) using sunlight, a number of things happened. The first was that OP cyanobacteria could out-compete their non-OP counterparts as they had access to a high-energy source and a metabolism to use it, while the non-OP species could only access lower-energy infrared and had a low-energy metabolism, not to mention finding oxygen poisonous. The second was that, provided they had nutrients in the sea, cyanobacteria-like OPs could colonise much of the sea's surface. Therefore some of the Earth's biomass changed: some non-OP biomass died off, being replaced by OP species. Secondly, it is very likely if not inevitable that the total sum of the Earth's biomass grew. But we need to be mindful that while the Earth's living biomass grew, it was still far off what it is today with the modern abundance of multicellular life (which back then had yet to evolve) and the mass colonisation of land.

Given all this you might be puzzled as to why the $^{13}C$ to $^{12}C$ record between 2.9 and 1.9 bya (see Figure 2.1) sees a scatter of points both above and below the long-term Earth average: surely if the Earth was seeing an increased biomass (even if it had not grown to the size of the present day) then the data points would all be on one side of this long-term average line given that the net effect of a larger biomass would be for carbon drawdown. Also how, according to Figure 2.1, could there be both carbon drawdown from, and carbon release to, the atmosphere both at the same time? Here we need to bear three things in mind.

First, because the Earth system was going through a critical transition, as we discussed earlier, the previous state was disrupted and so it oscillated back and forth before finally flipping. Second, the data points in Figure 2.1, especially to the left of the graph billions of years ago, have a large dating error. Remember, a dating error of just a couple of per cent in a billion years is an error of 20 million years. In fact if you look up the original source of the data you will see that the dating error for a number of the points from strata around 2 billion years ago was as much as 10%, covering a span of 200 million years during which a lot can happen especially with an oscillating Earth system. Consequently, it is not surprising, especially with thousands of data points, that many are not precisely dated so creating the illusion that overall the biosphere was seeing a net drawdown of $^{12}C$ from, as well as a release to, the atmosphere seemingly at the same time. Third, though an actively increasing in size global biomass would see increased carbon drawdown, the $^{13}C$ to $^{12}C$ record is actually a measure of carbon flow (draw down from, and release to, the atmosphere) and not biomass in itself.

Net carbon drawdown from the atmosphere, due to a larger-than-before biomass now photosynthesising using the new, improved oxygenic mechanism, resulted in less of the carbon dioxide greenhouse gas in the atmosphere and the decline of the potent greenhouse gas methane due to it being increasingly mopped up by oxygen, and both served to cool the Earth. Though for a while this methane oxidation actually replenished the carbon dioxide absorbed by OP photosynthesisers, ultimately there was net carbon dioxide drawdown and of course the decline in methane: cooling was inevitable. Indeed not only did the Earth cool, and experience a number of glaciations, but somewhere around 2.4 and 2.2 billion years ago it entered an extremely cold state: Snowball Earth I.

## Snowball Earth I (~2.4–2.2 bya)

With the new oxygenic photosynthesis and the (above) decline in greenhouse gas warming, the Earth cooled. With cooling, the Earth experienced a positive feedback effect where cooling served to reinforce further cooling as in Figure 2.2B: the ice caps expanded, reflecting more sunlight so enhancing the cooling. Normally, when this happens in an ice age's glacial, the Earth system runs into a limit before it gets too far. At that limit an opposite cycle of warming—as the ice reaches down into warmer latitudes so halting the growth of reflective ice-caps and preventing further cooling—balances matters. However, 2.2 billion years ago—according to the developing new narrative—such was the degree of life's greenhouse gas removal that the cooling continued and ice spread down towards the tropics.

It should, though, be noted that the presence of a near-equatorial supercontinent at that time with some snow-covered mountain ranges would have seen even more snow cover with this cooling and that too would have reflected heat away from the Earth. Additionally, there being significant equatorial land, as there was at that time, prevented ocean currents transporting heat away to higher latitudes: an equatorial supercontinent enhances cooling. Now, while a tropical latitude supercontinent may well—if not almost certainly—have been an additional cooling factor, under the new narrative it was the increase in oxygenic photosynthesisers drawing down carbon dioxide and its oxygen mopping up the potent greenhouse gas methane that was the critical one. (We will return to the debate between the old and new narratives later.)

Once the ice caps grew beyond the temperate zones and reached what today we would consider the tropical latitudes, somewhere between 25° and 30°, there was a point of no return: the temperature difference between temperate latitudes and the edge of the tropics is far greater than that at the edge of the tropics and the equator. The ice therefore grew to cover almost the entire planet and the Earth, from space, looked like a giant snowball (or 'slushball' if a significant band of ice-free water and land existed at the equator). During the growth of ice sheets, evaporation from the remaining warmer, equatorial seas enabled continued precipitation some of which was over land that away from the tropics that fell as snow. This compacted into ice so further developing, thickening, the ice sheets away from the tropics: in places, these sheets' thickness could be measured in miles.

The term 'Snowball Earth' was first coined in 1992 by Joseph Kirschivink at the California Institute of Technology in a paper contribution to a book, *The Proterozoic Biosphere* (*10*). Though here Kirschivink's Snowball Earth was actually Snowball Earth II, a different event, which we will come to later on (in Chapter 5).

Now, it has to be emphasised—because some scientists have exercised themselves over this point—that there is some debate, as mentioned, as to whether ice entirely covered the Earth or whether there was a small, ice-free band around the equator. The geological evidence is not conclusive. Some say that the balance of evidence leads one to conclude that the entire Earth was covered in ice with perhaps a few exceptional places such as where there were hot springs; others that there had to be a small but significant amount of open water for life to continue, so that the Earth was not a snowball but a 'slushball' with open seas in a belt around the equator. Having said that, would open water really be necessary for life to survive? After all, today cold environments beneath ice, such as in Antarctica, can harbour diverse prokaryote and eukaryote communities, including simple animals. However, in terms of this narrative, such delving into nuance is purely an exercise in pedantry: either way the Earth was largely, if not almost completely, covered in ice *and* life did survive!

With ice over much (if not nearly all) the ocean and land, greenhouse gas exchange with the oceans was curtailed. Similarly, carbon was not only frozen in reservoirs such as permafrosts *in situ* but the ice provided a physical barrier between the oceans and the atmosphere. Yet where, and while, there was remaining open water, this became colder, so up to a point facilitating the absorption of carbon dioxide, whose atmospheric concentration lowered, so further enhancing the cooling.

So what was the Earth like during the Snowball? Well, signs of glaciation have been found in strata dating from this time that were then (due to plate tectonic movement) in and around the tropics. (Remember, there was a tropical supercontinent.) In fact, such evidence was among the first to indicate that a later Snowball Earth existed.

As to the temperature, estimates vary, but it is likely that the *average* temperature of the Earth was somewhere around −50°C; this compares with a global mean temperature of roughly around 14°C across much of the mid-20th century. At the poles temperatures would obviously have been colder as with ice covering the oceans there would have been next to no heat carried by sea currents from the tropics: polar temperatures might have been −70°C and at winter could even have fallen to −100°C (*7*). With carbon dioxide freezing at −78.5°C, there would have been wispy carbon dioxide clouds above and in the coldest places solid carbon dioxide (dry ice) on the ground, which is what we find at Mars' poles today. This would have provided a further planetary sink for carbon dioxide so helping lower its atmospheric concentration. Today Antarctica provides such a sink for water which is trapped there as ice. Consequently, today, with a significant proportion on the planet's liquid water converted to solid ice, the Earth's seas are lower by roughly 80 metres than it would be in a warmer, largely iceless Earth (*11*) as has existed throughout much of the past 66 million years since the dinosaurs died out or indeed most (but not all) of the half billion years or so of multicellular life.

Today, in cold weather, when we breathe out we generate a cloud of water vapour. If we were at the poles during the Earth's snowball events then not only would we breathe out a visible cloud of water vapour but also a visible cloud of freezing carbon dioxide vapour. Today at the poles, explorers find that the water vapour they breathe out freezes on their scarves wrapped around their faces (and, if facially haired men, in their beards). So back in a Snowball Earth event, at the poles it was so cold that we really might expect, not just water ice but, dry ice (frozen carbon dioxide) to form as we breathed out. Temperatures would be so low that frostbite would virtually be a rapid certainty unless explorers took extreme protective measures. Here, a fully charged spacesuit would be the order of the day. Indeed, a space suit would be appropriate attire given the view. With the oceans capped by ice for thousands of miles to the equatorial zone, there would be little water vapour in the air hence few clouds apart from, possibly, some of the aforementioned wispy clouds of carbon dioxide. This would mean that the skies were largely clear and so—especially without today's industrial particle pollution and light pollution—the night time starscape would have been spectacular! Furthermore, given a uniform ice cap, there would be little thermal turbulence and this would make a Snowball Earth away from the equatorial zone an ideal place for astronomy. The only thing impeding astronomy at the poles would have been the aurora displays.

The oceans away from the equatorial zone would have been frozen and the land covered by ice sheets with only high mountain peaks piercing the ice (nunataks). The range of landscapes across much of the Snowball Earth would not be that different to those found in across Antarctica today, from its ice sheet domes through to its heavily glaciated mountain ranges.

Moving to the tropics, what things were like much depends on whether the Snowball Earth I was a true 'snowball' or more of a 'slushball' with considerable open water near the equator. If the Earth were a true 'snowball' then clearly at the tropics the temperature would have been at or below 0°C. Some computer models have suggested that this was likely. However, we need to remember that models are just that—models: they are not a complete capture of reality. If the Earth back then was a true snowball then there must have been a considerable number of warm refuges for life: there was no complete mass extinction event (nor was there one with the subsequent Snowball Earth II). As said, such refuges would consist of things like hot springs. But can we believe that there were enough of these of sufficient size and duration for life to ride out such an event? As noted earlier, others say that the answer is that the Earth at this time was more of a slushball with some open sea in an equatorial belt. This would have provided plenty of space for life to ride out the event. (Though, as there were no multicellular species for us to see overt signs of life, and with surviving geological strata from this narrow window in distant time sparse, we simply do not know how much life there was during this time.) If there was plenty of space for life, then we might expect Snowball Earth's tropical temperatures to be at least just a few degrees centigrade with conditions—in terms of temperature—a bit like parts of Alaska or Scandinavia today at the time of the autumnal equinox.

Of course, as noted above, any open seas during the Snowball Earth events would be much lower than the sea levels today as much of the planet's water would be in the form of giant ice caps miles deep covering most of the land outside of the equatorial zone. Were it not for the ice, the continental shelves would be certainly completely exposed (all be they covered with ice) with what remains of the seas confined to areas that are today those of abyssal depths.

Consequently, the first Snowball Earth in the run-up to 2.2 bya, and the second Snowball Earth within the Cryogenian (the geological period from 635 million years ago to 720 mya—see Appendix 3), saw the planet in uniquely cold states not found at any other times in its history. These two Snowball Earths were outstandingly distinct, individual events (all be they both featuring a number of severe glacial episodes), and they mark the times when the Earth—crossing a critical transition—flipped between states.

We will eventually look at the currently accepted explanations as to the different causes of both Snowball Earths, so as to show the advantage of the new explanatory narrative being explored by this book. For now we just need to note that the above is still a greatly simplified description (covering the principal factors only), as there was more going on. For instance, while the evolution of oxygenic photosynthesisers was key to oxygen's rise in the atmosphere after 2.2 bya, there was also chemistry going on that further helped oxygen's establishment. There is no space here to detail matters, but it is worth briefly mentioning some of the atmospheric circumstances around the time of the critical transition and touch upon some of what might else have been going on, especially as this too involved the interaction between life and planet.

'Before' the critical transition, the Earth's atmosphere had much methane and carbon dioxide and this methane mopped up any of the first traces of oxygen from the very first photosynthesisers (a biological factor).

'After' the critical transition the new oxygen reached possibly up to as much as ~4% oxygen of present atmospheric level (PAL) (*12*), but some say less (*13*), some more (*14, 15*).

To put this into context, today the Earth's atmosphere has a proportion of oxygen of nearly 21%. So up to 4% PAL equates to just 0.84% $O_2$ in the atmosphere. However, some estimate the atmospheric oxygen immediately following the initial oxygenation 2.2. bya was as low as 0.1%: we are talking about very low levels of oxygen, but still enough to affect atmospheric chemistry. As noted, some say that there was more oxygen than 4% PAL $O_2$ (0.84% of atmosphere) after the flip. This vagueness as to the exact concentration of oxygen a couple of billion of years ago is due to our not being able to measure it directly (we have no time machine) and so we have to infer it from the likely chemistry taking place at the time from the surviving geological record, which itself is sparse.

Whatever the exact figure, clearly between the methane richer 'before' and the oxygen richer 'after' the flip, there must have been a time when the falling concentration of methane roughly equalled that of the rising concentration of oxygen. The production rate of the two would have cancelled each other out, and so we got the lack of UV protection mentioned earlier—but also something else.

It was this unique—equal methane and oxygen—atmospheric mix that David Catling, Kevin Zahnle, and Christopher McKay of NASA's Ames Research Center in California chose to consider. Published in *Science* in 2001, their conclusions were that a combination of *both* biological action and atmospheric chemistry helped change the Earth from being one of having a reducing surface environment to being oxidised (*16*).

They postulated that not only did oxygenic photosynthesis produce oxygen but that, through biological fermentation, methane was also produced; much like methane is in the bottom of stagnant ponds and in marshland in a process that evolved a little time after microbial life first arose. However when the Earth was going through its first flip from an anoxic world to a lightly oxygenated world, and at the time when the lowering concentrations of atmospheric methane roughly equalled that of the rising concentration of oxygen, this methane was then photo-dissociated (photolysis) by ultraviolet light in the upper atmosphere to carbon and hydrogen. The carbon was then oxidised by the oxygen to carbon dioxide (which could then once again be available for photosynthesis) while the hydrogen escaped into space. Bringing these reactions together and simplifying them (see Figure 3.3), the net reaction is one of water being split into its constituents of oxygen and hydrogen with the latter—being light in weight—escaping into space and so *irreversibly* oxidising the Earth.

It needs to be stressed that this chain of reactions *only* works when oxygen and methane are in roughly equal concentrations and these can only co-exist if both are continually being generated by life (otherwise they would cancel each other out—one would oxidise the other). Remember, in the pre-oxygenated, pre-flipped world there was no oxygen (well, maybe a few parts *per* trillion before the evolution of OP), and in the post-flipped world very small amounts of

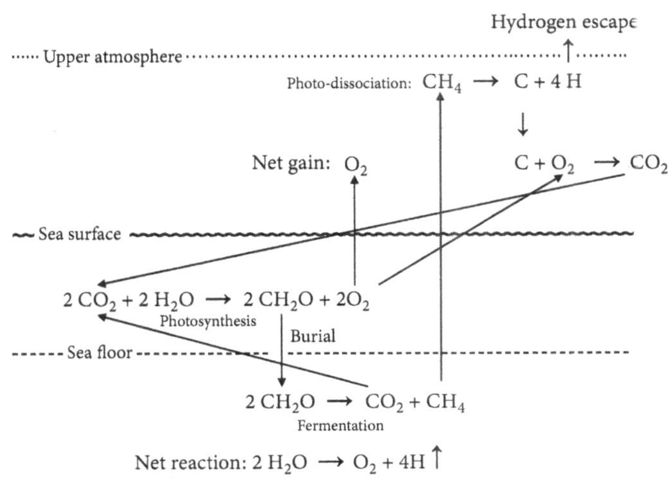


**Figure 3.3** Possible atmospheric chemistry at the critical transition helping raise oxygen from many parts *per* billion to up to 4% PAL. Based on Catling, D. C., Zahnle, K. J., & McKay, C. P., Biogenic methane, hydrogen escape, and the irreversible oxidation of early Earth. *Science*, **293**, 839–843 (2001). But also the core of this process was speculated much earlier by Harold Urey of the University of Chicago in 1952. Urey, H. On the early chemical history of the Earth and the origin of life. *Proceedings of the National Academy of Sciences*, **38**, 351–363 (1952).

methane (measured in parts *per* billion). Furthermore, as said, the reaction *also* requires substantial photo-dissociation: the breakdown of molecules by high-energy sunlight (typically ultraviolet). This could not have happened earlier, in the reduced, methane-rich, pre-flipped Earth as there was likely a high-altitude methane haze of heavier carbon molecules (tholins) protecting the lower, thicker atmosphere. Conversely, as we also previously covered, in the post-flipped, oxygenated world there was an ozone layer protecting the lower atmosphere and surface from ultraviolet (UV) radiation much as it does today. Both these UV shields (tholin and ozone, respectively before and after the flip) inhibited photo-dissociation. Also, remember our transitory MIF of S: this is further evidence that there was no UV protection around the actual time of the flip. In short, this hydrogen escape mechanism could only happen at the time of, and around, the flip when both the chemical reactants were more or less in equal concentrations and when there was no upper atmosphere protection from UV. This mechanism was important to additionally boosting oxygen levels and carrying the Earth system, pushing it through the critical threshold and to flip and change its state: the impact of oxygenic photosynthesis by life on the atmosphere was augmented by purely chemical reactions in the atmosphere that effectively split water, allowing hydrogen to escape and leave behind oxygen.

And so planetary biology *together with* atmospheric chemistry together drove the first major flip of the Earth state from a reducing one to a slightly oxygenated one and also caused Snowball Earth I.

Now, though this covering (possibly near covering) of the Earth with ice is called Snowball Earth I (as there was a second major planetary glaciation to come), Snowball Earth I itself, as previously mentioned, was not a single snowball (or slush-ball) event.

During the critical transition, the Earth system oscillated: it was not a smooth, or non-critical, transition and so there was more than one severe glaciation and very likely more than one Snowball or near Snowball/Slush-ball Earth following one after the other in succession. Alas, dating the geological record from so long ago is not accurate enough for us to discern the fine detail in the various geological clues and so does not allow us to deduce the exact causes and precise timings of the various oscillations. Variations in greenhouse concentrations almost certainly played a part, as did the presence, or absence, of a methane haze, but nothing can be deduced with certainty other than after 2.9 bya there was the first, Pongola glaciation. This glaciation was likely associated with a change of atmospheric chemistry. Then after 400 million years later there were among others the Huronian glaciation and then the Makganyene Snowball Earth. (The terms Pongola, Huronian, and Makganyene all being geological ages.) In 2021 a small international collaboration of researchers with Simon Poulton as the lead author conducted a high-resolution analysis of 2.4 to 2.2 bya geological strata using sulphur isotopes and the strata's iron chemistry. It was a sophisticated elucidation of what was going on back then in the distant past. Without going into detail—the science is complex—the researchers concluded that there was continued oscillation in atmospheric oxygen levels after about 2.32 bya and then much variation over 200 million years before a final oxidation event around 2.22 bya. This last oxygenation was some 100 million years later than researchers had, prior to 2021, previously thought (*17*). In short, there were a number of cold, superglacial episodes over many millions of years. The Pongola glaciation was sufficiently early as to be distinct from the Huronian and Makganyene. The latter two and the subsequent oxygenation at 2.22 bya were essentially part of a single transition event—albeit with multiple components—with the Earth changing state.

With reference to the graphical representation of a critical transition of Figure 2.4B, this number of oxygen (as well as carbon and climatic oscillations) is denoted by the 'zone of transition'.

Because there were a number of variations in atmospheric oxygen over this period, and the time over which they took place is so long (hundreds of millions of years), Simon Poulton and colleagues consider that the term the 'Great Oxidation Event' is something of a misnomer: 'Event' suggests a single occurrence, instead they suggest 'Episode' would be more appropriate.

It should also be noted that if life was expanding with the evolution of oxygenic photosynthesis (OP) then we would expect more $^{12}C$ (carbon-12) drawdown (remember photosynthesis prefers $^{12}C$ to $^{13}C$) and this would change the carbon-13 ($\delta^{13}C$) ratio to $^{12}C$ in the atmosphere, hence in the carbonate sediments being laid down at the time. This drawdown would see less carbon dioxide greenhouse gas in the atmosphere and so would cool the planet. In short, the suggestion is that the evolution of OP was involved in the cooling the planet with the early Pongola glaciation as well as with Snowball I. (As we shall see, the latter saw even more OP carbon drawdown due to nutrient runoff from a fragmented supercontinent. Here, there is also the possibility of the rise of the eukaryotes playing a role, but we must not get ahead of ourselves . . . .)

However, following each glacial phase there would, *ipso facto*, be warming: volcanoes would continue to return carbon dioxide to the atmosphere with a $^{13}C$ rich mix. There was also a return of carbon to the atmosphere from $^{12}C$-rich organic sediments in the form of both carbon dioxide and methane (again much as happens from the anaerobic breakdown of organic sediments in ponds, wetlands, *etc.*, today). This released carbon in the form of $^{12}C$ would in turn would change the $^{13}C/^{12}C$ ratio the other way. So these events—the OP-driven drawdown of $C^{12}$ and its subsequent return along with the volcanic re-setting of the $^{13}C/^{12}C$ ratio—easily explain both the high and low $^{13}C$ ratios at the times of Snowball events seen in Figure 2.1. Here, as previously mentioned, the highs and lows seem to happen at the same time, but this is an artefact of our not being able to date the rocks with the sufficient resolution to see the swings in carbon isotope ratio. (We will return to this again when discussing Snowball Earth II and the Earth's second big flip in Chapter 5.)

Having said all this, there were other mechanisms than the evolution of OP that could further explain the high-low $^{13}C$ ratios. A cooling Earth would see less anaerobic photosynthetic drawdown of $^{12}C$, and a (largely) ice covered Earth would see the further addition of $^{13}C$ from volcanic carbon dioxide. Equally, given that we know that there were algal mats before and after this time, we can still envisage a volcanic carbon dioxide warmed Earth following the cold phases to release $^{12}C$ to the atmosphere where ultimately it would end up as carbonate rock. I mention these biological dimensions now because the idea that it was the biological evolution of oxygenic photosynthesis that helped trigger the Earth's Snowball events is a recent one that has yet to become widely promulgated in the geoscience community even if it is gaining some traction among some specialising in Earth system science: earlier ideas have non-biological factors dominate and even stimulate the evolution of OP. These older alternative explanations will be further explored for comparison later (see Chapter 5).

But life was not the only thing evolving. That the Earth itself was geologically evolving, even deep down, is evident in a number of ways in addition to the previously mentioned crustal thickening. An example of recent research was published in 2015 when an international group, led by Andy Biggin from Liverpool University, published in *Nature* their work on Earth magnetic field fluctuations over the past 2 billion years that suggested that the Earth's core was changing and indeed had changed roughly around the time of the Earth's first big flip: the Earth is not a static system, not even deep down (*18*). Yet, this similar timing could simply be coincidence, but even if this was so, that reasonably large continents existed at least a little before the first big flip is relevant: a supercontinent—made up of a number of former smaller continents—did play a part (even if the albedo dimension is not as important under the new narrative scenario) Furthermore, as the Earth itself has evolved over time, it has gone through many changes. The question therefore arises as to whether some changes coincident in time are in fact genuine coincidences or whether there is some causal relationship, or even whether a coincident change was simply a contributing factor. (We will return to this later.)

Leaving the core and turning to the Earth's crust, its evolution was likely a factor contributing to the Earth's oxidation providing the necessary elements for life as well as various different types of carbon reservoir.

From its formation, over time, the Earth's crust had thickened and eventually the first large continents formed. Previously, many volcanoes in the early Earth were underwater but as the continents formed through colliding plates, more volcanoes were found on land too. These terrestrial volcanoes are hotter and have a more oxidised output of gases than their marine counterparts. Furthermore, volatiles in the crust would also have declined over time unless they were recycled (just as carbon as carbonate is subducted) back into the mantle. This they were through the now established tectonically driven continental drift.

What may well be relevant was that the time of the Earth's first big flip was also the time of the breaking up of a supercontinent: Kenorland. (Remember Chapter 1: with continental drift we get the Wilson cycle of supercontinent formation and fragmentation.) Was this also a coincidence? As we shall see again later on (Chapter 5 and 'Merging the old and new narratives' section) the standard explanation of Snowball Earths I and II's cause centrally involves the presence of a supercontinent. That not one but both Snowball Earths took place around the time of a supercontinent break-up makes the 'it's-a-coincidence' hypothesis less certain: once may be a coincidence, but twice?

The reason this is important is because if oxygenic photosynthesis evolved roughly 2.9–3.1 bya, why did not these photosynthesisers draw down carbon dioxide and plunge the planet into a Snowball Earth over half a billion years previously? The answer could be that a 'proper' supercontinent was required, but not necessarily for the albedo reasons the traditional narrative has it. (True, this factor would have played a part, but it was not as central a role as that of life as the new narrative promulgated here has it.)

As noted in Chapter 1 towards the end of the section 'The Earth and early biosphere', the Earth's first continents were both smaller and lacking the tall mountain ranges of today's continents: a greater proportion of these smaller continents would be beneath the sea. Indeed today, our continents do not end at the coast but under the sea at the continental shelf's edge. In fact, the notion that continents could be repositioned to join each other—like South America could with Africa—gets a better fit when looking at the continental shelf boundaries, not just the present-day coastline.

Over time, the Earth's continents have increased in size—both area and altitude—and Kenorland was smaller than an assemblage of today's continents would be into a supercontinent (which, by the way, palaeoclimate modellers should take into account when they try to recreate a tropical supercontinent-induced snowball Earth). Not only the Earth's cooling enables a thicker crust, but continents colliding sees the build-up at the collision point of crust both vertically (forming mountains) and in depth. These areas of thicker crust tend to be more stable and better able to withstand the rigours of plate tectonics. Today such stable areas are known as 'cratons' (from the Greek '*kratos*' meaning 'strength').

If modern-type plate tectonics did not properly get going until somewhere around 3.3–3.0 bya (see Chapter 1) then continental collisions and the building of bigger and higher-altitude continents could only have taken place after that time. The question then becomes, when did this happen?

Ilya Bindeman from the University of Oregon (US) had an idea as to how to get a handle on this and began working with a small team of colleagues from the United States, Switzerland, and South Africa. It was all to do with rain falling on land.

Rain is water that for the most part originates from ocean evaporation. Water itself consists of various isotopes of hydrogen and oxygen in molecules of $H_2O$. Just looking at the oxygen isotopes, the most common is oxygen with the atomic weight of nearly 16 is known as O-16, written as $^{16}O$. This isotope of oxygen makes up over 99.7% of the oxygen atoms in the crust. Conversely, there are other heavier isotopes of oxygen: $^{18}O$ makes up 0.2% and $^{17}O$ less than 0.04% (which is a small but still easily measurable proportion). The thing is that a cloud water vapour tends to lose water made up of the heavier isotopes first. So when water evaporates from the ocean and falls as rain just inland near the coast of a continent, the rainwater's oxygen molecules have proportion of isotopes quite similar to that of ocean water, known as Standard Mean Ocean Water (SMOW), or the often-used VSMOW (mentioned here for those following up the reference below—see also Appendix 1). There is a temperature complication in that a warmer Earth will see proportionally more evaporation of the heavier isotopes but the use of all three oxygen isotopes when analysing the geological record helps get around this problem. The thing is, when continents are large then far inland there is greater fractionation, as more of the heavier isotopes rained out nearer the coast. *Only* when continents are large does greater fractionation occur!

Such fractionation, especially with regards to the $^{18}O$ isotope in water, also occurs with altitude, or the presence of tall mountain ranges.

What Ilya Bindeman and colleagues did was to look at the oxygen isotope concentration of shales from many samples spanning from the present to 3.7 billion years in the past (before Snowball I some 2.2–2.4 bya). Shales are rocks formed from clays, and clays are formed from sediments carried down by water. Some of the minerals in clays are affected by the water that carried the clay particles. A consequence of this is that the sediments that go on to form shale contain the isotopic signature of the water that transported them.

Ilya Bindeman and colleagues found that the triple isotope signature of shales that formed before 2.5 bya—those that formed between 2.2 and 2.5 bya (the period that includes Snowball Earth I)—and more recent shales, each formed a distinct grouping of oxygen isotope signatures. Their interpretation is that those that formed before 2.5 bya are less fractionated because there were no large continents or continents with significantly tall mountain ranges. Conversely, those that formed 2.2 and 2.5 bya included those related to a large continent that featured tall mountains. This signature became even more pronounced after 2.2 bya, suggesting continued, gradual growth in continental size with time (*19*). This fits in with the Patrice Rey and Nicolas Coltice finding, discussed at the end of Chapter 1, that it was unlikely that the early Earth's thin crust would have become thick enough with cooling to support major mountains above 2.5 km until at least 2.5–2.8 bya.

All this is relevant to the idea of life being very much in the driving seat for this early (partial) oxygenation of the Earth and indeed, as Ilya Bindeman and his colleagues themselves speculate, a large landmass being eroded provides extra nutrients for life in the waters around a fragmenting supercontinent. It was this OP life, blooming due to—the argument goes—the nutrient release, that boosted oxygen generation to a level that it overcame the anaerobic Earth's reductive sinks. Further, where life thrives, so does speciation, hence evolution.

Rain on land erodes in a number of ways. In addition to the physical mechanical wearing of water on rock, water on land freezes much more easily than sea water. You don't even need tall mountains to see this: those living away from the tropics will not be unfamiliar with winter's frosty morning. Water freezing in small cracks and fissures in rock expands and so leads to fracturing. Also and importantly, rainwater is more acidic than seawater, which is buffered by salts. More acidity aids the erosion of more alkaline rock. Today, ocean seawater is slightly alkaline with a pH of around 8.1 at its surface (in pre-industrial times a couple of centuries ago surface ocean pH was a little more alkaline at around 8.2, which may not seem much of a change but, pH being a logarithmic scale, it does represent around 30% an increase in hydrogen ion concentration which is the key factor determining acidity). Back in the Archaean (2.5–4.0 bya) with less erosion from land it has been estimated that the oceans' pH was somewhere around 6.3–7.7 (though this estimate is subject to discussion, it is still alkaline) (*20*). Today, rainwater away from urban and industrial areas has a pH of ~5 which is decidedly acidic: this is due to both carbonic and sulphur acids (*21*). This is much more acidic than the estimated pH of the Archaean oceans.

Submerged 'land', such as that of the continental shelves of our present continents, is protected by the buffered sea from the sort of erosion undertaken by acidic rainwater: its acidity is due to it reacting with carbon dioxide in the air to form the aforementioned weak carbonic acid. (And this is also behind the ocean acidification we see today that add to corals' troubles: the extra carbon dioxide we humans add to the atmosphere causes increased carbonic acid production.)

So, rain erodes rocks on the surface of the land to release minerals, some of which are nutrients for life. Here, some of these nutrients are more important than others. One of the more important ones is phosphorous.

Phosphorous is an essential element for life on Earth. It is a constituent of nucleic acids (DNA and RNA) as well as fundamental to the energy biochemical adenosine triphosphate (ATP). As you may have realised from its name, ATP contains three phosphate groups. Life gets energy to drive virtually all (other than processes such as chemosynthesis and photosynthesis) the metabolic processes requiring energy from ATP. This happens when ATP loses a phosphate group to become adenosine diphosphate (ADP). As such, phosphorous is a fundamental element for life.

Yet, despite it being a vital element to life as we know it, it only makes up 0.1% of the Earth's crust: it is not one of the crust's top ten elements. It is even rarer in seawater at 0.1 parts *per* million. That it is a rare element and that all life uses it speaks to it being essential: despite its rarity, life has not been able to employ the more common alternative elements available to it.

Yet, despite it being rare, most of the Earth's crust has phosphorous albeit at these low levels. However, concentrations are a little higher near thermal vents as the hot water rising through geological strata helps release it. Evaporation ponds on land subject to occasional rainfall and then evaporation also see phosphorous concentrations build up. Further, some terrestrial thermal springs are also places where phosphorous concentrations are higher. (This combined with the presence of UV light to generate handed biochemicals, and the concentration of ions in cells is another argument for life starting in terrestrial geothermal ponds—see Chapter 1) (*22*).

So let us add this factor into the picture. This then is a summary scenario of what possibly happened around the time of Snowball I.

Plate tectonics properly became established somewhere around 3.0–3.2 bya. Oxygenic photosynthesis possibly evolved in cyanobacteria 2.9–3.1 bya (some say a little later) at a time where there were no large land masses sporting tall mountain ranges. Carbon drawdown from these photosynthesisers could have been responsible for the cooling of the planet that led to the Pongola glaciation.

At that time, there were continents, but they were smaller and much of these were submerged: surrounded by shallow seas. Then, through plate tectonic movement, the first large supercontinent formed around 2.5 bya and more substantive rainwater erosion began releasing some nutrients, including phosphorous. A little later, 2.2–2.3 bya (perhaps as early as around 2.4 bya as there is an oxygenation event at 2.43 bya), this supercontinent fragmented. This allowed seas to penetrate into the supercontinent and so made the supercontinents' climate more wet. It also enabled rivers from the supercontinent to carry more nutrient-rich sediments to these semi-compartmentalised, small inland/inlet seas that helped contain the nutrients, which in turn enabled the oxygenic cyanobacteria (and—to be discussed in the next chapter—possibly photosynthesising eukaryotes) to bloom. Further, thermal vents along the fractures would also have provided additional nutrients. (As we shall see in the next chapter, they are also a likely site for the evolution of eukaryotes and they too could have played a part.) Such thermal vents are not common between land masses coming together to form a supercontinent, but are when a supercontinent fragments. This bloom of oxygenic photosynthesisers provided a surge of oxygen that mopped up much of the atmospheric methane and inhibited the surface and near-surface anaerobic methanogens. It took us towards, if not across, the point of equal methane and oxygen atmospheric concentrations that enabled the chemistry to allow hydrogen escape and the further rise of oxygen which took the Earth system across a critical threshold hence transition.

The Earth had crossed a threshold in a critical transition from an anaerobic atmosphere to one with oxygen, albeit an oxygen concentration of up to just a couple of per cent.

## The end of, and time after, Snowball Earth I

But why did oxygenation stop at an atmospheric concentration of just a couple of per cent and not continue to grow to today's level of around 21%? Here there were a number of key reasons.

First, the creation of an ozone layer re-established UV protection in the upper atmosphere and this reduced the photo-dissociation of water vapour looked at earlier this chapter.

Second, with the concentration of oxygen greatly dominating the now minority gas, methane was easily mopped up by the oxygen and converted to carbon dioxide and water. (Due to both the first and second reasons, the [planetary hydrogen escape] reactions described in Figure 3.3, would be greatly diminished.)

Third, with the rise of oxygen, anaerobic methanogens were poisoned and so forced to retreat to anoxic refugia, reducing global methane production. Here, it is true that methanogens would survive away from oxygen in deep-sea sediments and even thrive on the rain of organic detritus of dead living matter from above. However, any methane produced would have to bubble up through the water column and so would have at least in part been gobbled up by other bacteria that metabolise methane (methanotrophs). And of course, most of any methane reaching the surface, hence atmosphere, would be mopped up by the oxygen now present in sufficient quantity: this took out a proportion of the oxygen.

Fourth, there is only so much simple prokaryotic oxygenic photosynthesisers can do. Two billion years ago, there were not the complex, multicellular organisms with roots and so forth that greatly enhance the additional release of even more nutrients from minerals that much of the living biomass today requires.

And so the further growth in atmospheric oxygen concentration through the photo-dissociation mechanism ceased and there was also not enough oxygenic photosynthesis taking place to continue its building up.

The critical transition was a temporary stage (albeit lasting many millions of years) through which the Earth system traversed and, with Snowball Earth I's series of planetary glaciations that were the most dramatic characteristic of this transition, the transition had to come to an end.

With the Earth largely (if not nearly completely) covered in ice, the oceans were physically absorbing less carbon dioxide and, with most (but not all) life killed off by the cold, there was little photosynthetic drawdown of carbon dioxide. Yet volcanoes continued to release carbon dioxide into the atmosphere.

This volcanic release, on the global scale, provided a slow but nonetheless steady trickle of carbon to the atmosphere. To put this into present-day context, today volcanoes release around 100 million tonnes of carbon a year to the atmosphere. This is equivalent to 1%–2% of the amount of carbon we humans are currently adding burning fossil fuels (*23*); it is likely that the volcanic release back then was of a similar order of magnitude if not greater (the inside of the younger Earth had not had so long to cool and it is possible that some deep, residual volatiles were still migrating towards the crust). This volcanic carbon dioxide is still a trickle globally speaking but, at the time of Snowball, as it is now, it was a trickle that persisted over millions of years. It enabled carbon dioxide once more to accumulate to unprecedented levels in the atmosphere. Eventually, with so much of this greenhouse gas, the planet warmed. Indeed, because Snowball Earth's ice prevented much ocean uptake of this carbon dioxide, and because of Snowball Earth's low level of life drawing down atmospheric carbon, it is quite likely that carbon dioxide levels rose considerably which in turn conferred a substantive greenhouse effect. And once melting kicked in (which inevitably it did with the greenhouse forcing) there was a positive feedback effect that was the reverse of the feedbacks during Snowball Earth's establishment (or each of the Snowball's major glaciations): ice melt resulted in a reduced reflective ice cap which encouraged more ice melting and so on (again see Figure 2.2b). Indeed, that there was a very high atmospheric concentration of carbon dioxide in the atmosphere at the end of Snowball I is evidenced by cap carbonate strata seen in the geological record. This much warmed Earth would also see the return of carbon to the atmosphere from some organic sediments.

This cycle of snowball creation and dissolution continued a number of times as reflected in the Huronian and Makganyene (among other) glaciations that together comprised the time of Snowball Earth I. This caused the carbon isotope fluctuations.

The carbon cycle then reconfigured between each snowball glaciation before stabilising with different amounts of carbon in the biosphere's different carbon reservoirs than in Earth's purely anaerobic time. This ensured that the planet's first great flip, ended.

Life, though, continued to evolve.

## References

1. Tribe, M. A., Morgan, A. J., & Whittaker, P. A., *The Evolution of Eukaryotic Cells*. Studies in Biology no. 131. Institute of Biology & Edward Arnold: London, Great Britain (1981).
2. Gaucher, E. A., Govindarajan, S., & Ganesh, O. K., Palaeotemperature trend for Precambrian life inferred from resurrected proteins. *Nature*, **451**, 704–707 (2008).
3. Boussau, B., Blanquart, S., Necsulea, A., Lartillot, N., & Gouy, M., Parallel adaptations to high temperatures in the Archaean eon. *Nature*, **456**, 943–945 (2008).
4. Robert, F., & Chaussidon, M., A palaeotemperature curve for the Precambrian oceans based on silicon isotopes in cherts. *Nature*, **443**, 969–972 (2006).
5. Hren, H. T., Tice, M. M., & Chamberlain, C. P., Oxygen and hydrogen isotope evidence for a temperate climate 3.42 billion years ago. *Nature*, **462**, 205–208 (2009).
6. Ono, S., Beukes, N. J., Rumble, D., & Fogel, M. L., Early evolution of atmospheric oxygen from multiple sulfur and carbon isotope records of the 2.9 Ga Mozaan Group of the Pongola Supergroup, Southern Africa. *South African Journal of Geology*, **109**, 97–108 (2006).
7. Lenton, T., & Watson, A., *Revolutions That Made the Earth*. Oxford University Press: Oxford, Great Britain (2011).
8. Farquar, J., Bao, H. M., & Thiemens, M., Atmospheric influence of Earth's earliest sulphur cycle. *Science*, **289**, 756–758 (2000).
9. Papineau, D., Mojzsis, S. J., & Schmitt, A. K., Multiple sulphur isotopes from Paleoproterozoic Huronian interglacial sediments and the rise of atmospheric oxygen. *Earth and Planetary Sciences*, **255**, 188–212 (2006).
10. Kirschvink, J., Late Proterozoic low-latitude global glaciation: The Snowball Earth. In Schopf, J. W., & Klein, C., eds., *The Proterozoic Biosphere: A Multidisciplinary Study*. Cambridge University Press: Cambridge, Great Britain, pages 51–52 (1992).
11. US Geological Survey, *U.S. Geological Survey Professional Paper—Estimated Present-Day Area and Volume of Glaciers and Maximum Sea Level Rise Potential*. US Geological Survey: Reston, VA, USA (1999).
12. Blättler, C. L., Claire, M. V., A. R. Prave, A. R., *et al.*, Two-billion-year-old evaporites capture Earth's great oxidation. *Science*, **360**, 320–323 (2018).
13. Rye, R., & Holland, D., Paleosols and the evolution of atmospheric oxygen: A critical review. *American Journal of Science*, **298** (8), 621–672 (1998).
14. Lane, N., *Oxygen: The Molecule That Made the World*. Oxford University Press: Oxford, Great Britain, pages 43 & 53 (2002, 2016 revised).
15. Holland, H. D., The oxygenation of the atmosphere and oceans. *Philosophical Transactions of the Royal Society B* **361**, 903–915 (2006).
16. Catling, D. C., Zahnle, K. J., & McKay, C. P., Biogenic methane, hydrogen escape, and the irreversible oxidation of early Earth. *Science*, **293**, 839–843 (2001).
17. Poulton, S. W., Bekker, A., Cumming V. M., *et al.*, A 200-million-year delay in permanent atmospheric oxygenation. *Nature*, **592**, 232–236 (2021).
18. Biggin, A. J., Piispa, E. J., Pesonen, L. J., *et al.*, Palaeomagnetic field intensity variations suggest Mesoproterozoic inner-core nucleation. *Nature*, **526**, 245–248 (2015).
19. Bindeman, I. N., Zakharov, D. O., Palandri J., *et al.*, Rapid emergence of subaerial landmasses and onset of a modern hydrologic cycle 2.5 billion years ago. *Nature*, **557**, 545–548 (2018).

20. Krissansen-Tottona, J., Arney, G. N., & Catling, D. C., Constraining the climate and ocean pH of the early Earth with a geological carbon cycle model. *Proc. Natl. Acad. Sci.*, **15** (16), 4105–4110 (2018).
21. Charlson, R. J., & Rodhe, H. Factors controlling the acidity of natural rainwater. *Nature*, **295**, 683–685 (1982)
22. Mulkidjaniana, A. Y., Bychkovc, A. Y., Dibrovaa, D. V., *et al.*, Origin of first cells at terrestrial, anoxic geothermal fields. *Proc. Natl. Acad. Sci.*, **109** (14), 5,156–5,157 (2012).
23. Intergovernmental Panel on Climate Change, *Climate Change 2013: The Physical Science Basis. Contribution of Working Group I to the Fifth Assessment Report of the Intergovernmental Panel on Climate Change.* Cambridge University Press: Cambridge, Great Britain (2013).

# 4

# A Billion Years towards the Second Flip

The Earth 2 billion years ago, despite having flipped from a reduced, anoxic state to an oxidised state with free oxygen in the atmosphere, was still very unlike our world today with roughly a tenth of the atmospheric oxygen we currently enjoy (on average today close to 21% by volume at sea level). There were no plants or animals: microbes were still the only life forms. Life was still largely confined to the seas. However there was some life on land, and not just in freshwater pools but also in moist cracks in stone. Oxygenic photosynthetic life could even be found in seemingly dry deserts beneath small fragments of semi-transparent rock such as quartz that trapped moisture beneath it while allowing just enough light through for photosynthesis. Nonetheless, the total amount of terrestrial biomass was still minute compared to its marine counterparts, but even in the seas there was far less back then than there is today.

There were a number of reasons for this constrained biomass size. The principal ones relate to the (lack of) supply of nutrients. Nutrients, such as phosphorous (see Chapter 3), in the ocean originate either from land runoff or from stirring sediments. With runoff, rain erodes rock which releases mineral nutrients that get carried to the ocean. With sediment stirring, embedded nutrients are released from the sediment back into the ocean.

Yet well before a billion years ago both these nutrient release mechanisms were muted, even with the evolution of a more modern water cycle following the creation of the first large, land area supercontinent sporting significant mountain ranges: the supercontinent whose breakup was associated with Snowball I (again, see the previous chapter). Today we have multicellular plants with their roots penetrating, and widening, cracks in rock, hence enhancing erosion. Today we also have large (that is to say *not* microbial scale) burrowing animals digging their way through sediments. A billion to 2 billion years ago there were neither. On land, nutrient release from rain falling on rocks relied on the chemical and physical action of the water: this last included freeze-thaw erosion processes and the former some acidic dissolving mechanisms. In the sea, nutrient release from stirring sediments could only commonly happen where the sediments were either close enough to the surface for wave action to stir things up or through happenstance such as an undersea landslide along the continental shelf.

Yet life was evolving—genes were being shuffled and mutated with those conferring the greatest benefit surviving to the next generation—but life still had to

*Earth's Flips of State*. Jonathan Cowie, Oxford University Press. © Oxford University Press (2026)
DOI: 10.1093/9780197825402.003.0004

develop certain key traits before we could get multicellular plants. Also, larger, multicellular animals could not evolve until there was even more than 4% of the present atmospheric level of oxygen (PAL) in the atmosphere and, as that was as yet absent, so the potential for multicellular animal evolution was also constrained to very small-sized species.

Nonetheless, following the first big flip to 0.5%–4% (or possibly, some say, more) PAL of oxygen, evolutionary developments were to take place . . .

## The un-'boring' billion

The period roughly stretching between 2 and 1 billion years ago (but actually covering a longer period) has been described by some geoscientists as 'the boring billion'. This is somewhat understandable as the geological record is sparse with few biological developments found compared to more recent times. Still, the description 'the boring billion' is unfair and is, as we shall eventually see, also somewhat of a misnomer.

The supercontinent, Kenorland, whose breakup coincided with the rise of dominance of oxygenic photosynthesis (OP) and the living Earth's first big flip of state, had split into a number of smaller continents. Eventually, as hundreds of millions of years passed and as the Wilson cycle (of formation and disruption of continents and supercontinents—see also the next chapter) progressed, a new assemblage of continents, called Nena-Columbia, formed. Eventually, this also split and once again continental drift spread the continents across the globe. Regardless, the Wilson cycle continued with yet another iteration, and by a billion years ago yet another supercontinent formed, called Rodinia. A few hundred million years later, this too broke up and at this point—over a billion years after the first big flip—the Earth went through its second great flip of state. (Though we must remember that, while this is the second great flip, this is the third transition—the first being more gentle with the transition of a lifeless Earth system purely in chemical equilibrium to one in which life disturbs this equilibrium.)

Even if one could describe a billion years that saw two iterations of the Wilson cycle that included the formation and breakup of two supercontinents as 'boring', on the biological front there was also a lot going on.

One reason why this billion-year period has been described by some geoscientists as 'boring' is because it is difficult to discern the detail of life's evolution: microbial fossils lack detail and, of course, are small. Consequently, one needs other evidence. Fortunately there other signs of what was going on in the form of biomarkers (distinctive biochemicals) and various isotopic evidence to be found in the geological record. This isotopic evidence can, for instance, suggest how much oxygen there was in the ocean at a given time as well as the state of various chemical element global cycles.

In addition to this, we have present-day life forms providing evidence to past developments as these contain genetic sequences that have been formed over the many millions of years. As each generation passes, so genetic sequences are shuffled and mutations creep in. So, by comparing the similarity of specific genomes and genetic sequences for specific genes (often for specific proteins) in different but related species we can see how one species is related to another and how that species in turn relates to yet another, and so even how this first species relates to the last. This in turn means that we can deduce the order in which they evolved. Furthermore, as the DNA sequence for a gene for a specific protein changes with time, and as that protein may be used by a number of species, we can pull all this evidence together to create a kind of biological or molecular clock.

Now, while we might not be able to exactly tell the time with such a molecular clock, we can compare different genomes of present-day species to estimate roughly when a species speciated (evolved and branched off into another species). Finally, we can also compare this genome evolution evidence gathered from present-day life with that of past isotopic evidence, from the geological record, and putting these together see if a coherent narrative emerges as to what might have taken place long ago.

The other thing about the 'boring billion' was, it is thought, that the atmospheric concentration of oxygen largely hovered between less than 1% up to somewhere around 4% (or possibly more) PAL for much of this time (see Chapter 3). Remember, hydrogen escape (Figure 3.3) was key—in addition to the evolution of oxygenic photosynthesis—to changing the atmosphere from having next to no free oxygen up to possibly very roughly 4% PAL oxygen and (as noted at the end of the last chapter) this non-biological process driving the increase in concentration stopped when oxygen concentrations were substantially greater than that of methane as any methane generated by life was mopped up by the said oxygen. This meant that (as is today) vanishingly little entered the upper atmosphere to be photo-dissociated and the 0.5%–4% or more PAL of oxygen conferred an ozone shield protecting the lower atmosphere from the higher-energy ultraviolet. Further to this, as Tim Lenton and Andrew Watson point out in chapter 13 of their *Revolutions That Made the Earth* (*1*), there were various nutrient cycles conferring positive and negative feedbacks (somewhat analogous to the climate feedback cycles in Figure 2.2 but involving nitrogen, iron, and molybdenum) that too helped stop the further rise in atmospheric oxygen concentration.

The above explains why the increase in atmospheric oxygen stopped following the transition and the boost conferred by the atmospheric chemistry outlined in

Figure 3.3. But why was there no subsequent further rise due to the new oxygenic photosynthetic microbial species?

Well, there may have been some variation across the boring billion, as estimates from the geological record are not precise. However, there is general agreement that despite some likely variation, oxygen concentrations were more or less constant. There were feedback mechanisms that prevented further rise. Here, quite likely is that there was ongoing methane release from anaerobic life that would have reacted with oxygen, so preventing further marked release. Remember, atmospheric concentrations of oxygen were still low during the boring billion and it is certain that the oceans were not as oxygenated as they are today so they still provided a significant habitat for anaerobic life.

To get an increase in oxygen to present-day levels would require a kick to the Earth system enabling it to cross another critical threshold. This did not happen during the boring billion.

Finally, before continuing with the co-evolution of life and planet narrative, some of you with some knowledge of the Earth system in the deep-past will be wondering how this new narrative reconciles with, or even supplants, the traditional view of Snowball I. Alas, just now is not the space to explore this in detail in this chapter but, be assured, it will be discussed later. Meanwhile, you can check the specific works as well as reviews cited in this book and so readers can follow up various avenues of interest as they wish. For students used to reading science papers, with the growth of open access policies, let alone online access, there has never been a better time, or easier, to track down the academic literature. For now, the important thing to note about the so-called boring billion is that it was a time that saw a new, turbocharged form of life flourish and that this would lay the ground for—following another flip of state—the Earth to become much more like the world we recognise today. And that is anything but boring.

## Turbocharging life

Backtracking a bit, up to somewhere around over 2 billion years ago, life consisted of simple cells: cells that were far simpler to the cells that make up your own body or the cells in the plants and (unless you are vegetarian or vegan) the animals you ate with your last meal.

Life of the early Earth consisted of simple, prokaryotic cells that include bacteria and eventually the cyanobacteria (which, remember, are sometimes confusingly called blue-green algae even though they are most certainly not plant algae): it was the prokaryotic cyanobacteria that first used oxygenic photosynthesis that—according to the new Earth system view—was integral in triggering Snowball Earth I and the Earth's first flip of state ~2.4–2.2 bya. And you will recall that these oxygenic cyanobacteria only managed to generate oxygen by bolting together two photosynthetic systems so as to be able to raise an electron's energy level to the degree necessary to split water. (See Chapter 3.)

These early, prokaryote cells consisted of a cell wall, a cell membrane, and an internal loop of genetic material (DNA); they did not have a distinct membrane-bound nucleus and neither did they have other membrane-bound components.

Conversely, the cells of your body all have—at least in some stage in their life cycle—a nucleus and internal membrane-bound specialised components. These more modern cells are called eukaryotes.

The term 'eukaryote' relates to the Greek '*eu*' meaning 'good' and '*karyon*' meaning 'nut' or 'kernel': 'nut' and 'kernel' standing in for the biological term 'nucleus'. (Not to be confused with the Latin word for 'kernel' which is the basis for the term 'nucleus'.) Hence 'eukaryote' in terms of its Greek roots means 'good nucleus'. Figure 4.1 consists of a very simple depiction of a prokaryote and beneath that a eukaryote.

Looking at Figure 4.1, and its simple cartoon of some prokaryote and eukaryote features, you can see that eukaryotes and prokaryotes are very different: it is not just size, there are a lot more structures inside eukaryotes. Of great relevance to us, there are also structures within a eukaryote cell that themselves have striking similarities with prokaryotes. These structures include chloroplasts (that undertake photosynthesis in plants) and mitochondria (that undertake energy metabolism in plants and animals). Prokaryotes are, being cells, bounded by a wall. Yet chloroplasts and mitochondria are also bounded by a membrane, all be they already within a eukaryotic cell. Prokaryotes have genetic material, and so do mitochondria. Indeed in animals (including humans), while nuclear DNA gets shuffled between generations with contributions from each parent, mitochondrial DNA (often written as mtDNA) in the mother's egg gets passed down with relatively little change. This is why scientists seeking to trace the earliest humans often say that they are seeking 'Eve', the earliest female human, because they are doing this by looking at mtDNA passed down through the generations by mothers.

Though eukaryotes and prokaryotes are different in size, prokaryote cells and mitochondria and chloroplasts (within eukaryotic cells) are more similar in size. Other similarities include that all three—prokaryotic cells, mitochondria, and chloroplasts—have inner-membrane infolds. A prokaryotic cell's infold is called a mesosome while an infold in a mitochondria is called a crista (the plural of which is 'cristae'): mitochondria have many cristae. Also, prokaryotic cells, mitochondria, and chloroplasts contain ribosomes (responsible for protein synthesis). Again, see Figure 4.1.

The question then arises, were these similarities are purely coincidental, or is there some sort of causal relationship between prokaryotes and mitochondria or chloroplasts?

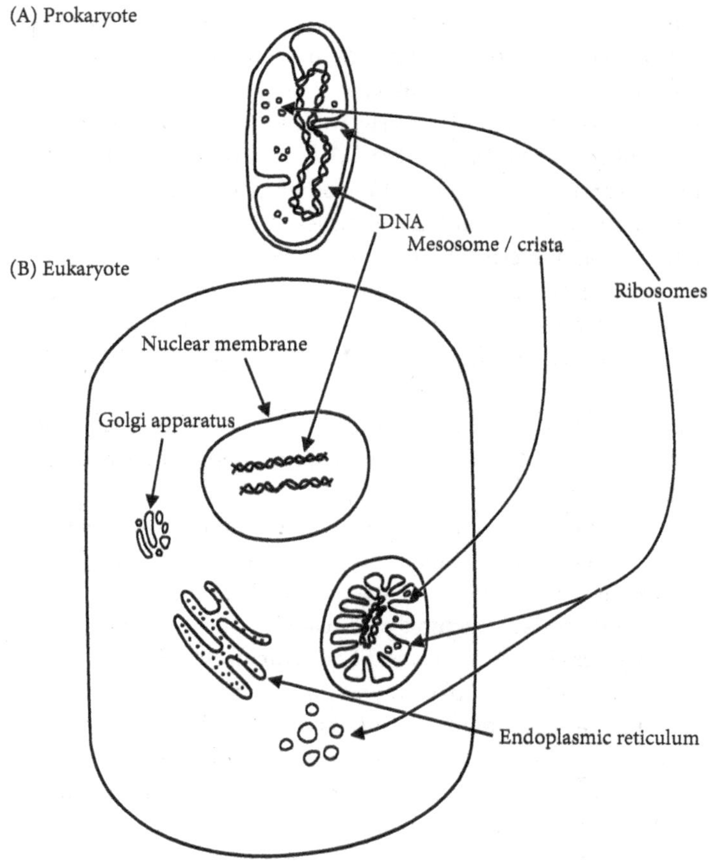


**Figure 4.1** (A) A simple depiction of an idealised prokaryote. (B) A eukaryote.

In science, this sort of coincidence-*vs.*-causal question has arisen before a number of times and arguably most famously in the geosciences with Alfred Wegner's hypothesis of continental drift in 1912 (we mentioned continental drift and plate tectonics in Chapters 1 and 3). Wegner noticed that the western coast of Africa was similar to the eastern coast of South America: could this be coincidence? Despite some early support—notably from Arthur Holmes in 1920—the idea of continental drift was largely dismissed (indeed, ridiculed by some) and it did not gain widespread currency until nearly half a century later in the latter half of the 1950s and early 1960s. Yet Wegner was right, the coastal similarities were not purely coincidence, as South America and Africa were once joined as Gondwanaland, that in turn had been part of a supercontinent called Pangaea.

Likewise, the notion that the similarities between prokaryotes and the organelles (such as mitochondria) of eukaryotes are not simply coincidental is now an established biological view.

This prokaryote cell and eukaryote organelle (such as the chloroplast and mitochondrion) similarity was first meaningfully articulated by the Russian biologist Konstantin Mereschcowsky in two papers in a German biological journal in 1905 (*2*) and then enlarged on in 1910 (*3*). However, as with Wegner and his notion of continental drift, the idea of prokaryotes somehow being related to mitochondria and chloroplasts in eukaryotes did not gain traction for many decades until the American biologist Lynn Margulis (at the time called Lynn Sagan) proposed a mechanism for the evolution of eukaryotes that involved prokaryotes (*4*) and subsequently popularised it in her book, *Origin of Eukaryotic Cells*, in 1970 (*5*).

While we will explore this further in the next section, the idea was that prokaryote cells—or rather a vestigial form of them—live within eukaryotic cells. This relationship, called endosymbiosis, has since gained significant traction, albeit that today there are a number of variations of the central theory as to how this came about. Furthermore, since Margulis's work in 1970, new biological tools (such as DNA sequencing) have enabled additional evidence to come to light and support the core theory even if some of the detail remains debated.

Both chloroplasts (which capture light energy to split water and create carbohydrates and oxygen) and mitochondria (which produce the ATP high-energy molecules) enable eukaryotic cells to channel more energy: you could say that they are kind of turbocharged.

So, how did the new eukaryote cells likely arise from prokaryotes?

## The evolution of the eukaryotes 2.5–1.5 bya(?)

As noted, the new advanced cells, the eukaryotes, are very different from prokaryotes. With regards to looks (Figure 4.1), the latter are smaller, they lack nuclei, they lack internal, membrane-bound structures, and their DNA is a single circular chromosome, as opposed to eukaryotes with multiple linear DNA chromosomes. There are other differences too. Prokaryotes reproduce through binary fission with genetic recombination by lateral gene transfer, whereas eukaryotes use a process called mitosis for cell division and meiosis (the process underpinning sex) for genetic recombination (genetically shuffling the pack). Furthermore, the larger more complex eukaryotes have genomes typically with around a thousand (or thousands) of times more base-pairs than the prokaryotes.

The other big difference between prokaryotes and eukaryotes is that the former have a strong cell wall but eukaryotes have a thinner, flexible one. Prokaryotes had a thick wall that prevented osmosis (the process by which smaller water molecules will seek to dilute a solution of chemicals on the other side of a membrane) from bursting them. This is particularly problematic for the thinner-walled eukaryotes (especially in freshwater environments) that require a pumping mechanism to bail out the incoming water: it most likely why the earlier prokaryotes had their thick wall. (We will also return to this in Chapter 5.)

As we previously noted, oxygenic photosynthesis is only carried out by cyanobacteria in prokaryotes as well as eukaryotic plants (including single-celled eukaryotic algae). Here, within eukaryotic plant cells, oxygenic photosynthesis is conducted by chloroplasts: the internal, membrane-bound structures that Konstantin Mereschcowsky observed that had some striking similarities to prokaryotes.

For many years it was thought that eukaryotes somehow evolved directly from prokaryotes. However, as noted in the previous section, Konstantin Mereschcowsky in 1905 speculated that eukaryotic plant cells acquired the ability to undertake oxygenic photosynthesis from cyanobacteria. They did this through some form of symbiosis whereby the eukaryotic plant cells somehow took onboard (engulfed) some prokaryotic cyanobacteria and these then became the chloroplasts found in plant cells today. Mereschcowsky's idea, while intriguing to the scientific community, did not become accepted in his lifetime though it is now firmly entrenched in biology.

The story as to how this understanding came about has recently (2015) been summarised in an open access paper by William Martin, Sriram Garg, and Verena Zimorski of Germany's Dusseldorf University in the *Philosophical Transactions (B) of the Royal Society* (6). They note that a German biologist, called Andreas Schimper, had actually previously considered Mereschcowsky's idea back in 1883, but he only mentioned it briefly in a footnote: Konstantin Mereschcowsky was the first to consider it in detail in his 1905 paper.

In 1925 the American biologist Ivan Wallin also posited the idea that mitochondria—the organelles responsible for key aspects of energy metabolism within eukaryotic cells of both plants and animals—were originally prokaryotes (7). He would later work with Konstantin Mereschcowsky.

There were a number of reasons why at first Mereschcowsky's and Wallin's ideas did not gain much traction with the rest of the scientific community at the time. The suggestion that somehow free-living bacteria became incorporated within, to become an integral part of, eukaryotic cells seemed too outlandish.

The then-predominant view of Darwinian evolution was that successive generations of an organism would experience mutations and that only the beneficial mutations (making the organism 'fit') would survive to be handed down to further future generations, while those with deleterious would die out: the idea that two different microbial species could merge to become a very different, synergistically greater, new species (strictly speaking 'domain' or 'superkingdom') was simply too radical; that two different organisms could come together to form a single entity—a chimera—was unbelievable and too exotic for many.

The criticism was so strong and vociferous that Wallin himself discarded the idea. One unresolved question (of a number) that made the Mereschcowsky–Wallin hypothesis difficult to accept was, if it was true, then what happened to the original prokaryote's DNA? The accepted wisdom in the first half of the 20th century was that *all* eukaryotic cell's genetic material was in the membrane-bound nucleus, the nucleus that was quite separate to chloroplasts and mitochondria. It was not until 1962, when Hans Ris and Walter Plaut (1962) demonstrated the presence of DNA in chloroplasts (*8*) and then in 1963 with the discovery of mitochondrial DNA (mtDNA) by Margit and Sylvan Nass by electron microscopy (*9*), that the idea of genetic material outside of the nucleus was accepted. This would have lent credence to the Mereschcowsky–Wallin hypothesis were it not that their hypothesis—such was its reception's disdain—had been largely forgotten. Yet, among a few, the idea that these organelles might once have been free-living prokaryotes quietly gained credence. The partnership of the ancestors of eukaryotic cells having prokaryotic cells living within them became known as 'endosymbiosis' and the evolutionary mechanism for this 'symbiogenesis'.

In the late 1960s, the aforementioned Lynn Margulis began to play a leading role in re-popularising, and indeed further developing, the hypothesis so that today with regards to the evolution of cells it is now considered as accepted theory. One of her earliest significant papers on the topic was her paper written in 1968 under her married name of Sagan (*4*).

Acknowledging the earlier work of Wallin, Mereschcowsky, and others, as well as drawing on more recent understanding of microbiology, Lynn (Sagan) Margulis marshalled a cogent argument. Her suggestion was that a prokaryote with flexible cell wall (she considered a spirochaete as a likely candidate species) ingested a cyanobacterium through a process called 'endocytosis'. Here the endocytosis involved the spirochaete engulfing the cyanobacterium and ingesting it whole. (This is a particular type of endocytosis called 'phagocytosis'.) The process is one you might have seen in school biology with an amoeba ingesting a particle. However, in this instance, instead of then digesting the ingested cyanobacterium, the idea is that one spirochaete kept the cyanobacterium bacterium alive. The spirochaete therefore gained the benefits of photosynthesis while the cyanobacterium acquired a safe home within the spirochaete. A similar process, Margulis thought, was behind the development of mitochondria.

There is not the time here to go into all the detail of the various nuanced variations of endosymbiotic theory; suffice to say that in its broad terms it has become established within biology. It is, though, worth looking at four recent areas of research that are illuminating our understanding of the evolution of early eukaryotes.

First, in 2015 an international team of European scientists and their lead author Anja Spang, were looking at the genomes of bacteria found when they were surveying microbial diversity in the deep marine sediments near hydrothermals from the Arctic Mid-Ocean Ridge (*10*). What they found were members of the prokaryotic Archaea, or if not that then cousins to Archaea, to which they assigned the phylum name of 'Lokiarchaeota'. The Lokiarchaeota have a genome that has many of the genes that are common to eukaryotes and so they are therefore likely to be closely related ancestors of the first eukaryotes.

As Anja Spang and her team explain, the name 'Lokiarchaeota' was coined in reference to the hydrothermal complex sampling location, Loki's Castle, which in turn was named after the Norse mythology's shape-shifting deity Loki. The deity Loki has been described, in analogy to the ongoing debates on the origin of eukaryotes, as 'a staggeringly complex, confusing, and ambivalent figure who has been the catalyst of countless unresolved scholarly controversies'.

Not mentioned in Anja Spang and her colleagues' paper, but which will be of relevance later when considering the new notions of Earth flips and the evolution of life against past-accepted explanations, is that the hydrothermals from the Arctic Mid-Ocean Ridge are hydrothermals in a zone of plate tectonic spreading. As mentioned above, eukaryotes are thought to have evolved (if not earlier) around the time of the breakup of a supercontinent; that is to say, when a supercontinent was being fractured by spreading geological faults. In other words, hydrothermals would be common around and in the breakup, inland or intruding seas of supercontinents and life by a number of these vents would benefit from nutrient runoff from the said supercontinents and so thrive. Plenty of life means more chance for mutation, speciation, hence evolution.

The second piece of research was published in 2005 having been undertaken by German researchers from the University of Köln led by Birger Marin (*11*). They looked at the DNA of a freshwater amoeba, a eukaryote called *Paulinella chromatophora*. They found that the DNA of its chloroplasts differed from that of other eukaryotic oxygenic photosynthesisers and—molecular-clock analysis suggests—the ancestors of *P. chromatophora* acquired photosynthesising cyanobacteria much later than other photosynthesising eukaryotes. This indicates that the evolution of photosynthetic organelles through symbiogenesis to endosymbiosis from cyanobacteria was not an unique event as might be believed, but happened a number of times; it may even be an ongoing process. Indeed, we already know that endosymbiosis also took place with mitochondria, and so with symbiogenesis also happening at least twice with chloroplasts, this suggests that symbiogenesis is not that hard an evolutionary step.

The third of the four papers comes from two cousins working on either side of the Atlantic, in Wisconsin and London: David and Buzz Baum. It concerns the mechanism, symbiogenesis, by which endosymbiosis arose. All told, there are a score or so of theories as to how one cell engulfed another through phagocytosis, starting with Lyn Margulis's suggestion of a spirochaete engulfing the cyanobacterium. There is not the space here to go into all the Margulis-type endosymbiotic theories, but in essence they are all 'outside in' hypotheses. 'Outside in' in that the prokaryote started off *outside* and then through 'phagocytosis' was ingested so ending up *inside* what would then become an eukaryotic cell with the ingested prokaryote becoming a chloroplast or mitochondria.

What David and Buzz Baum came up with was the opposite: what they called an 'inside out' model.

Here, the 'inside out' model has the photosynthesising cyanobacterium (or the energy-harvesting prokaryote) outside—but next to—another larger prokaryotic cell that effectively confers protection and possibly some metabolites for the prokaryote next to it while getting the smaller prokaryote's own metabolic products. This is like the straightforward symbiotic relationship found elsewhere in biology, such as birds picking fleas off the back of buffalo: the bird gets a meal and the buffalo gets rid of its parasites. However, with the cyanobacterium and the larger prokaryote, what happens next is that a bit of the larger cell extends and protrudes (Figure 4.2A ii) through a pore in its own cell wall to touch more of the neighbourly prokaryote; in this way the larger cell gets more of the prokaryote's metabolites. This extension (or 'bleb') then enlarges and wraps itself around the prokaryote (Figure 4.2A iii and iv). Eventually the prokaryote is engulfed (Figure 4.2A v). The membranes exposed to the environment beyond the partnership toughen and finally the original larger cell's own cell wall is reduced in thickness leaving just a thin membrane that encloses the cell's nucleus (Figure 4.2A vi) (*12*).

One advantage of the Baum cousins' 'inside out' model is that not only does it explain how prokaryotes became organelles, it also explains the evolution of the eukaryotic cell's nuclear membrane and possibly other membrane structures found within eukaryotic cells. Another is that it also obviates the need to explain, with Margulis's 'outside in' theory, why the ingested prokaryote did not simply get digested as are food particles. The Baum cousins' 'inside out' model explains both more and also more easily.

The fourth paper on 'inside out' endosymbiosis for the evolution of eukaryotes relates to the first, Anja Spang and colleagues' paper on Lokiarchaeota

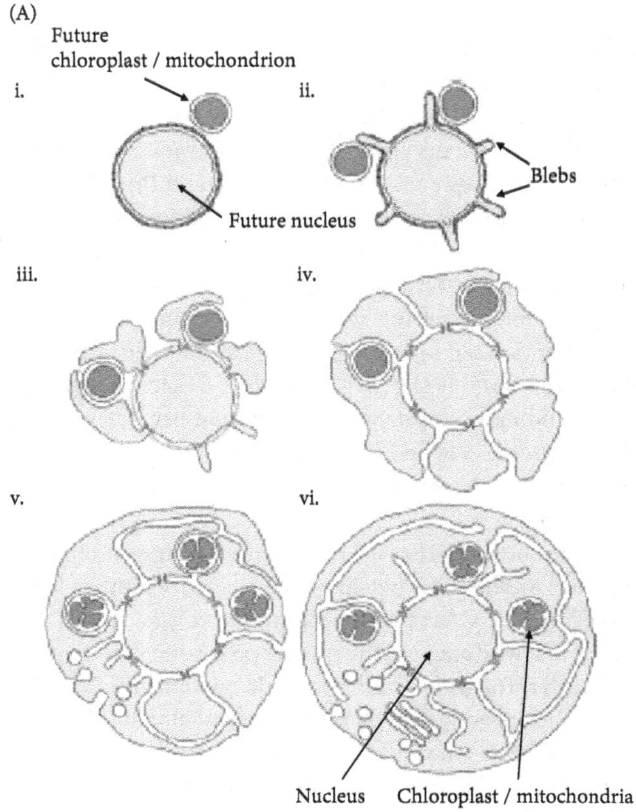


**Figure 4.2A** 'Inside out' hypothesis for the endosymbiotic development of chloroplasts and mitochondria and the evolution of the eukaryotic cell.
Adapted from Baum, D. A., & Baum, B., An inside-out origin for the eukaryotic cell. *BMC Biology*, **12**, 76 (2014). Reproduced with permission.

likely to be closely related ancestors of the first eukaryotes. This new work's lead authors were Hiroyuki Imachi and Masaru K. Nobu, microbiologists, with the latter working at the Japan Agency for Marine-Earth Science and Technology in Yokosuka. He and his team built a chamber that simulated conditions in methane-rich sediments of the type near the thermal vent where Anja Spang found the Lokiarchaeota, though the Japanese researchers obtained theirs from deep-sea methane seep sediment of the Nankai Trough, 2,533 m down. They cultivated and refined this microbe community from the vent until they had one that consisted of just Lokiarchaeota together with a different methane-producing archaeon. They were then able to take micrograph pictures of the

(B)

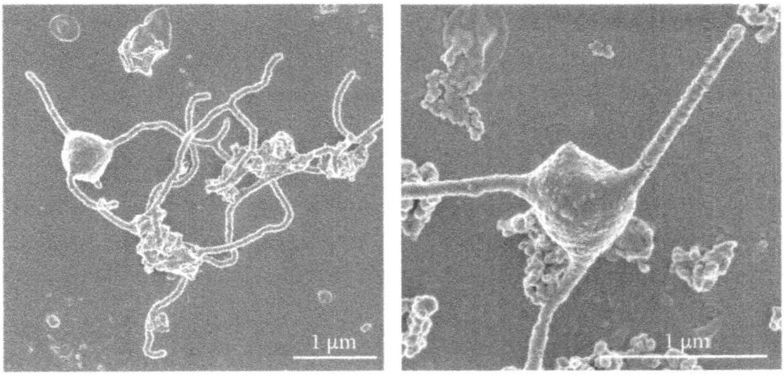


**Figure 4.2B** Micrographs of a Lokiarchaeota. (Please see the Introduction and Acknowledgements for an update as to this species' nomenclature.) Micrographs provided by Hiro Imanchi.

 As from Imachi, H., Nobu, M. K., Nakahara, N., Yuki Morono, Y., *et al.*, Isolation of an archaeon at the prokaryote-eukaryote interface. *Nature*, **577**, 519–525 (2020).

Lokiarchaeota. (Please see the Introduction and Acknowledgements for an update as to this species' nomenclature.)

What was striking, when seeing a picture of a Lokiarchaeota prokaryote, is that it has a central mass but coming from it are a few (three or four) long thin protuberances. A news article in *Nature* referred to them as 'wisp-like protrusions' (*13*). To my mind, it was easy to see these wrapping around smaller prokaryotes with which they may have a symbiotic relationship. If this is so, then it could be the first stage in the process of prokaryotes becoming eukaryotes. What do you think? See Figure 4.2B.

Hiroyuki Imachi, Masaru K. Nobu, and colleagues in their own paper themselves envision a variant process of David and Buzz Baum's (see above) 'inside out' model for the evolution of eukaryotes. This was explicitly stated in their preprint (*14*) but, for whatever reason, was excised in the subsequently published peer-reviewed version (*15*).

The evidence that prokaryotes became endosymbionts is more than compelling, even if there are plenty of details remaining to elucidate. Given that we have the University of Köln team's (and other) evidence suggesting that chloroplasts arose from prokaryotes more than once (*11*), so it could be that there was more than one endosymbiotic mechanism, albeit possibly nuanced variations on a single theme. Indeed there is other evidence (too much to go into here) that makes the idea of there being multiple endosymbiotic events convincing.

Finally, with regards to the evolution of mitochondria from prokaryotes, the Arctic Lokiarchaeota evidence that mitochondria evolved from bacteria-like prokaryotes near submarine thermal vents also explains how mitochondria handle electrons so well, if, that is, the prokaryotes' ancestors were involved in metabolising hydrogen.

To put this into the broader context of this book's narrative, do note that such vents are common near spreading plates, and that supercontinental fracture into more than one plate initially sees a confined marine area (inland waters/seas) that can accumulate run-off from the supercontinent, hence nutrients, so allowing life there to thrive.

The above broad, endosymbiotic idea as to how eukaryotes likely evolved is now firmly embedded in science even if some elements still have to be elucidated and confirmations needed: the notion of the endosymbiotic origin of mitochondria and chloroplasts has gone from hypothesis to accepted theory.

One final thought, the nucleus of prokaryote cells contains information in the form of genetic code (this was briefly touched on in Chapter 3). With endosymbiosis the eukaryote's nucleic acid (genetic) information is added to by the genetic information in the endosymbiotic prokaryote that is now a mitochondria or chloroplast: hence mitochondrial DNA. Could this additional information be a diagnostic indicator of a key evolutionary step? For now consider this increased information dimension a musing, but it's one we will return to later (in Chapter 8).

The rise of eukaryotes and their establishment is quite a story and, once they had evolved, much of it happened in the so-called 'boring billion'. Yet, given there were a number of endosymbiotic origin events, when did the evolution of eukaryotes first take place?

## When did the first and the more complex eukaryotes evolve?

The timing of the evolution of the first eukaryotes, let alone more complex ones, is debatable. In answering this section's titular question this observation is something of a disappointing start. Nonetheless, it is a critical question in terms of the biosphere evolution of the carbon cycle. For example, did eukaryotes evolve at the time of Snowball Earth I or after it? If eukaryotes evolved at the time of, or before, Snowball Earth I then perhaps the new dimensions to carbon cycling eukaryotes afforded enhanced the greenhouse, carbon dioxide gas drawdown and so contributed to the cooling with Snowball Earth I. (And this last does not alter the idea presented in the last chapter that the Earth's first large supercontinent's rain facilitated erosion releasing nutrients: oxygenic photosynthetic eukaryotes would equally benefit from nutrient release as would oxygenic photosynthetic, prokaryotic cyanobacteria.)

Alternatively, if eukaryotes evolved during the so-called boring billion that followed Snowball Earth I, then any eukaryote-caused changes to carbon cycling could not have lowered atmospheric greenhouse gas climate forcing in any very major (Snowball Earth-inducing) way. This last (late eukaryote evolution) might seem unlikely because surely, with the more advanced eukaryotes that photosynthesised, it would seem likely (with early eukaryote evolution) that there would be more photosynthesis globally hence more drawdown of carbon dioxide needed for a Snowball Earth. The counter to this argument is that single-celled eukaryotic species, such as some species of amoeba that feed off organic matter from their single-celled plant counterparts as well as prokaryotes, respire out carbon dioxide. So it could be that while there was more carbon drawdown from the atmosphere following the evolution of the eukaryotes, so there was increased carbon return to the atmosphere. The carbon cycle might have changed, but not in a way that changed the amount of atmospheric carbon, rather in that the carbon cycle speeded up.

Such questions are of great interest to biosphere and Earth system scientists and lend importance to asking, when did the first eukaryotes evolve? Saying it is 'debatable' is not particularly helpful.

So what is the evidence?

Fossils of single cells are obviously small and so harder to detect than, say, the big bones of dinosaurs (which only came into existence much, much later). Also, single cells, lacking bone and hard parts, do not readily fossilise. Yet there are some microscopic objects found in the geological record that could possibly reflect the remains of these early more complex single cells. Even so, some of what are claimed to be such microfossils could easily be coincidental artefacts rising out of pure chance: they just happen to look like they might have once been cells or colonies of cells. (Multicellular eukaryotic species are easier to identify but, clearly, they had to evolve *after* the rise of single-celled eukaryotes. Fossils of multicellular species are easily discernible in the geological record in the Ediacaran period that began 635 mya, and so eukaryotes must have first evolved well before then.) Finally, on the fossil front there is less accessible rock that dates from further back in time.

This means that we have to rely on other, secondary evidence (and return to the question of eukaryotic fossils shortly). Such alternative evidence includes molecular evidence.

An example of indicative molecular evidence includes sterols. Eukaryotes have some unique biochemicals in their cell membranes, and these break down to sterols. (You may not know that you have likely come across sterols before, as they are a class of chemicals that includes 'cholesterol', a substance of which you will surely have heard, that all animals make.) In 1999 a team of researchers led by Jochen Brocks discovered sterols in oil from strata in Pilbara, Australia. These shale strata date from 2.7 billion years ago. Could this, as the researchers suggest, be a sign of early eukaryotes (*16*)? Here, one problem is that sterols are also made by a few species of bacteria and so this evidence is not as conclusive as would be liked but, it can be argued, it is nonetheless at least suggestive.

So, while the above work has been replicated by others, not everyone is convinced and even some of those who made the original discovery have their doubts.

In 2015, US geoscientists Katherine French and Christian Hallmann led a small team who repeated the sterol analysis but worked at being as scrupulously clean as possible to avoid contamination of their samples. They could not find such sterols (*17*). They also supported their conclusion with their argument that eukaryotic cells need free oxygen and that 2.7 billion years ago was *before* the Great Oxidation even and the Earth's first big flip around 2.2 billion years ago changing state from an anaerobic world to a 0.5%–4% PAL aerobic one: so there was no free oxygen 2.7 billion years ago. (Remember, at the most there were only the faintest of traces measured in parts *per* million or even billion.)

Having said this, there is a counter to the above argument. True, there may not have been free oxygen in the atmosphere 2.7 billion years and earlier, but there could hypothetically, just possibly, have still been eukaryotic plants (algae) generating their own oxygen internally for their own consumption with any leakage being mopped up by methane in the atmosphere. Such life would be a bit like a human carrying their own oxygen tank when scuba diving. Nonetheless, 2.7 bya might be considered a little early if cyanobacteria oxygenic photosynthesisers evolved 3.0 bya: in the great scheme of things 0.3 billion years (the difference between 3.0 and 2.7 bya, 300 million years) is not that long.

And so the debate continues. Where we are in the decade beginning with 2020 is waiting to see if others using ultra-clean techniques also fail to find sterols from 2.7 billion years ago.

Nonetheless, those favouring the idea of an early evolution of eukaryotes can also draw some (limited) support from bio-clocks, or molecular clocks, based on the degree of genetic variance between species mentioned earlier this chapter. (Remember, as time progresses so mutations [changes] accumulate and this can be used as a very rough dating mechanism.) This method is far from exact and the range of estimates for the evolution of eukaryotes made this way extends back well beyond 2.7 bya. Yet as the number of genomes analysed increases, and more experience is gained, accuracy will increase.

So where does this leave us? British Earth system scientists Tim Lenton and Andrew Watson (whose work was mentioned above as well as in

Chapters 2 and 3) in 2011 consider the biomarker evidence as reasonably strong for some sort of early eukaryote (a 'proto-eukaryote') being between 2.45–2.2 bya (*1*). This is around, indeed overlaps with, the time leading up to the Great Oxidation and the first Earth Flip, which, if true, makes one wonder whether the rise of single-celled plant eukaryotes—a kind of proto true algae—was the (or one of the) causal factor(s) behind the first great Earth flip with Snowball Earth I, Though it is important to remember the possible alternative that this Great Oxidation Event could have been down to cyanobacteria alone, and that eukaryotes rose later during the boring billion. Tim Lenton and Andrew Watson suggest that true eukaryotes did not rise until around 2.0 bya (*1*). This was after Snowball Earth I, and so their view is that it was likely that proto-eukaryotes were involved with that event rather than true eukaryotes.

Here, for the record, I should perhaps nail my own flag to the mast. My personal contention is that carbon drawdown by prokaryotic cyanobacteria undertaking oxygenic photosynthesis (OP) was responsible for the early Pongola glaciation (2.9 bya) but that eukaryotic OP (together with cyanobacteria) were implicated in Snowball I some 400 million years later (~2.4–2.2 bya). So eukaryotes (or their precursors, perhaps some Lokiarchaeota-type species in a symbiotic relationship) possibly arose after 2.9 bya (maybe around 2.4 bya at the earliest?) but certainly before 2.2 bya. However, let's look at the range of estimates others have given.

Elsewhere in the 'eukaryote evolving date' debate, ranges of estimates vary. As mentioned, there are those that think they could have evolved quite late: eukaryotes could have arisen around 1.5 bya in the midst of the so-called boring billion. (And of course, remember that we are talking about when eukaryotes 'first' evolved as, as we have previously noted, symbiogenesis leading to eukaryotes occurred a number of different instances and times. So some eukaryotes may have arisen as late as in the boring billion.)

What certainly did happen later in the boring billion was the diversification of eukaryotes. Both biomarkers and bio-molecular clocks suggest that eukaryotes were diversifying *after* the Earth's first great, big flip and its 'Great Oxidation', so diversifying during the boring billion. For example, it is likely that the big 'split' between most plants and fungi (or, for biologists, the split between bikonts and unikonts, respectively) took place somewhere at the very least between 900 mya and 1.2 bya, but quite possibly a little earlier. This split is but one component to the diversification of eukaryotes. (Though 'split' is not really accurate, as it is more likely that unikonts and bikonts arose independently of each other in separate endosymbiotic events.) Clearly though, for eukaryotes to diversify they had to evolve before then, and here there is still that question, exactly when?

Returning to fossils, some have been found that are thought to possibly be of one of the first early eukaryotic aerobic species and originally considered to date from 2,100 mya. They were similar to a species of multicellular alga living today, *Grypania spiralis*. But it is possible that the dating of this was overly early. Also, though this looks like it could be eukaryotic, this is not certain (*18*). Another good candidate for a modern descendant of an early eukaryotic oxygenic photosynthesiser has been suggested by a large international team of researchers in 2008 led by Robert Moore and Miroslav Obornik. It is a microbial marine brown ball related to present-day apicomplexian parasites (one of which you may know, *Plasmodium falciparum*, the malaria parasite). This brown ball is called *Chromera velia*, and its relation to other species came as a surprise, for though related to apicomplexian parasites, which live within animals and so don't use sunlight, this one is fully photosynthetic (*19*). The trouble is that, while this may be a good candidate for this species being a modern descendant of an (there was more than one) early eukaryotic photosynthesiser, the evidence is purely what we see in its descendents today, and so we have no strata to date. Furthermore, the evidence suggests that strictly speaking these hypothetical early eukaryotic photosynthesisers were precursor eukaryotic species and not true eukaryotes.

If it really is difficult to pin down a date for the origin of eukaryotes from fossils, then what else do bio- or molecular clocks tell us?

One of the more recent estimates came in 2014, when Laura Eme of Dalhousie University, Canada, was the lead author of a paper that compared the differences of 159 proteins as they mutated/changed with evolution. To do this they looked at 85 species and calibrated the rates of change with 19 known evolutionary divergences within the fossil record. This method is sound and would have given a clear answer were it not that the fossil record of evolutionary change is most accurate in more geologically recent times, and not at all detailed, or amenable to accurate dating, a billion or more years ago.

To understand this, let's take a clear but geologically recent example, we might look at a protein in a brown bear and a polar bear today. We note a certain number of differences. We also know that brown and polar bears diversified from each other 400,000 years ago. And so we can say that this amount of protein change took place over that time and use this as one instance of calibrating bear protein and genetic change with time. However, the divergence of eukaryotes took place very much prior to 600 mya (as by then multicellular species had evolved), and possibly it happened as much as 2 bya, if some of the tentative fossil conjecture is to be believed. Even if the assumption that the rate of protein molecule change remained the same over this time is sound, the further back into this unknown territory we go the greater the uncertainty.

Laura Eme and her colleagues have tried to get around this by correlating the results from many molecular clocks. Even so, the margins of error are large.

Just how large is made clear by the Eme team's study (*20*). They estimate the age of divergence from the last eukaryotic common ancestor to be between a recent estimate of 1,007 mya (1.007 bya) to an older estimate of 1,898 mya. However, both these recent and older estimates have their own margins of error. The recent 1,007 mya estimate itself has an upper and lower limit of 943 mya and 1,102 mya, respectively. Whereas the older 1,898 mya estimate has an upper and lower limit of 1,655 mya and 2,094 mya, respectively. This means that their estimate for divergence from the last eukaryotic common ancestor is somewhere in the envelope of 943 mya and 2,094 mya. This effectively covers much of the boring billion period. Indeed, it almost gets us back to the end of Snowball I (2,200 mya or 2.2 bya).

Fortunately, there is a mix of evidence and we do not have to rely on just one category, or even one study within a category of evidence, when forming an opinion as to when the first eukaryote evolved: though clearly this must have been at least a little before this ancestral eukaryote diverged. So, biologists can come to an opinion even if it is only an informed guestimate. If Snowball Earth I, around ~2.4–2.2 bya and its 'Great Oxidation', was associated with prokaryotic cyanobacteria having a global impact following having evolved the ability to undertake oxygenic photosynthesis (which must have originated at least a little before then, as said, possibly around 2.9 bya if not a little earlier), then the next step of endosymbiotically bringing cyanobacteria into the proto-eukaryotic cell must have been after that time. (This at least almost fits in with the Eme study's oldest estimate for eukaryotic divergence from the initial ancestral cell of 2,094 mya.)

Yet, some others feel that eukaryotes arose somewhere around 2 bya or a little earlier. This is not that long, in the grandest, deep time, geological terms, after Snowball Earth I (though you might be forgiven in feeling that a couple of hundred million years was indeed a long time) if not during it. Others consider that eukaryotes first rose a little later. (Indeed, as time was needed for evolution to take place and given that cyanobacteria were an endosymbiotic partner in the evolution of photosynthetic eukaryotes, you might ask yourself, how much time was needed following the evolution of cyanobacteria for eukaryotes to first evolve?) What is likely, taking different views into account, is that this eukaryote evolution at the *very latest* first happened during the so-called boring billion. Here we can sensibly hypothesise (or 'guess' as the Nobel-winning physicist Richard Feynman would say) that, if at the latest it first took place earlier in the boring billion than later, then it is possible that it actually took place earlier: certainly proto-eukaryotes evolved earlier, as it is unlikely that the major key steps in evolution all occurred together—time was needed for diversification and survival of the fittest. We can, though, take comfort in that as research continues more fossil and molecular clock evidence will come to light. When it does we will be able to date the age of eukaryotes' rise more precisely.

If the evidence suggests that at the latest eukaryotes first arose early in the boring billion then it is not a huge stretch to consider that the earliest they could have evolved was at the time of, or before, Snowball I (~2.4–2.2 bya). If they had, then their photosynthetic capability would have made Snowball I taking place more likely. This, I contend, is what happened. Though remember, some others (but not all) do disagree.

Notwithstanding the above eukaryotes-evolved-early-in-the-boring-billion estimate, we must recognise that that is still only a 'guestimate', albeit a reasoned one with some supporting evidence. We also need to recognise that such reasoning might still be wrong; we need to at least recognise that if it is wrong that the reality still must lie within the wider ranges that the evidence dictates. Here, the bottom line of this wider range is that eukaryotes seem to have first evolved within an envelope (or boundary) of time bounded earliest around, or a little before, the Earth's first really big flip of state ~2.4–2.2 bya and bounded at the latest was 1.5 bya in the midst of the so-called boring billion. And if eukaryotic evolution and its oxygenic photosynthesis was not involved in instigating Earth's first big flip, Snowball Earth I, then cyanobacteria's oxygenic photosynthesis most certainly was.

Either way, life—according to the new Earth system science thinking—was involved in Snowball Earth I taking place.

As said, my personal thoughts are that within the envelope of estimates given above, eukaryotes evolved earlier rather than later. Here, I wish to make a further point regarding the evidence. I am mindful of what is known as the Signor–Lipps effect. Named after Philip Signor and Jere Lipps, this is the idea that any sampling of the fossil record will not hit upon the remains of the very first (or last) of a particular species. This is for two reasons: (1) the fossil record is incomplete, and (2) a sampler is unlikely to happen upon the very first (or last) fossil of a particular species (*21*). This a bit like opening at random early on in the dictionary you are unlikely to hit the very first page of words beginning with the letter 'c'. True, you may get close and if you are already in the 'c's you will turn back, and if in the 'b's you'll flip pages forward: the thing is, you are unlikely to hit the page containing the very first letter of the 'c's by simply opening the dictionary more or less at random. (Unwittingly, astrophysicist Richard Gott III independently, and subsequently, followed this line of thinking and created a 95% confidence simple model to suggest the possible future lifetime of our human species among other exotic questions [*22*].) I am therefore inclined to go along with the likes of the previously mentioned Lenton and Watson (*1*), of an early estimate for the first eukaryote (or at least proto-eukaryote) at 2.7 bya and perhaps even a little earlier than that: would 3 bya be too much of a stretch? Possibly (it is too close to the rise of cyanobacteria which needed to be there first). Specifically, with regards to true eukaryotes (not some proto-eukaryote), Lenton and Watson estimates their first rising later than I at somewhere around 2 bya: as I said, there is some debate.

Remember, in the very loosest sense (such is the difficulty of hypothesising the dating of event so long ago), this is somewhere around the time of the 2.4–2.2 bya Snowball I and, given the evidence weak as it is, it is hard not to think that there is an association between the rise of OP eukaryotes and that event.

We also need to bear in mind that others argue for a much later evolution of eukaryotes. In 2023 a small, largely German-based group of researchers lead by Jochen Brocks and Benjamin Nettersheim, suggest that eukaryotic life appears to have flourished surprisingly late in the history of our planet. (Their words 'surprisingly late' not mine.) Their view is based on proto-sterol evidence in sediments dating from 1,640 to around 800 mya. However they do not attribute the source of these proto-sterols to true eukaryotes but 'probably' (again their word) to ancient protosterol-producing bacteria and deep-branching stem-group eukaryotes. These last are now extinct, early eukaryotic relatives. The team are of the view that modern eukaryotes started to appear in the Tonian period (1,000 to 720 million years ago) (*23*). This really is very late, as less than a hundred million years later we most certainly had multicellular life that flourished. But if you are of this view, and as the Brocks and Nettersheim team themselves say, this 'Tonian transformation' emerges as one of the most profound ecological turning points in the Earth's history. Having said that, they do also hypothesise that stem group eukaryotes could have existed as early as nearly 1.7 billion years ago . . .

Now, readers might have found this section a little confusing, such have been the range of estimates as to when eukaryotes arose and holding these in your mind to compare them with events. To make it easier, these are visually laid out in Figure 4.3 together with the suggested rise of cyanobacterial OP discussed in Chapter 3.

So, was Snowball I associated with the rise of OP eukaryotes?

Snowball I took place at a time of the breakup of a supercontinent with narrow seas penetrating the newly fractured continent. These seas spreading inland had thermal vents, and we do see Lokiarchaeota associated with those events today. These inland seas would have been fed by considerable in-continent rainfall hence nutrient runoff. Add this nutrient supply to the volcanic activity driving the fracturing, meaning plenty of submarine thermal vents, and you could consider this proto-eukaryotic and prokaryotic heaven: there would have been both sea vents' infrared for the proto-eukaryotes and a sea vent nutrient-rich environment for cyanobacteria near the sea surface to access solar energy.

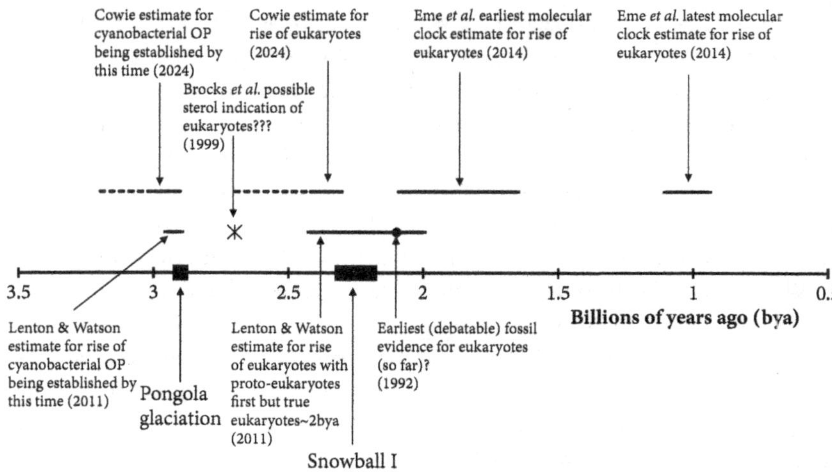


**Figure 4.3** When did eukaryotes evolve? See text and text references. Note to be clear: the horizontal lines depict the rough range of each estimate (the vertical arrows do not denote exact dates but point to the range lines). (I should point out a detailed nuance in my estimate for *cyanobacterial* OP 'being established *by* this time'. My personal opinion is that OP arose well *before* this time.)

I would add to the mix that in places there was a great tidal range—which you get off continents bordering a large ocean—that facilitate landform inlets cycling through periods of isolation and seawater mixing. (Remember, the Moon was closer to the Earth over 2 billion years ago and days were shorter. Both would facilitate tidal erosion of shores, hence nutrient addition.) It was therefore a time of interconnected, diverse environments with a good nutrient supply. One might therefore reasonably suspect that along with the oxygenic cyanobacteria benefiting there would equally be large populations of a diverse number of other prokaryotes. This provides an increased opportunity or chance for symbiotic relationships to form and significant evolutionary change (such as endosymbiosis) to take place. These nutrient supply and tidal range matters I will return to at the end of the next chapter when discussing the second Snowball Earth.

Irrespective of when—within the boring billion or earlier—eukaryotes evolved, by the middle of the boring billion, Earth had both cyanobacteria and eukaryotic life. And though atmospheric oxygen had not reached present-day levels of 21%, it was still a few per cent: the Earth was a very different place from what it was before Snowball Earth I—the flip of planetary state and the 'Great Oxidation'. Indeed, life on Earth was now equipped to move further on.

## Going multicellular

Most of the species you most likely refer to on a daily basis are things like mammals (cats, dogs, cows, you, *etc.*), birds, trees, garden plants (be they ornamentals for garden decoration or vegetables for eating), and fungi. *All* these species are multicellular: they have many cells of different function making up the individual. The rise of true multicellular life took place during the boring billion but after the rise of eukaryotes and so most likely in the boring billion's second half.

At this point it is briefly worth noting the afore term 'true multicellularity', as it is possible to have intermediate forms between single-celled species and those that are properly multicellular. These can consist of many cells that form a single connected, clearly structured, whole organism. For instance, there can be colonial organisms that are effectively single cells capable of independent existence that have come together to form a (less clearly) structured form. An example of this would be those species that belong to the phylum Porifera, the sponges. They are among the immediate ancestors of the simplest of multicelled animals, or Metazoa. But they are not true Metazoa: they are not Eumetazoa (remember from the Greek '*eu*' meaning 'good') which include fish, birds, mammals, and all that you would consider as being proper animals. Nonetheless, sponges do consist of a conglomeration of many cells and so are multicellular.

The way biologists often classify species, multicellular fungi, animals, and plants all belong to different kingdoms. ('Often', because over the years there have been different species classification systems: there's a lot of baggage in biology.) So you might think that a clue as to what was the first multicellular species could be obtained by mapping the genes of representatives of each of these kingdoms and then see which genes all these have in common. Such an analysis, you equally might think, would give us a set of genes that would also be present in their common ancestor (albeit in a slightly different form due to mutation over generations). Indeed, when we look at the genomes of fungi, animals, and plants, we do find that there are some genes that all these have in common. However, these in-common-genes are not connected to multicellularity functions; instead they have more individual cell housekeeping functions such as DNA replication, repair, and so forth. The multicellularity genes of those in the kingdoms of Fungi, Animalia, and Plantae—such as intercellular communication—are all different. The conclusion is that multicellularity (as did the earlier evolution of eukaryotic cells) arose independently a number of times to give us the distinct kingdoms that respectively ended up as fungi, animals, and plants (24).

This conclusion of the evolution of multicellularity multiple times has been arrived at by various independent studies even if estimates as to the number of times it arose differ. For example, Laura Parfrey and Daniel Lahr of the universities of Colorado and São Paulo, respectively, said in 2013 that multicellularity has arisen more than 25 times across the eukaryotic tree of life and in all of the major clades (*24*). Indeed, some estimates of the number of times multicellularity arose include those that exceed 40 times! The Laura Parfrey and Daniel Lahr analysis (which does chime with a number of others) looked at genes across a number of evolutionary lines and suggests that multicellularity arose as a process of convergent evolution. To put simply, different species at different times, evolved multicellularity independently and that multicellularity gave them each a very great evolutionary advantage; in short, different species 'converged' on this advantage. Indeed, some evolutionary lines seem within them to have seen multicellularity evolve a number of times: Laura Parfrey and Daniel Lahr note that it likely evolved three times among unikonts. Given there was a single, *first* origin of multicellularity—given that some evolutionary lines contain both genomically different multicellular and single-celled species—then the ability to be multicellular must have been repeatedly lost in a line in order for it to arise again. This may seem highly unlikely but there is evidence that multicellularity has been lost on some occasions, especially in the Fungi.

True (proper), complex multicellular life did not arise from single-celled life quickly. The Plantae (plants), and especially Animalia (animals), have a number of quite distinct cell types, yet they all these cell types to work together while doing their own specialised jobs and this necessitates the evolution of intercellular communication. Furthermore, as noted earlier, the leap from single-celled eukaryotes to multicellular organisms (such as your good self) went through a number of intermediate stages.

One such intermediate stage was the aforementioned evolution of colonial species such as sponges (Porifera). These consist of individual cells that live together in a single colony in which many cells arrange themselves to form a number of entrances and spaces within the colony and so give it a structure.

That the colonies are composed of individual cells can be remarkably demonstrated by forcing the sponge through a sieve so that the sponge is broken down into its individual, constituent cells. However, leave these cells in the water and slowly they will combine to re-form a whole sponge.

It is not difficult to imagine how such colonies could evolve more specialist cells to become a single, true multicellular organism. This is the 'Colonial Theory', and was proposed by Haeckel in 1874 (*25, 26*).

Another such intermediate stage to true multicellularity is represented by species whose individuals can be quite large, for example a few metres across, and with specialised parts such as a root attachment or fronds. These types of species are actually composed of a single cell but these species have amongst the largest of cells. Surprisingly for many, these species are not that rare and it may very well be that you will have seen some without realising it.

One example is a *Caulerpa*, a genus of aquatic weed, species that is fairly common in aquaria. You may have eaten members of *Caulerpa* species: they make up dishes such as sea grapes or green caviar.

Given their size and the specialist areas of individuals, you might wonder how they could not be multicellular. However, they all have a special property. Although individuals truly are a single cell, they have many nuclei: it is just that each nucleus is not separated from its neighbour by a cell wall, rather the entire individual has one encompassing cell wall. Such a cell with many nuclei are known as a 'coenocyte'. Again, it is not difficult to imagine how such a species could evolve cell walls between each nucleus to become a single, true multicellular organism. This is known as the 'Coenocytial Theory' of multicellular evolution, or sometimes Syncytial Theory: the former is when many nuclei arise within a cell and the latter when cells come together and the cell walls between them go (*27*).

There are other theories as to how multicellularity evolved. But given that multicellularity arose a number of times, it is not that surprising that there may have been more than one route to multicellularity rather than different evolutionary lines following the same route.

## Multicellular life and the world

Other than in deserts, often today life is clearly visible to us; plains of grass, shoals of fish, valleys of woodland, forest, swamps, and farmland, all provide vistas in which life is clearly the major feature. Pick up any magazine to look at the photographs and the odds are that pictures include images of life. Yet the life in these images, and the life with which you are most familiar, is almost invariably multicellular. The single-celled world of the prokaryotes before the Earth's first big flip (and Snowball Earth I), and the planet after an early 'boring billion' world of prokaryotes and single-celled eukaryotes, was very different to that of the subsequent multicellular time that has continued to the present.

About the only 'visually' common feature of the Earth's single-celled worlds before and millions of years after Snowball I, and much later the subsequent multicelled world, was that the sea's surface would have looked largely the same. In places without much sunlight and nutrients, the sea's colour would reflect that of the sky; elsewhere, with plenty of sunlight and nutrients, the sea has a distinct green tinge. (The oceans of the very early Earth—pre-Snowball I—nearly 4 bya would also have had a green tinge but only a faint green tinge due to the reduced [ferrous] form of iron.) This new, post-Snowball I, distinct green is the green of chlorophyll, the big molecule at the heart of photosynthesis This, to a human sightseer from the future visiting the single-cell world eons ago, would be arguably the greatest, cursory visual clue that life abounded in those long-gone days. To discern more, a closer inspection would be required to actually—if our time traveller had a microscope—see the individual living cells. However, in places even without a microscope there would be signs of life. Bacterial and algal films might cover some habitually wet rocks. In addition to these biofilms, the biggest living things one might come across would be assemblages up to a metre or two across lying in very shallow seas. These would be the 'stromatolites': colonies of bacteria, mentioned in the previous chapter. Other than that, there was little else living to see with the naked eye. There was hardly any but the scarcest amount of life on land and these would primarily be confined to aquatic and damp environments. This was largely a three-dimensional world in water as a thin soup of microbes, while on solid surfaces it would for the most part be a two-dimensional, biofilm, world of sheets of cells.

With the rise of multicellular life that would change.

Multicellular life has the potential to create structures and has physical presence to burrow into sediments and so extract nutrients that would otherwise be inaccessible. Multicellular life in the aquatic environment has the potential to evolve physical augments to more easily propel themselves through the water and so more easily access floating nutrients, ingest organic fragments, and even eat other living cells. Multicellular life has also the potential for structures to penetrate soils on land and small cracks in rock. And, as with its aquatic cousins, multicellular life has the potential to evolve physical augments to traverse land in search for food. Some multicelled species, such as birds and insects, even take to the air. This is the world of multicelled plants and animals, and it is a fully 3-D world.

As with the evolution of oxygenic photosynthesis enabling life to access the considerable resource of solar energy bathing our planet, as opposed to the sparser and lower-grade thermal energy and/or chemicals from vents or chemicals from rock, the evolution of multicellular life opened up access to far more nutrients. In both instances, the global carbon cycle ratcheted up manyfold.

So, following each major flip of state there was a change in the way nutrients were accessed, and concomitantly a change the global carbon cycle; there was also change in the way life used spatial dimensions.

We can get a sense of this change today with food chains (or what ecologists call trophic levels, with a plant at a base trophic level, an herbivore the trophic level above that, and a carnivore in turn one above that). As a very loose rule of thumb, very roughly a tenth of the energy in one trophic level gets transferred to the next (*28*). So a carnivore burning $x$ joules of energy a year as well as transforming some of its prey's biomass into its own tissues, will need very roughly $10x$ joules from the herbivore on which it preys, which in turn will need to feed off very roughly $100x$ joules of plant material: it is energy cascading from one ecological (trophic) level to another. Today's ecosystems, with multicelled species, are far more complex than the earlier ones from which they evolved.

While the above is a concise summary of the benefits of multicellular life, there were a number of stages necessary to take life from being a solely single-celled world largely in the seas, to a world that also has multicellular life both in the seas and also on land. Along the way to a planet with considerable multicellular life, the Earth and its carbon cycle would go through another (following Snowball I in which the rise of OP was a critical factor) major critical transition and a flip of state driven by biological innovation.

## References


1. Lenton, T., & Watson, A., *Revolutions That Made the Earth.* Oxford University Press: Oxford, Great Britain (2011).
2. Mereschkowsky, C., Uber Natur und Ursprung der Chromatophoren im Pflanzenreiche. *Biologisches Centralblatt*, **25**, 593–604 (1905).
3. Mereschkowsky, K., Theorie der zwei Plasmaarten als Grundlage der Symbiogenesis, einer neuen Lehre von der Ent-stehung der Organismen. *Biol Centralbl.*, **30**, 353–367 (1910).
4. Sagan, L., On the origin of mitosing cells, *Journal of Theoretical Biology*, **14**, 225–274 (1967).
5. Margulis, L., *Origin of Eukaryotic Cells.* Yale University Press: New Haven, Connecticut, USA (1970).
6. Martin W.F., Garg S., & Zimorski, V., Endosymbiotic theories for eukaryote origin. *Philosophical Transactions (B) of the Royal Society*, **370**, 20140330 (2015).
7. Wallin, I. E., On the nature of mitochondria. IX. Demonstration of the bacterial nature of mitochondria. *Am. J. Anat.*, **36**, 131–139 (1925).
8. Ris, H., & Plaut. W., Ultrastructure of DNA-containing areas in the chloroplast of Chlamydomonas. *Journal of Cell Biology*, **13**, 383–391 (1962).
9. Nass, M. M. K., & Nass, S., Intramitochondrial fibers with DNA characteristics: I. Fixation and electron staining reactions. *The Journal of Cell Biology*, **19** (3), 593–611 (1963).
10. Spang, A., Saw, J. H., Jørgensen, S. L., *et al.*, Complex Archaea that bridge the gap between prokaryotes and eukaryotes. *Nature*, **521**, 173–179 (2015).
11. Marin, B., Nowack, E. C. M., & Melkonian, M., A plastid in the making: Evidence for a second primary endosymbiosis. *Protist*, **156**, 425–432 (2005).
12. Baum D. A., & Baum, B., An inside-out origin for the eukaryotic cell. *BMC Biology*, **12**, 76 (2014).
13. Lambert, J., Cells hint at roots of complex life. *Nature*, **572**, 294 (2019).
14. Imachi, H., Nobu, M. K., Nakahara, N., Yuki Morono, Y., *et al.*, Isolation of an archaeon at the prokaryote-eukaryote interface. Preprint on *bioRxiv* (2019). http://dx.doi.org/10.1101/726976.
15. Imachi, H., Nobu, M. K., Nakahara, N., Yuki Morono, Y., *et al.*, Isolation of an archaeon at the prokaryote-eukaryote interface. *Nature*, **577**, 519–525 (2020).
16. Brocks, J. J., Logan, G. A., Buick, R., & Summons, R. E., Archaean molecular fossils and the early rise of eukaryotes. *Science*, **258**, 1033–1036 (1999).
17. French, K. L., Hallmann, C., Hope, J. M., *et al.*, Reappraisal of hydrocarbon biomarkers in Archean rocks. *Proc. Natl. Acad. Sci.*, **122** (19), 5915–5920 (2015).
18. Han, T. M., & Runnegar, B., Megascopic eukaryotic algae from the 2.1 billion year-old Neguanne iron formation. *Science*, **257**, 232–235 (1992).

19. Moore, R. B., Obornik, M., Janouškovec, J., *et al.*, A photosynthetic alveolate closely related to apicomplexian parasites. *Nature*, **451**, 959–963 (2008).
20. Eme, L., Sharpe, S. C., Brown, M. W., & Roger, A. J., On the age of eukaryotes: Evaluating evidence from fossils and molecular clocks. *Cold Spring Harbor Perspectives in Biology*, **6** (8), a016139 (2014).
21. Signor, P. W., III, & Lipps, J. H., Sampling bias, gradual extinction patterns, and catastrophes in the fossil record. In Silver L. T., & Schultz P. H., eds., *Geological Implications of Impacts of Large Asteroids and Comets on the Earth*. Geological Society of America Special Publication, no 190, pages 291–296 (1982).
22. Gott, J. R., III, Implications of the Copernican Principle for our future prospects. *Nature*, **363**, 315–319 (1993).
23. Brocks, J. J., Nettersheim, B. J., Adam, P., *et al.*, Lost world of complex life and the late rise of the eukaryotic crown. *Nature*, **618**, 767–773 (2023).
24. Parfrey, L. W., & Lahr, D. J. G., Multicellularity arose several times in the evolution of eukaryotes. *Bioessays*, **35**, 339–347 (2013).
25. Haeckel, E., The gastraea-theory, the phylogenetic classification of the animal kingdom and the homology of the germ-lamelle. *Quarterly Journal of Microscopical Science*, **14**, 142–165 (1874).
26. Haeckel, E., Die Gastrea Theorie, die phylogenetische Klassifikation des Tierreichs, und die Homologie der Keimblätter. *Jenaische Zeitschrift für Naturwissenschaften*, **8**, 1–55 (1874).
27. Ruiz-Trillo, I., Kin, K., & Casacuberta, E., The origin of metazoan multicellularity: A potential microbial black swan event. *Annual Review of Microbiology*, **77**, 499–516 (2023).
28. Lindeman, R. L., The trophic-dynamic of ecology. *Ecology*, **23**, 399–417 (1942).

# 5

# The Second Flip and Other Snowballs

It was, according to the new narrative, multicellular life that was key to triggering the Earth's second big flip and its own episode of dramatic climate change with the second Snowball Earth. ('Second big flip', as for now. We will ignore the first, less marked, change of state from a lifeless Earth to one where the Earth system was perturbed by life.) But—as with the evolution of cyanobacteria's oxygenic photosynthesis and the timing of eukaryotes in the last chapter—this did not happen immediately, which raises the question as to when multicellular life arose.

## When did multicellularity arise?

Many species of animals, land plants, and fungi are multicellular, as are many true algae. So, when asking the question as to when multicellularity arose, we first need to consider the question touched on at the end of the last chapter, whether or not it rose just once—hence in an evolutionary ancestor of *all* multicellular animals, land plants fungi and algae—or whether it did a number of times? You may have guessed from the latter half of the last chapter that the quick answer is that it seems to have risen independently a number, likely more than a score, of times (*1*). After all, as noted in the previous chapter, it was after the proto-eukaryotes had evolved that there was an evolutionary emergence of new domains (or taxonomic supergroups) of life. One domain was that of the bikonts (that includes plants) and another the unikonts (that includes fungi and animals). Subsequently, multicellularity arose separately in both these domains a number of times. (See Appendix 2 for definitions.)

As you will also recall from the previous chapter's discussion, it is difficult to discern from the fossil record exactly when the more eukaryotic-like cells (precursor eukaryotes) evolved from the simpler prokaryotes. The situation is a little better in ascertaining when multicellular life evolved; yet, it is still not easy. The earliest multicelled organisms had no skeletons as we commonly think of them and, being soft bodied, do not leave readily discernible fossils even if they are easier to see than single-celled life unless it is clumped. Furthermore, we are still talking of events that took place roughly a billion years ago and, even though this is over a couple of billion years later than life's early days, there is still not a huge

*Earth's Flips of State*. Jonathan Cowie, Oxford University Press. © Oxford University Press (2026).
DOI: 10.1093/9780197825402.003.0005

amount of accessible strata surviving today from those times: most of the rocks we see today mainly date from the past few hundred million years.

Starting with a perspective on which all researchers on palaeoecology agree: early ecosystems featuring many multicellular species existed in the Ediacaran, the geological period some 635 to 539 million years ago (mya; see Appendix 3). A number of these species were large in size compared to microscopic single-celled species, and some even measured a few centimetres in length so their remains are far easier to discern. Though this is still far smaller than birds, cats, and dogs with which you are familiar, it is nonetheless many, many times bigger than single cells that are just micrometres across. Such early Ediacaran ecosystems of multicellular species are exemplified by a discovery in 2010 by Xunlai Yuan and colleagues from China, of fossils representing an ecosystem in southern Anhui Province of South China (2). Here, around 3,000 fossils were identified. They grouped these into five shape categories: 16–40 mm sized and fan-shaped; a group similar to the afore but with a single, broad ribbon that then splits into a fan; centimetre-sized conical fossils; those with a thin stalk (1.3 mm long, 0.4 mm wide) and a cylindrical tube attached to a base that would be attached to rock; and finally a species shaped somewhat similar to those in the previous group but with a longer stalk and not-so-developed a base.

Many of the fossils showed branching patterns and other features typical of today's macroscopic (large) fleshy algae, some of them comparable with modern kelp. Others, the researchers muse, could be worm-like animals. These species, the researchers consider, would require more than up to a few per cent of oxygen in the atmosphere that existed after Snowball I and before Snowball II and the second flip.

This is in line with the accepted view that Ediacaran life, in its near present-day level of oxygen world after Snowball II, was multicellular *because* the higher oxygen levels allowed them to exist; they could only have the complex Ediacaran ecosystems because the Earth had earlier gone through another (a later event after Snowball I) phase change, a critical transition, in which saw the atmosphere's oxygen levels rise very roughly tenfold. Therefore, this traditional view has it that multicellular animals (metazoans) needed far more oxygen than single-celled eukaryotes, hence only evolved as a result of a rise in oxygen levels to near modern values with another flip of the Earth's state.

However, if, with the new narrative, it was life's evolution itself and the rise of multicellularity that caused the Earth to flip states, then multicellular animals and in other biological kingdoms would need to have evolved *before* the Earth's second great flip of state and the second increase in atmospheric oxygen to near today's levels. In short, this is because—the new idea goes—multicellular oxygenic photosynthetic species critically contributed to this tenfold rise of oxygen levels.

It should be once again stressed that there are a number of thoughts as to exactly when true multicellular life arose and not everyone buys into the afore view. The problem is that the geological evidence can be confusing. For instance, is that shape found in a rock a fossil of a former living species or something else non-biological in origin that just happens to look like it might be something that once lived? This is a bit like the shapes some sometimes see in a pizza that miraculously appear similar to the image of a famous celebrity or religious icon ('pareidolia').

To give an illustrative, recent example, in 2016 a group of largely Chinese geoscientists led by Shixing Zhu published a paper that claimed the discovery of fossils of multicellular species in geological strata some 1.56 billion years ago (bya) in the Gaoyuzhuang Formation, Yanshan area, North China (3). This was dated from *after* the 'Great Oxidation Event' and the Earth's fist big flip ~2.4–2.2 bya that took the Earth's atmosphere from next to nothing to possibly somewhere around 4% PAL (of the near 21% present atmospheric level) oxygen or more (see Chapter 3) but dated well *before* the second flip and its rise in oxygen levels to near present-day concentrations.

As first glance, the Gaoyuzhuang fossils look almost like stains in the rock, but they have statistically regular linear to lanceolate (lance-head like) shapes up to 30 cm long and nearly 8 cm wide. This, and their carbon content, is likely explained if they were the remains of living creatures. At the time of publication, these fossils provided the strongest evidence yet that multicellular eukaryotes many centimetres in size, and with a regularity of form, possibly populated the marine biosphere getting on for a billion years before the Cambrian Explosion.

Having said that, there are other possibilities that cannot be wholly discounted. For instance, these life forms could possibly be 'coenocyte': they could be a single, large cell but having many nuclei, as discussed in the last chapter. Yet, even if this was so, coenocyte species are still a halfway house to true multicellularity.

With regards to early true multicelled animals, the traditional view generally accepted (at the time of writing but equally one that is questioned), is that these did not properly evolve either until Snowball Earth II (Earth's second big flip) in the Cryogenian period that spans 720–635 mya: perhaps even a few million years towards its end before the Ediacaran and its clear assemblage of multicellular life. (We will talk more of the Snowball II timing in the section after next.)

As said, this commonly accepted view has it that it was this second Snowball Earth, and importantly its rise in atmospheric oxygen to near present-day levels, that spurred the evolution of multicelled animals. This is because it had been thought that multicelled animals simply could not live in a world of under 4% PAL (present atmospheric level) of oxygen. This last is the upper end of atmospheric oxygen concentration at the end of Snowball I ~2.4–2.2 bya. And if oxygen concentrations were lower than that then there was no chance that multicelled life could exist.

Such was the thinking just a few years ago, yet come the 2010s and there would be firm evidence that multicellular life could theoretically exist before Snowball II in a world with much less atmospheric oxygen that we have today.

Some, in line with the new alternative Earth system scenario, did consider the possibility that multicelled animal life evolved (after Snowball I but) earlier than Snowball II ~720–635 mya and its substantive rise in oxygen levels. This is a hypothesis with which I have sympathy, though I must stress that it is still debatable and (currently) many vehemently disagree. Fortunately, some decided to experiment to see at how low a level of oxygen multicellular life could—at least in theory—exist.

Time for a personal aside. Both this debate and discovery was exemplified in 2012 when I attended a life and planet symposium at the Geological Society (London) that looked at Snowball Earth II and the oxygen rise from 2% (2% of an atmosphere pressure), or ~4% PAL, to around present-day levels of nearly 21% of the atmosphere (*4*). At this meeting, I came across a poster by a post-graduate student and his colleagues from the University of Southern Denmark. (Posters are literally that, paper 'posters' mounted on a notice board. They are means by which scientists communicate work in progress at symposia prior to their peer-review and formal publication.) This postgrad and his supervisor sought to see if—at least in theory—sponges could thrive in water beneath an atmosphere of less than 4% PAL of oxygen (the oxygen level the Earth had in the period between the end of Snowball I and the beginning of Snowball II, ~2.2 bya and 720 mya). These experimental sponges were kept in the dark, deliberately so, so that there was no possibility of their getting oxygen from photosynthesis. And yet, the researchers discovered, they thrived!

This was clear experimental evidence that Metazoans (multicelled animals) could in theory live on the low-oxygen Earth as it was between the two Snowballs (*5*).

All well and good, but a little later at that symposium's Q&A session the question of when the Earth's first multicelled animals appeared and I opined 'possibly before 720 mya' only to be firmly chastised by the symposium session moderator that animal life could not survive in a world of just 4% PAL atmospheric oxygen. Not only was I put in my place but the postgrad student authoring the poster kept quiet: clearly the session's moderator had not done a complete round of the posters at that event and those that had were a little intimidated from the podium to silence. The point here is that, even in science, dogmatically firm beliefs can block the route to potentially viable, yet radical, new thinking and hypotheses. It also illustrates that in science new ideas are (rightly) challenged and need more than passing supporting evidence. It also demonstrates the value of posters at symposia.

Now, I have to say that the poster's researchers had gone one step further than I had dared in my considerations back then of the evolution of life and planet: my own thinking at the time was more modest. I then only thought a possibility for multicelled life back then in an oxygen-poor Earth would be if it symbiotically partnered with an oxygenic photosynthesiser. We know that oxygenic photosynthesis (OP) evolved prior to Snowball I (see Chapter 3), and I recalled my school zoology studies of the coelenterates (most of which these days are called cnidarians) and living *Hydra*.

*Hydra* are small animal species, a few millimetres long, commonly found in ponds, and are related to jelly fish. I was looking at one of these animals through a microscope. The thing I noticed was my specimen's colour. I was told that this was due to algae that lived between its cells in symbiosis. Could—I decades later mused—some animals have a symbiotic relationship with an OP cyanobacterium or algae so as to live in a low 4% PAL oxygen world? If so, it would be like they're carrying around their own portable oxygen-generating kit with them. Years later, this notion seemed to me quite likely and indeed this possibility was one that the University of Southern Denmark researchers had sought to exclude (remember they kept their sponges in the dark). What the researchers had done was to go further and demonstrate that sponges could live in a low-oxygen environment even without an OP symbiont!

Jumping forward from 2012 a couple of years and the poster's work was subsequently published in the peer review literature by Daniel Mills, Lewis Ward, and colleagues. In their paper, they showed that a modern sponge (*Halichondria panacea*)—serving as an analogue for early animals—can survive under low-oxygen conditions of 0.5%–4.0% PAL. Because the last common ancestor of multicelled animals likely exhibited a physiology and morphology similar to that of a modern sponge, its oxygen demands may have been met in the low oxygen environment earlier in the Cryogenian and Snowball Earth II well before the enhanced oxygenation of the Ediacaran Period that began 635 mya or so (6).

The bottom line is that this is experimental evidence that the origin of multicelled species may not have been triggered by a rise from ~4% PAL of oxygen content in the atmosphere but they might have evolved before then in the 0.5%–4% PAL of oxygen world prior to Snowball II. Indeed, I would say that this was even likely given the possibility of some metazoans having oxygenic photosynthetic symbiotic partners, even if not all of them needed such.

Given this, the idea that the fossils from 1.56 bya that Shixing Zhu and colleagues discovered in the Gaoyuzhuang Formation, Yanshan area, North China, could be of multicellular species (3) becomes a distinct likelihood. And even if these were coenocyte species, it does demonstrate that life could have been at the least on the road to multicellularity well *before* the onset of Snowball Earth II (~720 mya).

Now, you may be wondering, why all this emphasis on the rise of multicellular species? In addition to how they process carbon, the importance of such species to the Earth system is that some of those are much better at releasing nutrients from rock and sediments: much better than rain (which is slightly acidic) on rock (see Chapter 3). For instance, as we shall see, multicellular plants' roots can explore and enlarge tiny cracks in rock so speeding up rock erosion and the release of nutrients. Meanwhile, multicellular animals can burrow into sediments, churning them up and also releasing nutrients. These nutrients can then facilitate the development of more life and increase the size of the Earth's biomass. More biomass, in turn, can mean more carbon burial. This last is important, as it leads to a net accumulation of oxygen in the atmosphere.

This oxygen increase occurs because carbon burial disrupts the balance between oxygen generation by photosynthesis on one hand and metabolic respiration using oxygen that releases energy (which life uses to power its metabolism) on the other: organic matter is removed from the equation by burial, so leaving an oxygen surplus.

The simple reaction is this . . .

$$\underset{\text{water + carbon dioxide}}{6H_2O + 6CO_2} \underset{\textit{respiration}}{\overset{\textit{photosynthesis}}{\rightleftharpoons}} \underset{\text{organic matter + oxygen}}{C_6H_{12}O_6 + 6O_2}$$

With organic burial we get this...

$$6H_2O + 6CO_2 \xrightarrow{\textit{photosynthesis}} \underset{\downarrow \text{ burial}}{C_6H_{12}O_6} + \overset{\text{release } \uparrow}{6O_2}$$

Having said that, true multicellular plants did not colonise the land until sometime after 450 mya (the earliest fossil evidence so far is for true plants in the Devonian 420 mya) which was well after Snowball Earth II and the Earth's second big flip. However, there is evidence of precursor species to true plants dating from 480 million years ago and this helps reconcile molecular clock evidence for the rise of plants at around 500 mya (*7, 8*). Nor did multicellular animals have a significant presence on land until after Snowball II. But other multicellular species could have colonised the land before then and so helped erode rock and release nutrients. This last was key to the new hypothesis regarding Snowball Earth II.

## The move to land

*If* multicellular species had evolved *prior* to the rise of oxygen from around 2% to close to today's (21% of the atmosphere) value (at Snowball II), an increase equivalent to a change from around 4% PAL to near 100% PAL, then this oxygen rise could not be involved in the evolution of multicellular life: multicellular life was already there! (We will once more return to this question of which came first—the further increase in oxygen or the evolution of multicellularity—later in this chapter.)

It is likely, from what we know from both present-day biology and geological evidence (such as $^{13}C$) from around 850 mya and earlier, that there was some biological activity on land, albeit in freshwater environments and damp areas. These would have included bacterial and algal communities. Some, as Tim Lenton and Andrew Watson point out in chapter 14 of their *Revolutions That Made the Earth* (9), make a case for significant colonisation of the land by proto-lichens beginning just prior to Snowball Earth II (~ 720–635 mya).

Lichens are a result of the symbiosis between a fungus and a photosynthesising plant partner, and often—it has been recently been indicated—a yeast that may help confer on some lichens their overall shape (*10*).

Lichens (with their algae and fungi) can insert hair-like structures (hyphae) into rock to access nutrients. The important outcome of this is that it increases the erosion of rock. This, in turn, has two further essential consequences: first, increased atmospheric drawdown of carbon dioxide reacting with exposed rock and, second, increased nutrient release into streams, rivers, and ultimately the seas. This last increases marine algal growth hence also further increases carbon dioxide drawdown. It also overcomes the nutrient limitations that had since 2.2 bya kept atmospheric oxygen concentration hovering somewhere around 2%.

The former, exposed-rock drawdown of carbon dioxide, essentially occurs due to the following summary reaction:

$$CaSiO_3 + CO_2 \rightarrow CaCO_3 + SiO_2$$

Calcium silicate + carbon dioxide → Calcium carbonate + silicon dioxide

The actual reactions are a little more complicated. For example, the carbon dioxide actually reacts with water in the atmosphere to form carbonic acid, which attacks the rock (as it began to do in greater earnest around the time of Snowball Earth I—see Chapter 3). Nonetheless, the above equation captures the essence of what is happening, and it is a key Earth system reaction. Indeed, such carbonate formation is so common in water that you will have noticed it furring up your kettle.

That lichens are good at colonising bare rock is evident with many examples about us: lichens grow on roofs, the stone walls of buildings and gardens, and quite simply growing on many types of rock including (as you can see at your local church) a number on some types of stone used to make tombstones. So while the early Earth's land may have been inhospitable to life away from wet and chronically moist areas, stretches of bare rock occasionally wetted by rain were ripe for colonisation by lichens.

Under this scenario, the colonisation of land by species more advanced than single-celled microbes and their colonies (such as variations on the shallow water stromatolite microbial mats) will result in enhanced silicate weathering and its concomitant lowering of atmospheric carbon dioxide, hence helping (along with increased biomass and some of its concomitant burial) planetary cooling. This would continue until various biosphere reservoirs of carbon fill and then carbon recycling takes place. This scenario in essence defines the scene of the Earth going through another flip, changing state, and the process of going through a critical transition that again involved another extreme climate change event as happened with the first flip (Snowball Earth I ~2.4–2.2bya).

## The second flip and another Snowball

Before exploring some further detail behind Snowball Earth II, it should be stressed that as with Snowball I (Chapter 3) there is much debate as to whether the Earth became a true snowball with ice stretching from poles to equator, or whether there was a band of equatorial water that never disappeared during the event. In this last case, the Snowball would have been more of a 'Slushball'. Once again, there is little point in exploring this particular detail as it is largely irrelevant to the overall notion of co-evolution of life and planet narrative being presented here: I am not going to take sides in that debate. So, once again, consider the term 'Snowball Earth' as a stand-in for 'Snowball or Slushball Earth'.

Again, as with Snowball Earth I, Snowball Earth II consisted of a number of distinct cold events that spanned a period of very roughly 90 million years: each of these events could be considered a distinct snowball, but the term 'Snowball Earth II' here refers to all these collectively.

Snowball Earth II principally consisted of three significant individual cold episodes that we know of; there could well be more, let alone minor ones, but we have not (yet) discerned them all from the geological record if they did indeed take place. These three—two major and one minor—glaciations were: the Sturtian that ended ~700 million years ago (mya), the Marinoan that ended 635 mya (both coldest), and the Gaskiers (which was milder and not global) that ended 585 mya. The Sturtian and the Marinoan could have been proper Snowballs (firm Slushball theory adherents notwithstanding) and they occurred in the geological period of time called the Cryogenian that spanned 720 mya to 635 mya (see Appendix 3): from the beginning of the Sturtian to the end of the Marinoan. On our arm's-length journey of the Earth's history we are now a long way from our start at the shoulder—we are near the knuckles of the hand.

The milder Gaskiers ending 585 mya occurred within the geological period known as the Ediacaran that spanned from 635 mya to 541 mya.

The notable thing about the Ediacaran is that it was a time following a rise in atmospheric oxygen to near present-day levels (the Earth's second great flip of state) and a time (as mentioned at this chapter's beginning) in which much multicellular life was clearly evident including early multicelled animals. This in turn led to the Cambrian boom (which is also known as the Cambrian explosion or Cambrian diversification) of species that began after the start of the Cambrian period 541 mya.

The question that arises is whether there is a relatively simple, and most certainly a logical, explanation that provides a coherent narrative for all these events?

Again, it needs to be stressed that there are a number of competing ideas and I will return to some of these later (see the section later this chapter 'Alternative Snowballs *vs.* the new narrative'), but the narrative I present here falls into the camp of 'it was the evolution of life that triggered the flip of the Earth's state', as opposed to the direct opposite view of the Snowball Earths spurring life's evolution.

As discussed in the previous chapter, the 'boring billion' saw the Earth in a comparatively stable state with atmospheric oxygen hovering somewhere around 0.5%–4% PAL (some say possibly more, see Chapters 3 and 4) or 0.1%–0.84% of total atmospheric concentration (again some say more, ~2%). So by what mechanism, mostly during the Cryogenian geological period (720–635 mya), did oxygen concentrations rise to a value near present-day levels?

The answer, as alluded to above, may well involve the colonisation of land by more than microbial communities of species: the multicellular proto-lichens. This is the view that Tim Lenton and Andrew Watson present in their book *Revolutions That Made the Earth* (9).

As noted in the previous section, early lichens are good candidate species for such land colonisation. Consider their potentially colonising starting well before Snowball Earth II and its first (the Sturtian) major glacial phase. Other than microbial communities in freshwater and habitually damp places, these early colonisers would have had no biological competitors. So, on one hand it was a bit like being able to move into an empty house as opposed to one already occupied; moving in you were free to do what you wanted without having to worry about any fellow residents or rooms having already been taken. On the other, the early land colonisers would find life harsh: there was no proper soil just bare rock, some pools, streams, and rivers. Soil is a complex substance consisting of organic matter together with finely broken-down rock, and it needs life to make it, with fungal hyphae and plant roots eroding the surface of either the bedrock or the surface regolith. Life's organic matter produces carbonic acids (in addition to that in rain) that further erode mineral particles to a very small size, usually a couple of micrometres (μm) or less: most fine-particle clays need life to form, and prior to the colonisation of land there was markedly less clay to subsequently form shales (which we see today in the geological record). So there was less clay formation prior to the rise of terrestrial multicellularity and this helps inform of earlier events discussed in Chapter 3.

Aiding early soil formation, fine dust called loess (formed from blown rock flour created by glaciers grinding rock and also blown from deserts) was trapped by life so aiding soils' formation. Since before the early multicelled colonisers of land, other than bacteria and algae, there was *ipso facto* no life on land, so there was no soil (strictly speaking not soil as we commonly know it today): it was going to be tough for the new multicelled colonisers of land.

Indeed, it is this absence of soil from the time before the hypothesised colonisation of land, by the lichen and fungal colonisers, that is one piece of evidence that the land prior to the first major glaciation, the Sturtian, had no soil-making multicellular life: there are no major fine-particle clay deposits. Conversely, clays are far more common *after* Snowball II (*11*).

There is other evidence that there was the new life on land at this time and this comes once again from $\delta^{13}C$ to $^{12}C$ (changes in carbon-13 to carbon-12) isotope ratios (see Chapter 1). $\delta^{13}C$ evidence in non-biotic carbonate by itself only tells us whether or not there is carbon drawdown from (usually by photosynthesis), or release back to, the atmosphere of carbon from compounds (originally) created by life: non-biotic carbonate by definition is itself not life. (Conversely, $\delta^{13}C$ depletion in fossils, including in fossil fuel strata, is indicative of former life.) However, the $\delta^{13}C$ evidence in non-biotic carbonate does suggest $^{12}C$ drawdown elsewhere by life associated with the Snowball II glaciations. Further, it is also possible to see what was happening to other element isotopes at the same time.

Just as $^{12}C$ is the most common isotope of carbon, so $^{16}O$ (oxygen-16) is the most common oxygen isotope that makes up over 99.7% of oxygen (remember Chapter 3). But there is also a rarer oxygen isotope, $^{18}O$, that makes up 0.2% of oxygen. You will recall that water, being $H_2O$, contains oxygen and most of this is $^{16}O$, but those water molecules that contain $^{18}O$, being atomically heavier, are a little less likely to evaporate (see Chapter 3). This in turn means that rain water contains even less of the minority $^{18}O$ than water in the oceans. It is therefore possible to look for carbon in carbonate that has been photosynthetically enriched with $^{12}C$ and also has $^{18}O$ from depleted freshwater. The enriched $^{12}C$ suggests photosynthetically processed carbon (hence life-processed carbon), while the depleted $^{18}O$ suggests that the oxygen came from rainwater that fell on land (if the rain fell onto the sea it would be re-mixed with the standard oceanic cocktail of $^{18}O$ and $^{16}O$). This is what US geoscientists Paul Knauth and Martin Kennedy realised when they set about looking for life on land in the geological record. They presented their results in a 2009 *Nature* paper (*12*) that was also open access reviewed in the same issue of that journal by Michael Arthur of the Pennsylvania State University (*13*). ('Open access' means you can access it on the internet without having to have a subscription for the journal, so you can look up this review article should you be interested.)

Paul Knauth and Martin Kennedy noted that the sediments in estuaries today have both an enriched $^{12}C$ signature because they contain carbon from the land's biota and depleted $^{18}O$ because the carbon has been washed down to there by what was rain water. They compiled all published oxygen and carbon isotope data for marine carbonates from the time of Snowball II and found that some had a $\delta^{13}C$ and $\delta^{18}O$ signature that was similar to that from present-day sediments found near river outflows to the sea. Importantly, they did not find this signature in carbonates older than 850 mya. They inferred this as being due to an explosion of photosynthesising communities on late Precambrian land surfaces after 850 mya. This biologically enhanced weathering so generated carbon-bearing soils on a large scale, and their erosion sedimentation buried carbon (see the photosynthesis and organic burial equation towards the beginning of this chapter). This burial of carbon, be it *in situ* on the land or washed into the seas, facilitated the rise in atmospheric oxygen that was necessary for the subsequent expansion of multicellular life. The net oxygen gain was further enhanced by nutrient release into the oceans to the benefit of marine life and this resulted in even more carbon burial.

All these processes were taking place in the run-up to Snowball Earth II.

Let's now put this all together into a sequential, logical narrative.

What might have happened is this. Multicellular life in the form of fungi and early lichens (we will call them proto-lichens) began to colonise the land sometime from around 850 mya or later (we do not know exactly when, but the Paul Knauth and Martin Kennedy evidence suggests this). These fungi and proto-lichens lived on rocky surfaces and their hyphae found fissures in the rock which they could enlarge from which they could extract nutrients. On death, their decaying organic matter also provided carbonic acids that further aided the erosion of rock and the release of nutrients (at a level above that of previous erosion rates). Soil formed which trapped some carbon, while other carbon got washed down into rivers and out into the oceans with some carbon settling out in sediments around where the rivers entered the sea.

As noted, not all the rock nutrients released by land's new life remained on the land: some was transported with the carbon to the oceans. Here Tim Lenton and Andrew Watson hypothesise, in their *Revolutions That Made the Earth* (to which I have referred previously) (*9*), that the liberation of the nutrient phosphorous from the land into the seas may have played a key part in helping life in the oceans go through a boom.

Tim Lenton's and Andrew Watson's phosphorous liberation, enriching the seas, is supported by evidence from geological strata that dates from that time. Geoscientists Noah Planavsky, Olivier Rouxel, Andrey Bekker, and colleagues analysed around 700 samples through time. They found that samples dating from 750 to 635 mya featured unusually high dissolved phosphate concentrations in the aftermath of the widespread, low-latitude Snowball Earth II glaciations. They proposed that this phosphorous led to a boom in life that generated oxygen and enhanced carbon burial so elevating the Earth's atmospheric oxygen concentration. However, their theory is that this paved the way for multicellular species not the other way around as Tim Lenton and Andrew Watson suggest: the Planavsky, Rouxel, Bekker, and colleagues' notion is that the phosphorous came solely from Snowball ice eroding land and not additionally from multicellular life (possibly proto-lichens) colonising land, enhancing erosion so releasing phosphorous (*14*). If the Planavsky and colleagues' hypothesis were true we might have expected the previous Snowball I to have engendered similar phosphorous release, hence a similar magnitude of carbon burial and so a similar rise in atmospheric oxygen levels or alternatively at least a larger increase in oxygen than seen at the Snowball I transition. Yet assuming this was correct and that there was much oxygen generation at Snowball I but much was mopped up in overcoming the primordial Earth's huge reducing reservoirs, once these reservoirs had gone, why did the oxygenation not continue? Clearly, the oxygenation during Snowball I differed from Snowball II even if nutrient release from ice erosion was similar. Something else must have been happening with Snowball II and that something, as Tim Lenton and Andrew Watson suggest, may have been early multicellular terrestrial life releasing phosphorous and other nutrients.

Returning to Tim Lenton and Andrew Watson's scenario, with the first boom of life in the oceans as well as the new life on land, there was increased carbon burial, hence in turn there was a small accumulation of oxygen in the atmosphere. At the same time, the increased land erosion exposed silicate that drew down carbon dioxide from the air (see the previous section) in addition to life's drawdown of carbon dioxide, part of which ended up buried. With less greenhouse gas in the atmosphere, the Earth cooled and entered into a super-glacial period or a Snowball Earth: Snowball Earth II, or more accurately it was the first intense cold period of a number within Snowball II.

However, with the planet Earth frozen, life died out on land. With no longer any (or at least no significant level of) terrestrial life (which depends on whether your view is of this time being one of Snowball or Slushball), erosion of rock surfaces exporting nutrients declined and at the same time organic burial ceased. The Earth was in the first major glacial time, the Sturtian episode, of Snowball Earth II and ice covered much, if not all, of the planet. This ice swept the land clear of nearly all life.

(As in our previous discussion of Snowball Earth I, if you are a Slushball person then there would still be terrestrial life in a narrow equatorial band, or if the Earth was more a completely ice-covered Snowball, terrestrial life would only have clung on in small areas of refugia such as around hot springs or on mountain peaks that poked through the ice [nunataks]. Either way—as with the oxygenic photosynthesisers riding out Snowball I—life, and the newly evolved multicellular life, survived.)

Meanwhile, again as with Snowball Earth I (see Chapter 3), volcanoes continued to out-gas, which included venting a small (on the global scale) but steady stream of carbon into the atmosphere. Slowly, over a few million years, the atmosphere's carbon dioxide increased so warming the Earth and eventually, around 700 mya, it ended the first cold phase (the Sturtian) of Snowball II and the ice melted. But now, with the carbon burial, there was a little more oxygen in the atmosphere.

With the land swept clean by the former ice, terrestrial multicellular life—the fungi and proto-lichens—had to start again as did the whole slow business of: soil formation; organic burial; erosion of nutrients and carbon into rivers and then the oceans providing more carbon burial; the enhancement from these nutrients of marine life (and yet more carbon burial). This, all again, provided a further net gain of atmospheric oxygen combined with another period of carbon dioxide drawdown. This all ended with a second major cold period of Snowball II, the Marinoan (sometimes referred to as the Varanger, should you wish to do an internet search for further information). The volcanic carbon dioxide build-up and then re-colonisation of land between the two glaciation events took approximately 60 million years. Meanwhile, as before, volcanoes kept recycling carbon as part of the plate-tectonic-driven deep carbon cycle (see Chapter 2) providing a steady stream of carbon dioxide to the atmosphere that eventually warmed the Earth and ended the Marinoan 635 mya. So the Earth's atmospheric oxygen had increased yet again.

The Earth was to go through at least one more iteration with another cold phase roughly another 60 million years later, the Gaskiers at about 580 mya. This was a shorter and less intense an iteration than the Sturtian and Marinoan. By the end of Snowball II, the Earth's atmospheric oxygen had increased in a stepwise fashion with each major glaciation to end up somewhere nearly around today's (100% PAL) level of 21%.

During this process, and as with Snowball I, the times of photosynthetic drawdown of carbon (this time in the land colonisation phases) reduced the atmosphere $^{12}C$. Conversely, some $^{12}C$ was released after each cold glacial event ended.

This last saw the Earth was coming back through a critical transition (see Figure 2.3B). The Earth at its maximum Snowball state reflected more of the Sun's energy compared to when it was not covered in brilliant, white ice. This in turn meant that there needed to be mid-Snowball even more carbon dioxide than was otherwise needed to trap the necessary heat to melt the snowballed Earth. The steady stream of volcanic carbon dioxide (richer in $^{13}C$) provided this greenhouse gas, which built up in the atmosphere in the absence of much life to bring the level back down. In short, an overshoot of carbon was needed, and took place, to get back through the critical transition's zone of transition to make a threshold jump back to a non-snowballed Earth state. So after each major glaciation the Earth became very warm and this enabled fermentation of any carbon accessible to methanogens (remember a considerable body of organic carbon had been buried). This in turn released $^{12}C$ back into the atmosphere. Also, the overshoot of volcanic carbon dioxide ended up as cap carbonate strata laid down after each glaciation event and these strata are clear in the geological record.

What this all meant was that, as with Snowball I, there were great swings in the $^{12}C$ and $^{13}C$ ratio during the time of Snowball Earth II (see Figure 2.1 and the vertical lines between 850 and 640 mya) as the Earth crossed a threshold, going through a critical transition.

The rise in oxygen to near present-day levels further facilitated oxygen respiration: the oxidation with (appropriately here) oxygen, of carbohydrates. This is a metabolic process that releases considerable energy, far more than other metabolic mechanisms, such as fermentation, on which previously some life relied (see Chapter 3). Those species that could do respiration had more energy available to them (per unit of biomass metabolised), which made things like physical movement easier. The higher oxygen concentration, to somewhere around nearly 100% of the present atmospheric level (PAL), compared to the previous levels between Snowball I and II of 0.5%–4% PAL of the boring billion, also enabled the evolution of even larger multicelled animals (metazoans) than those smaller first ones that had already arisen. If simple multicelled animals, Parazoa, such as sponges, could at least in theory live in water under the low boring billion 0.5%–4% PAL (see the section 'When did multicellularity arise?' at the start of this chapter) then it would be possible for animal cells to be buried within layers of many cells also needing oxygen in the more oxygenated post–Snowball Earth II world: some of the cells deep in your body only have access to oxygen at these lower concentrations. And if smaller multicelled species already existed at the start of Snowball II, then they already existed to take advantage of the higher oxygen immediately after that flip of Earth state. These species could therefore evolve to be bigger and physically more active than the simpler Parazoans: if you like these could be Metazoa+, or good Metazoa. Indeed, most of the multicelled animals that you are likely to consider to be 'less primitive'* are of this type of metazoan which—as previously mentioned—biologists call Eumetazoa. (You may recall from earlier, '*eu*' from the Greek means 'well' or 'good'.)

To sum up, according to this new narrative, the Earth's second big flip was due to the colonisation of land by non-microbial, multicellular life (most likely significantly proto-lichens), that released nutrients from rock resulting in more photosynthesising life, increasing organic burial and nutrient transport to the oceans. (Here, recall the discussion in Chapter 3 on phosphate.) This in turn caused marine prokaryote and eukaryote life to bloom, further increasing organic burial. This carbon burial, together with increased silicate erosion (see the previous subsection), lowered atmospheric carbon dioxide, and the flip was marked with another Snowball Earth consisting of a number of cold phases. The organic carbon burial eventually resulted in a net increase in atmospheric oxygen to around present-day levels, and that in turn resulted in the expansion of simple multicellular life and paved the way for the less primitive multicelled animals.

This perspective is opposite to the view that the Snowball's rise in oxygen enabled the evolution of multicelled species; the new narrative has it instead that multicelled species were critical to Snowball Earth II and the rise in oxygen. (We will return to the alternative explanations shortly.)

## After Snowball II

Following the Gaskiers cold phase, the Earth's atmospheric oxygen composition stabilised to hover around 100% PAL. It should be noted that it did not hold firmly to 100% PAL and did vary. One notable variation occurred in the Carboniferous—a geological period 359 mya–299 mya see Appendix 3)—that saw the evolution of a woody substance called lignin and hence the evolution of trees more as we know them today. Lignin is harder to microbially decompose and initially there were no animals to eat it and fungi to decompose it: enzymes

* According to the Institute of Biology's and the Association of Science Education's *Biological Nomenclature* (see Appendix 2 references) terms such as 'lower', 'major', and 'advanced' are not recommended, as evolution is not a linear process. However 'primitive', hence 'less primitive', is advised to indicate as appearing in an early or later stage in evolutionary development. Having said that, as you may have already noticed, I find it difficult to avoid using terms such as 'more complex' species: please forgive me. For many this disdain of 'lower', 'major', and 'advanced' issue will be a point of pedantry, but to some readers working in biology it is understandably important.

had to evolve to access this new, potential food source. So lignin carbon persisted and so could be buried. It was also a time, due the converging of two plates, of the formation of a low, relatively flat continent called Pangaea. It had large low-lying equatorial areas prone to flooding. Both these factors increased organic burial. This caused a temporary (in geological terms) increase in atmospheric oxygen to around 150% PAL and a world, that among other things, included giant insects. But all this is quite another story.

One question you may have is, why were there not further cold phases within, so making a longer, Snowball Earth II and a further enhancement of oxygen's atmospheric concentration? Leaving aside that we may not have discerned all the glaciation events associated with Snowball Earth II, a number of factors limited the continual ramping up of atmospheric oxygen. First, the resulting high levels of oxygen at the end of Snowball II ironically impeded oxygenic photosynthesis (oxygen is a highly reactive element hence its metabolic value) and so a brake, or a negative feedback, began to kick in, impeding further oxygen generation. Second, with animals benefiting from each pulse rise of oxygen, they could provide greater ecological 'function' relating to the processing of carbon. Animals can move and burrow and they were better able to 'mine' organic (carbon rich) sediments for food and so thrive, consuming more oxygen: more animals means more oxygen consumption. Finally, at high levels of oxygen, the chances of vegetation fires increases. You may remember from your school science that a glowing splint of wood bursts into flame in the presence of a higher concentration of oxygen; so more vegetation fires consumed more oxygen, reducing its atmospheric composition. These factors combined prevented unduly excess oxygen, hence preventing atmospheric levels from increasing further. (During subsequent times, only when the carbon cycle was altered in some way—such as due to the aforementioned increased organic burial in the Carboniferous—did atmospheric oxygen levels fluctuate significantly and even then, with biological evolution, the Earth system adjusted.)

The Earth after Snowball Earth II was a world for animals and led to the Avalon boom or explosion of species. They are considered to have evolved in the period shortly after the most intense part of Snowball Earth II, in the Ediacaran between 635 and 541 mya (see Appendix 3). However in 2011 a group of Chinese researchers led by Xunlai Yuan, suggested that some of these new species evolved perhaps shortly after the Marinoan glaciation (*15*). The first of these new species were rather odd compared to later ones. Indeed some say that some of the species do not seem to properly fit into the animal (Metazoan) kingdom, or plants or fungi. These Ediacaran (or Vendian, or Vendobiota, as some used to call them) species often had a frond-like structure that caught organic particles drifting in the water. They had little need to move much, or have tough protection against predators, as eyes and jaws had not yet evolved.

However, eyes and jaws did eventually evolve and the Ediacaran period came to an end with the start of the Cambrian period 541 mya and the 'Cambrian boom' of species that wiped out the Ediacarans.

To date one of the first of animals, with a mouth and gut, and which moved leaving tracks, has been dated to 551–539 mya. This was before the end of the Ediacaran (541 mya). It was segmented, looking vaguely worm-like. The fossils found were ranged 7–25 mm in width and 76–608 mm in length. It is called *Yilingia spiciformis*: *Yilinigia* from the Yiling district of China where the fossils were found in 2019 and *Spiciformis*, from 'spiciform', with reference to the spiky appearance of the fossil that resembles ears of wheat (*16*).

And so life continued, evolving for over half a billion years to the present day. Yet, during this time, there were other perturbations in the carbon cycle. Some of these were due to what the science fiction authors Terry Pratchett and Stephen Baxter call 'jokers' in their 'Long Earth' sequence of parallel-Earth novels (*17*). (While most were recognisably similar, some of their parallel Earths had a markedly different history to Earth's.) These carbon cycle perturbation events were out-of-the-ordinary disruptions of our planet's carbon cycle and included episodes of near continental-wide seas of lava (large igneous province formation) such as that which took place at the end of the Permian 252 mya, or a large asteroid impact such as that at the end of the Cretaceous 66 mya. Both these were global mass extinction events, with the latter being famous for wiping out the dinosaurs. We will look at these in Chapter 6 purely for comparison purposes, but they were not life-evolution-driven changes of state. However, over half a billion years after Snowball Earth II, a single species finally arose that would deliberately short-circuit the deep carbon cycle with global consequences. We shall return to this in Chapter 8.

## Alternative Snowballs

It has to be emphasised that the aforesaid, new, Earth systems science narrative—of largely biological-evolution-driven changes of the Earth's state—is substantially different to the former standard view as to why Snowball Earth I and Snowball Earth II took place. I emphasise this debate again as readers need to be aware of it and (especially those who are students with some Earth system expertise) make their own evaluation. Notwithstanding this, can this new view be reconciled with the formerly established one?

The earlier standard view has been summarised by among a number of others, including not that long ago in a very readable way by the geologist Ted Nield in his book *Supercontinent* (*18*), which also contains many other fascinating geological evolution stories. The bottom line is that *Supercontinent* does a wonderful job of taking its readers on a whistle-stop journey of much of our planet's history, and geologists' standard narrative as revealed by the geological record and our understanding of it in the mid-2000s.

Not surprisingly, as his book is called *Supercontinent*, he includes the traditional Snowball Earth perspective as the two Snowball Earths took place at the times when the Earth sported a tropical supercontinent. He notes that supercontinent formation represents part of the previously mentioned (Chapters 1 and 4) Wilson cycle (a term coined in 1974) whereby continental drift periodically brings together the Earth's continents into a single supercontinent (such as Kenorland and Pangaea) before fragmenting into multiple smaller continents (as are the continents today). This cycles once (very roughly and variedly) every half billion (500 million) years or so.

The early stages of supercontinent fragmentation—at the end of an equatorial supercontinent's life—is thought to be key to Snowball Earth formation and, indeed—even though geological dating strata from, and elucidating events of, the first Snowball Earth is fraught with difficulty—it does seem as if both Snowball Earth I and II may well have taken place at a time of equatorial supercontinent breakup.

Ted notes that there are times when supercontinents straddle the equator and at these times the Earth is particularly susceptible to cooling. This is for a number of reasons. For one, water holds more heat, is mobile, and so can transport that heat about the planet; so ocean water from lower latitudes can carry heat to the poles. But if much of the equatorial zone is land, this heat is not available to the oceans but gets radiated back out into space: which is why lower-level cloudy nights are warmer as the clouds impede this re-radiation process. Also, snow-capped mountains straddling the equator reflect sunlight (as do the Himalayas today) so any ice forming there (even just a thin layer) particularly reflects sunlight and heat back out into space. This also gets exacerbated in a feedback effect (see Figure 2.2B): cooling leads to more ice formation, hence more cooling, and so on. Indeed land, especially desert, is generally lighter in colour than ocean and so reflects more sunlight. Supercontinents being super-large, would see rain only around the continent's margins with the vast inland area being a light-coloured desert, it being in a rain shadow. So if much of the tropics is covered by land then more heat is reflected back into space so cooling the Earth. All these factors (thermal heat capacity, heat transport and albedo) act together so that tropical supercontinents serve to cool the Earth.

Ted Nield affirms that once fully formed, supercontinents are largely dry places; only inward bound trade winds from the ocean carry moisture, and this falls as rain largely in a belt around the prevailing wind edge of the supercontinent. Once the rain has fallen out of the air, unless it is somehow replenished, there is none left to fall further inland. But when the continents fragment (though many of the equatorial mountains remain) sea penetrates into the interior of what was the supercontinent and then these waters feed the atmosphere, increasing rainfall on the continent hence leading to considerable silicate erosion. This has the effect of sucking carbon dioxide out of the air. (As noted in the calcium silicate to calcium carbonate equation earlier this chapter in the section 'The move to land'.) Such silicate erosion is more pronounced with the basalt-type rocks that emerge with spreading continental plates, as happens in the middle of supercontinents when they break up. This silicate erosion and carbon dioxide drawdown is also more pronounced in warmer environments and so is another reason as to why equatorial supercontinental breakup is important for Snowball Earth formation. Further, the tropics sees more ocean evaporation than near the poles, and the warmer the air the greater its water carrying capacity. So the tropics see more rainfall than, say, polar regions.

In short, a number of factors synergetically come together to make the tropics the place where markedly more silicate-mediated carbon dioxide drawdown occurs during times of supercontinent breakup than at cooler latitudes. Add in the dimension of tropical supercontinents heat transportation away from the tropics being reduced, and all these factors serve to reduce the overall planetary temperature.

Finally, the fragmenting of the supercontinent, that is caused by mantle heat that slowly accumulates beneath the supercontinent, means that when the supercontinent does first split into closely spaced, smaller continents there is much volcanic activity in the rifts and that this might at times inject dust and sulphates into the atmosphere causing even further cooling.

All these factors combined served to plunge the Earth into a Snowball (or Slushball). Sulphate in the atmosphere is known to reflect sunlight, so cooling the lower atmosphere, hence the planet. This sulphate injection provides—according to the view until now favoured by many geoscientists—a trigger as to why Snowball Earths occur at times of supercontinental breakup. Indeed, recently there has been additional evidence to support this perspective. In 2017 Francis Macdonald and Robin Wordsworth proposed that the first, the Sturtian, glaciation of Snowball Earth II was triggered by the injection of sulphate aerosols into the stratosphere as the supercontinent Rodinia broke up. They noted that a large igneous province (LIP) formed at the time of the Sturtian. This is what you would expect with the breakup of a supercontinent. What the researchers found was that this LIP (known as the Franklin LIP) encroached on an area of sulphur-rich minerals. Lava and sulphur-rich strata would undoubtedly generate sulphur gases, and there would be cooling (*19*). Add this cooling to the cooling already in train due to the silicate-induced carbon dioxide drawdown as well as the supercontinent's equatorial position (that as mentioned prevented ocean warming, hence ocean heat transport to higher latitudes), and if a Snowball Earth would not already have been likely, then it would with this additional cooling. Many volcanoes release sulphates and so it is not unreasonable to assume that the tectonic activity associated with supercontinental breakup would release cooling sulphates, and that the pulsed nature of Snowball Earth II (and presumably Snowball Earth I) could be put down to the pulsed nature of supercontinent breakup: indeed, it would be unreasonable to expect a supercontinent to fragment all in one go.

The above are the reasons many have accepted as being the causes of Snowballs I and II.

With this currently standard view, life was not responsible for the Snowball Earths: the Snowballs occurred because of the position of a supercontinent in the tropics. This standard view is perfectly logical and uses factors that are reasonably well understood. Here, life was along for the ride and these geological factors behind Snowballs I & II spurred the evolution of oxygenic photosynthesis and multicellularity respectively.

So how does it compete with the new narrative being developed?

Well, the above question is not entirely an 'either'/'or' situation. All the factors described in the previously accepted view still operate in the new narrative. The difference is how much does each factor individually, as well as sometimes acting synergically, cause the cooling resulting in the Snowball events?

The big difference between the formerly accepted and the new narrative being developed is the role life played!

The formerly accepted view has it that the major evolutionary events of oxygenic photosynthesis and also possibly oxygenic photosynthesising eukaryotes (associated with Snowball I) and multicellularity (associated with Snowball II) came about *as a result of* the Snowball events. The new narrative has it that these major evolutionary events played a major part *in causing* the two Snowballs. Both the old and the new narratives have silicate-mediated carbon drawdown, both have the slow accumulation of volcanic carbon dioxide ending the Snowball episodes.

So let's examine the two narratives in more detail to see if one both better explains the evidence and has more evidence supporting it than the other.

## Alternative Snowballs *vs.* the new narrative

There are arguably a number of shortcomings in the up-to-now standard model. Principal among these, as pointed out by Tim Lenton and Andrew Watson (9), is that while there have been a number of supercontinents in the Earth's history—Ur, Kenorland, Nena-Columbia, Rodinia, and Pangaea—only twice has there been a Snowball Earth (or, more accurately, twice periods of multiple Snowballs). Indeed, while both the Snowball Earths (supercontinent

Kenorland's ~2.4–2.2 billion years ago (bya) and the Rodinia 850–580 million years ago (mya), respectively) had a supercontinent breakup association, none of the others were (nor were many) of the comparatively more minor glaciations/ice ages, that have also taken place have had such an association. So supercontinents per se, it can be argued, are not contingent for Snowball Earths.

Now, there is debate and it is possible to counter the above argument quite simply by saying that these other supercontinents' breakups did not occur when their location was in tropical low latitudes at or near the equator. This allowed tropical waters to move to higher latitudes keeping the Earth warm.

Yet, such is the debate, it is possible to counter this counter, as the exact location of supercontinents is not precisely known; there are clues and likelihoods but not precision. Taking an extreme position of presumed complete ignorance, we could consider taking a random supercontinent in the history of the Earth and place it somewhere equally randomly on the Earth's sphere. (This is known as a 'Monte Carlo' approach, as it relies on a random gamble.) Now, you may think that a sphere is a sphere and that nowhere on it is particularly special. However, this is not so once the sphere is spinning about a specific axis, let alone illuminated along a plane more or less at right angles to the first. Such is the geometry of spheres rotating about a specific axis (as the Earth does) that most of the sphere's surface is located near the sphere's equator and not at its poles.

Now, this is may seem counterintuitive—surely a sphere is a sphere and so is the same no matter from whence it is viewed—but it can be explained simply. If you walked along entire the line of a high latitude (near 90°), around say the South Pole, indeed right next to it, then you would just simply walk around the Pole, covering all longitudes, in a second or two. Conversely, walk along the 0° latitude along the entire line of the equator and you would be facing around a 25,000 mile (40,000 km) hike. This geometry means that a band wrapped around the equator covers more of the Earth's surface than the same-width band along a latitude near the poles. In fact, 50% of the Earth's surface lies between 30°N and 30°S close to covering much of the tropics; not 45°N and 45°S which you might intuitively think as 45° since it is the halfway latitude between the 90° latitude poles and 0° equator. Pursuing this line of argument, then the purely Monte Carlo (random gamble) view as to a supercontinent's theoretical position at its time of breakup gives a marked likelihood that it would be located in the tropics or subtropics. Yet despite there having been several supercontinents and their breakups, only two (periods of) Snowball Earths have occurred. This suggests, whatever other arguments there are (and there are other arguments), that on the basis of pure chance alone, equatorial supercontinental breakup is not a *causal* factor for the creation of a Snowball Earth (even if it is implicated in other, lesser, ways).

Another problem, considering purely non-biotic theories, as to the cause of the Snowball Earths relates to the previously mentioned Francis Macdonald and Robin Wordsworth LIP sulphate suggestion that may work for triggering the Sturtian if volcanic activity encroached on sulphate rich strata (*19*). What it doesn't do is explain the other cold phases of Snowball II such as the Marinoan, or the cold phases of Snowball I let alone Snowball II's regular, ~60 million year, pacing. The chance of a sulphate strata encroaching LIP for all these cold phases seems rather unlikely. So while it may be a factor in the degree of cooling in the Sturtian glacial, it is not a causal explanation for Snowball Earths in general. Furthermore, while volcanic-related activity is certainly associated with continental breakups, it does not just stop with the breakup. Today the continents are fragmented, scattered about the globe, but we still have volcanic activity, and lava-spewing rifting events: you can go to Iceland or much of the length of the Pacific rim, for example, to see this.

Furthermore, if supercontinents were needed for an ice age, then how come we are in one today?

As mentioned in Chapter 2, at the moment we really are in an ice age right now; albeit in a warm interglacial part of what is called the Quaternary ice age. Yet, right now, the continents are widely distributed across most of the Earth (Pacific excepted) and not clumped into a recently fractured supercontinent. In short, ice age glacials, let alone Snowball Earths, need not necessarily be associated with supercontinents.

So if we are in an ice age right now, then why have not our ice age's coldest times (the glacials) of our Quaternary ice age become a Snowball Earth? The current explanation for this is that when the Earth's continents are distributed mainly in higher latitudes as they are now (as opposed to equatorially) and the Earth enters a time of cold, then the ice forming on rocks prevents atmospheric carbon dioxide being drawn down and so prevents further cooling through the silicate reaction discussed earlier this chapter (*18*). Conversely, during a Snowball Earth where the land is equatorially distributed, it takes a while for the ice to migrate down from the higher latitudes to the low latitude tropics which—remaining ice free—can continue silicate-mediated carbon dioxide drawdown. Here, the traditional view does score over the new life-is-in-the-driving-seat narrative.

It must be borne in mind that the above is a summary of some of the key arguments and that there is a fair bit missing. For example, Earth system scientists and bio-geoscientists will be quick to point out that from some 35 mya the rise and then spread of tropical grasses (with their new ability due to refinements of the photosynthetic process to drawdown carbon dioxide more efficiently in warm climates than some other plants) has helped lower carbon dioxide and so cool the Earth and help usher in the current Quaternary ice age of the past

2 million years. So, here no recently fragmented supercontinent was needed. But then this is the point, the new Earth system narrative has life very much as a key player in how the Earth has developed.

Life is simply not included in the previous standard model for Snowball Earths as—if not a causal—a key factor for instigating the Earth flips: life, under the currently standard model, simply reacts to the changes. So, for example, there is no hydrogen loss from biogenic methane in the standard model involving life's chemical reactions that, it is now conjectured, boosted atmospheric oxygen with Snowball I (see Figure 3.3). Conversely, the new narrative has life very much playing the central role for both Snowball I and II.

Having said that, life is included in the traditionally favoured explanation of Snowball Earths' effect on life's evolution. Indeed, it is not ignored in the standard view that Snowball I saw, if not a contemporaneous rise in atmospheric oxygen from a negligible level to a few per cent of present levels, a rise in oxygen after the evolution of oxygenic photosynthesis. Neither is it ignored that multicelled animals boomed in the Ediacaran at the end of Snowball II with the Vendian boom and then again, a few million years later with the further evolution of animal feeding parts, with the Cambrian boom. In the standard view, the Snowballs spurred life's evolution, not the other way around.

Yet recent evidence suggests that simple multicelled life possibly existed prior to Snowball II: here the standard view is silent on this and indeed denies it.

So if life—under the traditionally accepted view—did not cause the two Snowball Earths, then how does the traditional view have life react to the Snowballs to result in the evolutionary developments we see around those times?

Here the traditional, standard view explanation involves hydrogen peroxide.

Hydrogen peroxide is a reactive molecule similar to water ($H_2O$) but with the chemical formula $H_2O_2$. It can be formed a number of ways but principally in the early Earth by the action of ultraviolet light on water (radiolysis) that generates hydrogen and oxygen radicals (single atoms of oxygen that are very reactive and also involved in the creation of ozone—see Chapter 3). Some of these oxygen radicals in turn react with other water molecules to form hydrogen peroxide; others form oxygen molecules ($O_2$). Hydrogen peroxide itself is an oxidant whose molecules themselves quite easily split into water leaving an oxygen radical that can go on to oxidise other things.

Life has to deal with these radicals and the way it has done this is to evolve an enzyme called catalase. This splits hydrogen peroxide into water and oxygen in the following reaction.

$$2H_2O_2 \rightarrow 2H_2O + O_2$$

$$\text{Hydrogen peroxide} \rightarrow \text{Water} + \text{Oxygen}$$

Here, the standard view has it, that with Snowball Earths with its vast ice sheets beneath largely cloudless skies, there would be a slow trickle of hydrogen peroxide formation which, over millions of years of Snowball glaciation, would accumulate and release oxygen. This oxygen was poisonous to anaerobic prokaryotic life which would have to develop mechanisms so as to cope with it. One such mechanism is the enzyme catalase. The standard view notion is that the presence of oxygen from hydrogen peroxide from the Snowball ice sheets meant that anaerobes had to develop such mechanisms and with this protection it enabled the rise of oxygenic photosynthesis (OP). Once catalase evolves (in response to the presence of hydrogen peroxide from UV bathing Snowball Earth I, says the traditional view) it is a far easier step to get from that to OP's oxygen-evolving complex (*20*).

With catalase it is easy to evolve this key, oxygen-evolving complex, component of OP. This is because the structure of this enzyme bears a striking relationship with the oxygen-evolving complex component of OP prokaryotes. Indeed photosynthesis' oxygen-evolving complex can almost be considered as two catalase molecules joined together.

To sum up, it was this hydrogen peroxide from Snowball Earth I that—the argument goes—stimulated the evolution of the then-anaerobic life to evolve catalase, much like in our modern world the presence of antibiotics in low levels encourages microbes to become resistant (*21*). The melting of the Snowballs would release the hydrogen peroxide from the ice and into the small seas formed as a result of the beginning of the supercontinent breakup. Life evolved catalase to deal with this peroxide-generated oxygen and so came close to evolving oxygenic photosynthesis.

An alternative notion, that may also have happened simultaneously, is that the released hydrogen peroxide from Snowball (or Slushball) melting would itself decompose releasing oxygen and this oxygen would spur life: in the case of Snowball Earth I it would facilitate the evolution of OP and in the case of Snowball Earth II the resulting rise in oxygen would facilitate the evolution of multicelled animals. The hydrogen peroxide chemistry for this, with regards to Snowball I, has been modelled (*22*) and, in theory at least, it seems to work if you make some assumptions about the numbers.

The counter to all this is that hydrogen peroxide generation calculations used to support the above argument are based on assumptions that some might consider as almost wishful thinking and that much of the hydrogen peroxide produced would be diluted by the oceans and countered by sea salts and other chemicals in the marine environment.

Before we leave hydrogen peroxide and catalase, to further explore the standard traditional and new narratives, we need to address the antioxidant problem.

Strict anaerobes in an oxygen world, as noted, are poisoned by oxygen. So the question arises that if anaerobic prokaryote evolved into an oxygen-generating cyanobacteria-like prokaryote, then how did they deal with the toxic oxygen they were generating? It is a chicken or egg question: which evolved first, oxygenic photosynthesis or the antioxidant mechanisms needed to protect the organism from the oxygen? The answer is likely to be that there are a number of antioxidants and that the antioxidant mechanisms co-evolved with OP from already-existing precursors.

One possibility has been suggested by Reto Burn and colleagues at the University of Basel, Switzerland. They have found that the anaerobe *Chlorobium limicola* produces the antioxidant ergothioneine. (Ergothioneine is a sulphur metabolite that occurs in microorganisms, fungi, plants, and animals. And as previously noted—see Chapter 2—some early prokaryotes used sulphur as an energy source.) Could it be that there are a number of compounds that evolved in early anaerobic prokaryotes that were later re-purposed as antioxidants when oxygenic photosynthesis (OP) evolved (*23, 24*)? This would enable OP to evolve without the need for a Snowball ice sheet to generate hydrogen peroxide to stimulate the evolution of antioxidants and so protect OPs when they did evolve. This also fits in with the evidence that OP evolved substantially *before* Snowball I, which saw the Earth's atmosphere flip from an anaerobic state to a low-level oxygenated state (Chapter 3). Catalase evolution problem solved.

So now we can summarise comparing the old and new Earth system science narratives.

Remember, what we have with the standard traditional view is that the *causes* of Snowballs I and II were fundamentally nothing whatsoever to do with life but primarily, with both Snowballs, were due to the equatorial position of their respective supercontinents. The equatorial position affected heat transport from the tropics, its mountain ranges ice caps reflected light, as did the lighter coloured deserts in the supercontinent's interiors. Then the breakups allowed water, hence higher rainfall, into the splitting supercontinent so causing silicate-mediated carbon dioxide drawdown and both these cooled the planet with further cooling from sulphate released during the continental breakup, much like sulphate-rich volcanic eruptions today cool the Earth for a few years but here on a much bigger scale. With this view, life was not a principal factor in *causing* Snowballs, but the Snowballs did *influence* life's evolution.

Conversely, with the more recent view now being promulgated by some in Earth system science, new evolutionary stages of life—oxygenic photosynthesis with Snowball I and the colonisation of land by proto-lichens and other multicelled species with Snowball II—were key to Snowball Earth formation.

With this view life, if not in the driving seat, at the very least is very much up front as the principal factor driving cooling change along with the other non-life factors (such as silicate-mediated carbon dioxide drawdown of the traditional standard view). With the new view, the Snowball Earths did not so much spur evolutionary change, but biological evolution facilitated the Snowballs taking place.

At this point I do not wish to confuse matters but I must stress that the narrative I have been presenting and the former, standard traditional narrative are towards two ends of a spectrum of views. This whole question of life and planet is a contentious area (which is why studying it is so exciting) of what is called biosphere science, or Earth system science, and there are many researchers who have a different take and/or nuance, or even who draw upon different aspects of these narratives. This is fine: it is the way science works, challenging competing hypotheses. What I hopefully have done here is to present a possible outline of some of the recent discussion taking place and contrasting it with an outline of the traditional view: I have not had the time nor space here to go into the many complexities of much detail.

Nonetheless, while more casual readers of this book need not worry unduly about the finer points of debate, relevant researchers will, and relevant undergraduate students should. To these I offer but one example that is illustrative of such a variation of thought.

In 2017 Jochen Brocks, of the Australian National University, Canberra, and colleagues analysed geological strata from four locales in Australia as well as one in the United States and one in Sweden that together span the period from 541 mya to 820 mya, plus more recent times (*25*). They looked for a number of biomarkers such as sterols that indicate different types of life and what they found does fit in with the new narrative (and Earth system science undergraduates will want to seek this work out). Where it differs is that their idea still has it that the principal early phase of Snowball II was caused by non-life factors. This work suggests:

> The Sturtian glaciation may have perturbed this positive feedback, flipping the system into a new steady state. Subglacial weathering increased marine phosphorous concentrations.

What they are saying is that the huge planet-covering ice sheet of Snowball II ground the rock beneath it, so releasing nutrients (phosphorous) that then caused a bloom of oxygenic photosynthesisers.

It was this phosphorous that likely enabled true plant algae to overtake cyanobacteria as the principal oxygen generators on the planet (*25, 26*). Conversely, the new narrative (as, for example, championed by Lenton and Watson)

(9) has it that it was the colonisation of land by multicellular species such as proto-lichens that played a critical role in releasing nutrients from rock such as phosphorous.

Yet while the data Brocks's team have elucidated is very interesting, others disagree with aspects of their conclusions. For instance, the BBC, in covering this work, noted that Nicholas Butterfield of Cambridge University did not subscribe to Brocks's view that the new levels of phosphorous washing into the seas was the critical factor for the plant algae to bloom, rather that the earlier evolution of sponges reduced the numbers of cyanobacteria that cleared the way for eukaryotic algae to boom with the added phosphorous nutrients (*27*). The new narrative certainly supports an earlier rise of sponges, but that does not mean that phosphorous release from land did not play a part in the subsequent increased role of plant algae. Healthy debate is fundamental to good science.

Now, we do not have a time machine to go back and see exactly what happened before and after the Earth's first flip of state around 2.4–2.2 billion years ago and during its second big flip 720–635 million years ago. The evidence we have is disparate, painstakingly gathered, somewhat sparse (there is not a lot of old geological strata accessible to study), and fraught with a lack of precision (the accurate dating of such old evidence is difficult) plus some of the evidence is decidedly circumstantial and open to different interpretation.

It is a bit like considering court case testimony from unreliable witnesses. Yet there are an awful lot of these witnesses so that just a few scenarios seem reasonably plausible. Both the old traditional and newer camps of views, as to how and why the Earth flipped between states with each Snowball, have their merits. So the question that remains is, which scenario is the most convincing in terms of what evidence we do have, as well as being believable in terms of evidence interpretation and hence of narrative?

Nonetheless, on a personal basis (and I stress this being a personal view), I do wonder if it is possible to at least partly merge the old and the new narratives? After all, it seems quite a coincidence that life went through two major evolutionary developments at the time when a supercontinent was breaking up. Could it be that the rise of oxygenic photosynthesis and the rise of multicellularity were each necessitated by a respective supercontinent breaking up (it enabled maximal nutrient delivery to inland waters, boosting life, that nudged the Earth through a critical transition), but that life *previously* going through a major evolutionary stage was a key—if not 'the' key—factor instigating Snowballs I and II?

In balancing all the evidence and deciding between the 'Supercontinental fragmentation caused the Snowball Earths' on one hand and that 'life's evolution was the causal factor' (and I would add 'facilitated by supercontinent fragmentation') on the other, one needs to decide, which is more likely? We have a global carbon cycle and life is firmly carbon-based, processing carbon at key parts of this carbon cycle. Can we really expect major evolutionary developments affecting life's carbon processing not to have a profound effect on the global carbon cycle? Or do we expect abiotic (purely geological) processes to be primarily responsible?

Here an examination of the global carbon cycle today might offer a clue.

The IPCC summarised the carbon cycle in their 2007 Assessment Report (*28*). Its depiction of the modern global carbon cycle estimates that the marine photosynthetic drawdown (see the centre and centre right of Figure 5.1) is a few score times greater than drawdown due to silicate weathering (see the left of Figure 1.5). Clearly, life is playing a more significant role in processing carbon in the global carbon cycle than silicate weathering, and so surely, therefore, life processes are hard to ignore when considering the key factors leading to Snowball Earths I and II even if the carbon cycle back in those times was quite different?

Before I try to bridge the two hypothetical narratives, I should ask you to note that, apart from the following section (which is a synthesis of differing

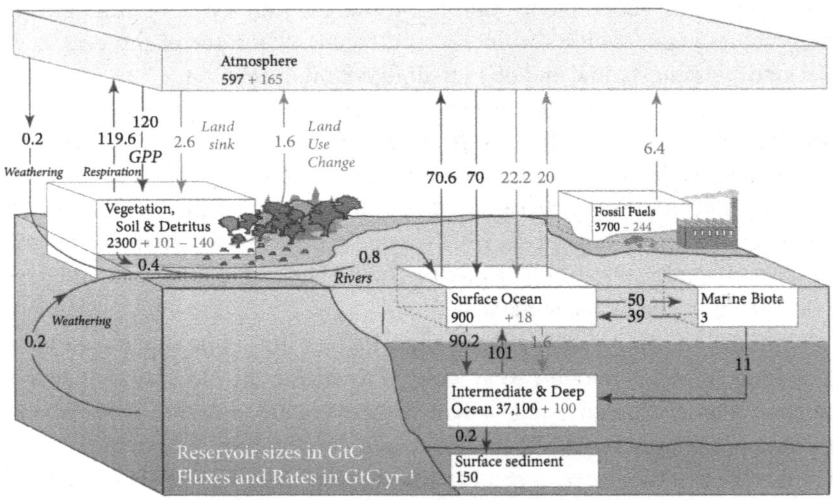


**Figure 5.1** The global carbon cycle for the 1990s, showing the main annual fluxes in GtC $yr^{-1}$: pre-industrial 'natural' fluxes in black and 'anthropogenic' fluxes are the numbers with '+' or '-' before them.

views previously covered) I have included key references throughout this book so that, should you wish, you yourself can explore the science papers and literature underpinning the way the Earth is considered to have flipped between states throughout its history. Alas there is no getting around it, if you wish seriously to decide between the two narratives—or come up with one of your own—you do need to read these papers (many of which are open access online) so as to be able to judge how the evidence fits in with each narrative. Meanwhile, the following section is a personal view (and I welcome it being challenged in an evidence-based way).

## Merging the old and new narratives

Any merger of the old (life was along for the ride) and new (life was in the driving seat) Snowball Earth and evolution of life narratives should, to have validity, at the very least explain *all* the evidence underpinning *both*.

First, stripping down the evidence it does seem that having supercontinent breakup when a supercontinent is *near the equator* seems important and not pure chance as the strong adherents to life-in-the-driving-seat argument as well as those in favour of life-along-for-the-ride would have it. As far as we can tell, an equatorial supercontinent almost certainly existed close to the time of both Snowball Earths I and II. They are needed to provide the nutrients if you are inclined to support the life-in-the-driving-seat argument; they are also needed to block ocean transport of heat away from the tropics and then albedo affects if you are inclined to the supercontinents-caused-the-Snowball-Earths-and-life-was-just-along-for-the-ride perspective.

Second, whatever the effect of each of these Snowball Earths on evolution, the evidence points to the biological machinery (hence an organism's genetic blueprint) of the respective evolutionary jump (oxygenic photosynthesis—and even OP eukaryotes—and multicellularity respectively for Snowballs I and II) being likely to have been in place *prior* to the respective Snowball. It is more likely so as to facilitate an immediate (geologically) biological boom associated with the Snowball event of the species involved in the evolutionary jump. This criterion plays to those supporting life-being-in-the-driving-seat hypothesis, as otherwise one has to argue that life responded to a global (or near global) ice sheet virtually immediately. This might be possible for small evolutionary change, but a big evolutionary leap (such as the rise of oxygenic photosynthesis or multicellularity), really?

Yet what of the former traditional narrative of a tropical supercontinent curtailing heat transport by sea away from the tropics being critical so triggering a glacial which—again due to supercontinent location—eventually led to a runaway Snowball. The traditional view has it that the supercontinent was necessary. Yet, some of the proponents of this narrative then go on to say that somehow—possibly through hydrogen peroxide formation on the global ice sheet dissociating into water and oxygen—this stimulated biological evolution, let us consider the supercontinents' role.

Let's assume that both Snowball I and II occurring at a time of supercontinental breakup (Kenorland and Rodinia respectively) is not merely coincidental but causal. Here a key question to ponder is, what was it about equatorial supercontinent breakups that facilitated the initial oxygen increase associated with Snowball I and the colonisation of land by multicellular life (possible proto-lichens among other species) with Snowball Earth II? Was it heat-transport-blocking with albedo kicking in, or was it life sucking greenhouse carbon out of the atmosphere being the key factor? Both factors were likely operational, but which one was fundamental to what was going on?

With the supercontinent heat transport/albedo argument, life's evolution of oxygenic photosynthesis (Snowball I) and then multicellularity (Snowball II) arose as a result of the respective Snowballs. With the life-in-the-driving-seat argument, it was life's evolution helping trigger the Snowball (not the other way around).

The thing is that both narratives do require a supercontinent but for different reasons.

With regard to Snowball Earth I, we have the Ilya Bindeman and colleagues water cycle evidence that Snowball I took place at a time of the first large supercontinent that additionally sported higher mountains as a result of the new larger continents colliding to form the said supercontinent (Chapter 3) (*29*). We also have the Patrice Rey and Nicolas Coltice argument of the Earth only having cooled enough for the crust to be thick enough to support tall mountains at the time of the first true supercontinent Kenorland (see Chapter 1) (*30*). Significant nutrient erosion of the land could not take place prior to then with less land and no extensive tall mountain ranges. These nutrients were required to boost the population of oxygenic species. This speaks to the life-in-the-driving-seat argument yet does still necessitate a supercontinent: the presence of a supercontinent was not pure chance!

Here then is a further personal thought. As already noted, there is ample evidence that oxygenic cyanobacteria evolved some considerable time *before* the Kenorland supercontinent, but what of oxygenic eukaryotes? Some researchers (such as Tim Lenton and Andrew Watson) suggest that the *earliest* eukaryotes (or proto-eukaryotes) could have arisen was around the time of Snowball I. Oxygenic eukaryotes could only add to the impact of the oxygenic prokaryotic cyanobacteria of that time: it might explain why the first big flip took place.

For what it is worth, this last is my contention too (though I would not rule out a slightly earlier rise of eukaryotes).

So (hypothetically at least) could a supercontinent breakup facilitate the rise of eukaryotes? The answer is (again hypothetically) yes, because it provided a diverse range of nutrient rich, multiple energy source environments perfect for these organisms for multiple reasons. (And multiple reasons increase likelihood.) A supercontinent breaking up is fractured tectonically and thermal vents are commonly (if not invariably) associated with such fractures. Some fractures would be more sheltered than others, there was some interconnectivity between them (the sea is a global common) and there were many fractures along with the vagaries of a fractured continental coastline. In short, there were a range of novel opportunities for organisms that already had developed a new evolutionary trick.

First, as previously noted, the first large supercontinent, with the first substantive mountain ranges altering the water cycle, hence increasing erosion of nutrients, would see nutrient delivery to the surrounding seas that would benefit both prokaryotes and any putative eukaryotes. This would be enhanced at a time when the continental fragments were both coming together and also just separating, as there would be water evaporating from narrow seas penetrating into areas of landmasses to provide more rain on land, hence erosion and nutrient runoff.

Second, the edges of a supercontinent—especially a tropical supercontinent—would at places (those with west facing east-west orientated funnels of land) for a number of reasons see very high tidal ranges: (a) because there would be the best part of half a planet of uninterrupted ocean for the tidal bulge to manifest, (b) because 2 billion (and indeed later two thirds of a billion) years ago the Moon was closer to the Earth than today (see Chapter 1). Further, tides would have been experienced more often as the Earth rotated faster in the past. Extremely high tidal ranges could only facilitate the rise of oxygenic photosynthesis *via* sunlight from non-oxygenic photosynthesis of communities associated with inland penetrating seas hypothesised with Snowball I. Also, extremely high tidal ranges could only facilitate the land colonisation by marine species hypothesised with Snowball II.

Third, especially at a time of supercontinent breakup there would be hot thermal vents in the narrow shallow seas of the fracturing. These would sport biological populations of their own and so add to the biodiversity of these inland-penetrating seas. (There would not be inland-penetrating-sea thermal vents during the supercontinent formation phase.)

Finally, and fourth, with regards to the eukaryotic colonisation of land, and with regards to Snowball I, not so much land itself but its freshwater systems, seas that are largely enclosed by land and fed by major continental rivers' freshwater, are less salty. This would have been important for the thinner-cell membraned eukaryotes that did not have the substantive thick cell wall of prokaryotes. This is because (as noted in Chapter 4) eukaryotes would be subject to osmotic pressures in freshwater environments. Having slightly less salty environments, as well as also substantively less salty environments, provides halfway houses—evolutionary stepping stones, if you will—to allow eukaryotes develop freshwater tolerating mechanisms.

Such a time of supercontinental breakup provides extensive nutrient rich, semi-compartmentalised, energy-source-diverse environments. The nutrients come from the supercontinental run–off: more so with Snowball II than I, but still taking place with I. The semi-compartmentalisation comes from the network of supercontinental-fractured inland seas. The energy source diversity comes from the thermal vents (common to, if not an essential part of, continental fracturing) and sunlight. And also there was the tremendous tidal range. As said, this last increases the chance of energy-rich sunlight for periods of time reaching areas of thermal vents, hence for species there to either use vent infrared as a source of energy, or sunlight, or even both. It also provides increased opportunity for symbiotic relationships to develop between species that do things differently, and this in turn increases the chance of the Baum-like inside-out endosymbiotic evolution of eukaryotes.

While in this last section I have summarised my own personal view, the debate continues as to whether life was along-for-the-ride and spurred into key evolutionary change by Snowball Earths I and II, or whether its evolution-was-in-the-driving seat.

The thing is that in both instances a supercontinent facilitated the *further* evolution of life.

Finally, I should again emphasise that the discussion within the Earth system science community is far more varied and there simply is not the space here to cover everything. For example, it is possible to argue that nutrient release with the Snowballs was due to their ice sheets grinding rock (as proposed by Jochen Brocks and colleagues discussed earlier this chapter). True, the ice sheets would erode and release nutrients, but this dimension speaks to *both* the former traditional view of supercontinental heat transport limiting and subsequent albedo causing Snowballs (life along for the ride) *as well as* that-life-was-instrumental-in-causing-the-Snowballs view: ice erosion and nutrient release by itself does not help us decide which of the views is more likely.

The question for those involved in the debate surely is which scenario best explains most of the evidence in a coherent, step-by-step way. Of course, in the process of doing that, assessment needs to be made as to the strength of each piece of evidence; remember, we are talking about events nearly a billion as well as billions of years ago.

Personally speaking, I am inclined to the new life-in-the-driving-seat perspective for a number of reasons. This includes that if it was the presence of a large, tropical supercontinent that caused both Snowball I and II due to both albedo and the prevention of heat transportation by sea to higher latitudes, then why did not the Snowballs occur as the supercontinent was forming? At the time of tropical supercontinent formation both albedo and heat transport factors would be in play, yet only when the supercontinents were breaking up do we see the Snowball Earths. There therefore had to be another critical factor and, other than substantive volcanic activity associated with the breakup (yet, as will be noted in the next chapter, well-known large igneous province—LIP—volcanic activity the past half billion years has never resulted in a Snowball Earth), the only factor that fits the bill is life.

Debatable, I accept, but arguably best fitting all the evidence is that life's evolution of oxygenic photosynthesis (and possibly OP eukaryotes) and multicellularity for Snowballs I and II, respectively, were critical for the Snowballs to take place; and not that Snowballs I and II spurred evolution.

Some may disagree. Either way, carbon-based life had evolved and, along the way, there was a profound change in the planet's carbon cycle that involved an episode of global climatic change.

There is one other thing that can be said to which both the old and new narratives hold. This is that one should not just view the evolution of the Earth simply as: our planet cooling following planetesimal accretion, then seeing chemical volatiles release and the creation of oceans and atmosphere, followed by plate-tectonics, the Wilson cycle commencement, and so forth, along with life interactions. Nor should biological evolution be considered simply as: biochemical synthesis from inorganics, the rise of genetic material, prokaryotes, oxygenic photosynthesis, eukaryotes, and multicelled species. Instead, we should adopt a 'co-evolution of life and planet' perspective.

Notwithstanding this and the Earth flipping state with evolutionary leaps, there were to be other disruptions to the planet's carbon cycle involving global climate change: there were the aforementioned jokers and wild cards, further evolution, and then intelligent design.

## References

1. Parfrey, L. W., & Lahr, D. J. G., Multicellularity arose several times in the evolution of eukaryotes. *Bioessays*, **35** (4), 339–347 (2013).
2. Yuan, X., Chen, Z., Xiao, S., Zhou, C., & Hua, H., An early Ediacaran assemblage of macroscopic and morphologically differentiated eukaryotes. *Nature*, **470**, 390–393 (2011).

3. Zhu, S., Zhu, M., Knoll, A. H., *et al.*, Decimetre-scale multicellular eukaryotes from the 1.56-billion-year-old Gaoyuzhuang Formation in North China. *Nature Communications*, **7**, 11500 (2016).
4. Fairchild. I., Condon, D., Lenton, T., & Shields-Zhou, G., (convenors), *The 2012 Fermor Meeting—The Neoproterozoic Era: Evolution, Glaciation, Oxygenation*. 19th–21st September. Geological Society: London Great Britain (2012).
5. Mills, D., Ward, L., & Canfield, D., Oxygen constrains on the origins of metazoans (poster abstract). In Fairchild, I., Condon, D., Lenton, T., & Shields-Zhou, G. (convenors), *The 2012 Fermor Meeting—The Neoproterozoic Era: Evolution, Glaciation, Oxygenation—Abstract book*. 19th–21st September. Geological Society: London, Great Britain (2012).
6. Mills, D,. B., Ward, L. M., Jones, C., *et al.*, Oxygen requirements of the earliest animals. *Proceedings of the National Academies of Sciences USA*, **111** (11), 4168–4172 (2014).
7. Strother, P. K., & Foster, C. A fossil record of land plant origins from charophyte algae. *Science*, **373**, 792–796 (2021).
8. Gensel, P. G., When did terrestrial plants arise? *Science*, **373**, 736–737 (2021).
9. Lenton, T., & Watson, A., *Revolutions That Made the Earth*. Oxford University Press: Oxford, Great Britain (2011).
10. Spribille, T., Tuovinen, V., Resl, P., *et al.*, Basidiomycete yeasts in the cortex of *ascomycete macrolichens*. *Science*, **353**, 488–492 (2016).
11. Kennedy, M., Drosser, M., Mayer, M., *et al.*, Precambrian oxygenation: Inception of the clay mineral factory. *Science*, **311**, 1446–1449 (2006).
12. Knauth, L. P., & Kennedy, M. J., The late Precambrian greening of the Earth. *Nature*, **460**, 728–732 (2009).
13. Arthur, M., Carbonate rocks deconstructed. *Nature*, **460**, 698–699 (2009).
14. Planavsky, N. J., Rouxel, O. J., Bekker, A., *et al.*, The evolution of the marine phosphate reservoir. *Nature*, **467**, 1088–1090 (2010).
15. Yuan, X, Chen, Z., Xiao, S., Zhou, C., & Hua, H., An early Ediacaran assemblage of macroscopic and morphologically differentiated eukaryotes. *Nature*, **470**, 390–393 (2011).
16. Chen, Z., Zhou, C. X., & Xiao, S., Death march of a segmented and trilobate bilaterian elucidates early animal evolution. *Nature*, **573**, 412–415 (2019).
17. Pratchett, T., & Baxter, S., *The Long Earth*. Doubleday: London, Great Britain (2012).
18. Nield, T., *Supercontinent: Ten Billion Years in the Life of Our Planet*. Granta: London, Great Britain (2007).
19. Macdonald, F. A., & Wordsworth, R., Initiation of Snowball Earth with volcanic sulfur aerosol emissions. *Geophysical Research Letters*, **44** (4), 1938–1946 (2017).
20. Lane, N., *Oxygen: The Molecule That Made the World*. Oxford University Press: Oxford, Great Britain (2002, 2016 revised).
21. McKay, C. P., & Hartman, H., Hydrogen peroxide and the evolution of oxygenic photosynthesis. *Origins of Life and Evolution in the Biosphere*, **21**, 157–163 (1991).
22. Liang M-C., Hyman Hartman, H., Kopp, R. E., Kirschvink, J. L., & Yung, Y. L., Production of hydrogen peroxide in the atmosphere of a Snowball Earth and the origin of oxygenic photosynthesis. *Proceedings of the National Academies of Sciences USA*, **103** (50), 18896–18899 (2006).
23. Burn, R., Misson, L., Meury, M., & Seebeck, F. P., Anaerobic origin of ergothioneine. *Angewandte Chemie International Edition*, **56**, 12508–12511 (2017).
24. Ruszczycky, M. W., & Lie, H.-W., The surprising history of an antioxidant. *Nature*, **51**, 37–38 (2017).
25. Brocks, J. J., Jarrett, A. J. M., Sirantoine, E., Hallmann, C., Hoshino, Y., & Liyanage, T., The rise of algae in Cryogenian oceans and the emergence of animals. *Nature*, **548**, 578–581 (2017).
26. Knoll, A., Food for early animal evolution. *Nature*, **548**, 528–529 (2017).
27. Pease, R., The algae that terraformed Earth. *BBC News* (2017). http://www.bbc.co.uk/news/science-environment-40948972.

28. Intergovernmental Panel on Climate Change, *Climate Change 2007: The Physical Science Basis. Contribution of Working Group I to the Fourth Assessment Report of the Intergovernmental Panel on Climate Change.* Cambridge University Press: Cambridge Great Britain (2007).
29. Bindeman, I. N., Zakharov, D. O., Palandri J., *et al.*, Rapid emergence of subaerial landmasses and onset of a modern hydrologic cycle 2.5 billion years ago. *Nature* **557**, 545–548 (2018).
30. Rey, P. F., & Coltice, N., Neoarchean lithospheric strengthening and the coupling of Earth's geochemical reservoirs. *Geology*, **36** (8), 635–638 (2008).

# 6

# Jokers and Wild Cards

Following Snowball II, multicellular life blossomed, first with the Avalon boom of Ediacaran biota and then later—with the evolution of true mouths and guts—the Cambrian boom (around 540 million years ago [mya]) and the extinction of the Ediacaran biota (see the last chapter).

On our arm's-length journey of Earth's history from the shoulder, we are now all the way down to the hand, just off the knuckles—leaving Snowball II behind—and onto the forefinger. From here on in you might think it plain sailing, mere centimetres, to a few hundredths of a millimetre from the fingertip and the evolution of anatomically modern humans possibly somewhere around 300,000 years ago (*1*). Not a bit of it: life was going to have to survive being thrown some wide curves. The Earth was going to see some (geologically speaking) short, sharp, highly disruptive episodes that would even, on occasion, lead to mass extinctions. Some such events as these undoubtedly happened prior to Snowball II (albeit with less impact to life that was primarily marine) but, in addition to these being so long ago, there were not the advanced, multicellular life forms with hard bits (shells, bones, and so forth) to leave a fossil record charting such events. However, post Snowball II, with the fossils of hard-bit, multicellular life, we can begin to see the impact of these wild cards and jokers in the geological record.

As shall be shown, though these jokers and wild cards did cause carbon cycle disruption and extinction events, despite some similarities *they are quite distinct from the evolution-of-life-driven Flips of Earth state* so far discussed. The jokers and wild card events are critical transitions, just not ones driven by a major new development in biological evolution. Nonetheless, these other critical transitions do impact on evolution even if they are not driven by a development in biological evolution. These events are more random than the equally non-biological-evolution-driven transitions such as the Milankovitch pacing of glacials and interglacials (see Chapter 2). But both these—joker events and Milankovitch pacing—involve critical transitions that do affect evolution (not the other way round). We will return to how Milankovitch-driven changes affected the evolution of our own species in Chapter 8, but for now we'll look at some of these joker/wild card events.

*Earth's Flips of State*. Jonathan Cowie, Oxford University Press. © Oxford University Press (2026).
DOI: 10.1093/9780197825402.003.0006

To greatly summarise matters, these jokers and wild cards can be principally categorised into two causal groups: asteroid impacts and, arguably more significantly, large igneous provinces (LIPs).

## Large igneous provinces (LIPs)

We have already seen that supercontinents were involved in Snowballs I and II (even if there is debate, between the new narrative explored by this book and traditional ones, as to exactly how). We have noted that supercontinent formation and then fragmentation occurs repeatedly as part of the Wilson cycle and that this takes place through plate tectonics that in turn is driven by the movement within the Earth's mantle: the mantle is the home of powerful forces.

The problem comes when these forces start to impact the Earth's surface in a major way, as opposed to the millimetre or centimetre *per* year movement of continents across the globe. One of the principal impacts is with what are called 'large igneous provinces' (LIPs). These form when molten igneous material from within the Earth burst through the surface over a large area or province.

Volcanoes, compared to LIPs, are minor features: LIPs are massive. Typically, LIPs are the size of some of today's nations and they involve thousands of cubic kilometres of lava flow on our planet's surface. Fortunately for us, there are no fully active LIPs today. However, we do today have an example of the remnants of activity of a formerly fully active LIP in the form of the Iceland hotspot. This sits at the northern end of the mid-Atlantic ridge between separating plates: Europe and North America, as well as Africa and South America are moving apart, part-driven by mid-Atlantic upwelling mantle convection. Back in the early Eocene this hot plume spot's centre was beneath Greenland, but it had hot material spreading beneath, and in some cases through, the crust to as far south as the northern coast of Scotland and the north-east Atlantic off of Norway. As we shall shortly see when discussing the IETM/PETM (an abbreviation to be explained later and also in Appendix 1) today's Icelandic hotspot had, back in the early Eocene, powered the North Atlantic Magmatic (or Igneous) Province.

The important thing is that LIPs can have a profound effect on the biosphere: they can release a pulse of gases into the atmosphere that cause planet-wide climatic disruption lasting thousands of years. Over the past half billion years, the Earth has seen an active LIP every few tens of millions of years (as well as smaller, less impactful, igneous provinces more frequently).

Figure 6.1 shows the approximate location on a map of the present-day Earth of some of the LIPs discussed in this chapter. Here you may have noticed that the

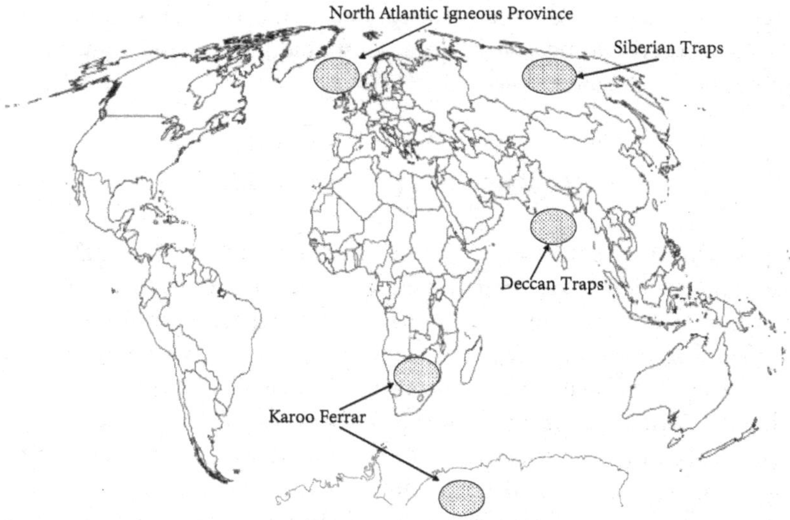


**Figure 6.1** A map of some examples of Large Igneous Provinces discussed in this chapter.

Karoo-Ferrar LIP is found both in southern Africa as well as part of Antarctica. The reason for this is that, back when that LIP was created and active, the Earth's continents (again due to the Wilson cycle of plate tectonic movement) were in different places: there was a time when Africa adjoined Antarctica, and the Karoo-Ferrar LIP straddled the boundary.

Now, you might think that the emergence of lava on such a scale as to create a small 'sea' of rock would have a profound impact on the planet. Yet for the most part of the LIP formation process, it is thought that their effect is less than one might think this formation period lasts for many tens or hundreds of thousands (sometimes even a million or so) of years. Indeed, as we shall shortly see, it has been postulated that the Deccan Traps (see Figure 6.1)—possibly through dust and sulphate injection into the atmosphere—lowered the global temperature by a couple or so degrees centigrade. As we shall shortly see, this (perhaps with alternating periods of warming through carbon dioxide release) may have been cause for the late Cretaceous decline of the dinosaurs before the Chicxulub asteroid impact 66 million years ago finally wiped them out. However, sometimes LIPs see a rapid release of gas: it is as if the Earth burps. Their effect can be globally profound.

The idea is this. A hot mantle plume heats the crust, and/or possibly previously subducted ocean crust, that releases gas ahead of the rising plume. This gas often raises the ground in an almost balloon-like way before—to continue the metaphor—the balloon pops, releasing the gas (2). Alternatively, some LIP-related events do not see such a rise in the land above the plume head. This can be explained (here very simply): the hot plume, containing subducted ocean floor, is so hot that it literally eats its way through the Earth's crust as opposed to pushing it up. Either way, there can be a massive degassing event at the start of the time when the LIP begins to form at the Earth's surface (3). This degassing can have a profound effect on the biosphere, including inducing mass extinction events.

Such LIP events are often associated with carbon isotope excursions (CIEs) and climate change. While these CIEs were neither as large as those of the Earth's two big flips, they still caused global temperature changes and did have a discernible impact on life and the fossil record.

Let's look at an example (and a little later another one).

## The end-Permian event (252 mya)

The end-Permian event was the largest of all the mass extinctions of the past 600 million years. As its name suggests, it took place at the end of the Permian geological period and the start of the Triassic (see Appendix 3); it is therefore also sometimes known as the Permo-Triassic event of 252 mya. (Students need to internet search separately both 'end-Permian' and 'Permo-Triassic' with 'event' and separately 'extinction' to capture more of the academic literature.) More colloquially, it is also known as the 'Great Permian Extinction' and even the 'Great Dying'.

This episode is of such biological significance in the geological record that geoscientists also use it to define the boundary between the Palaeozoic and Mesozoic eras (the latter being famous for being the era of the dinosaurs).

The end-Permian event, being the largest mass extinction for over the last half a billion years, saw a marked turnover of species: both fauna and flora, both marine and terrestrial. At that time, over 60% of animal species—both marine and terrestrial—disappeared. Overall (for both plants and animals) 90% of ocean species vanished, as did 70% of vertebrate families of species on land!

At the event's end, so much life was wiped out that of some of those species that survived were those whose lifestyle was associated with living off of dead biota. Indeed, immediately after the event, there is what is called a 'fungal spike' in the geological record.

Over the years, there have been a number of hypothetical explanations for the end-Permian event, including a giant asteroid impact. However, the hypothesis that has arguably gained the most traction with the geoscience and Earth system community in recent years is that it was due to the volcanic activity associated with the Siberian Traps LIP that dates from that time.

The Siberian Traps today are centred around the present-day Siberian city of Tura and also encompass Yakutsk, Noril'sk, and Irkutsk in Russia (see Figure 6.1).

To give you an idea of the scale of this LIP, the Siberian Traps has an area of just under 2 million km$^2$: that's greater than that of present-day Central and Western Europe combined. But area is one thing; it is the volume of the lava that more relates to the emissions of dust and gas injected into the atmosphere, affecting the globe. Estimates of the original volume of the Traps range from 1 up to 4 million km$^3$ (3 million km$^3$ is often cited). The Siberian Traps eruptions lasted at full intensity for about 1 million years, which—given the errors level of the dating methods—coincides with the mass extinction event. Yet it looks as though possibly there was more than one extinction episode within the end-Permian event. Without going into all the evidence (a book in itself), boron isotope data from the geological record of the time suggests a sharp drop in ocean pH—abrupt acidification—in a second phase of extinction. This is attributed to a massive injection of carbon dioxide into the atmosphere (*4*).

Here, the contention is that this injection of carbon dioxide, being massive, was critical in addition to whatever else was going on. Whatever else was going on may well have included a massive eruption within the LIP injecting dust high into the stratosphere to shield the Earth's surface from the Sun, so plunging the Earth into a freezing nuclear-winter-like state devoid of sunlight and so significantly shutting down photosynthesis (infrared prokaryotic photosynthesis by thermal vents, *etc.*, obviously excepted).

Lee Kump, of Pennsylvania State University, considers the end-Permian injection of carbon into the atmosphere was so massive that it could have overwhelmed the Earth system's buffering mechanisms. For example, unlike later LIP events (and we will come to an example shortly), at the time of the end-Permian there was no extensive sea floor carbonate to buffer the acid from the carbon dioxide released (*5*, *6*). This would have globally warmed the Earth to such a degree that the oceans would have lost nearly all its oxygen and made life on land in many places unbearable for the species that we see from the geological record go extinct.

However, there is other evidence to suggest that instead of multiple events much of the extinction, as far as land-based life was concerned, could have occurred in a single bad day. Peter Ward of Washington University, Seattle, and colleagues looked at the geology dating from the time of the extinction at seven sites scattered across 400 km of South Africa. They noted that the ancient beds of rivers in all these places suddenly changed but had, up to the time of the extinction, seen an orderly deposition of sediment. They attribute this dramatic difference to all the vegetation dying in each of the respective river basins and so, with their roots not holding the soil together, resulting in the rivers filling with a mass of sediment and detritus (*7*). They also suggest that the extinction event was first felt on land before it took hold in the oceans.

It is hard to grasp the impact of the end-Permian event, it being such an extreme mass extinction. Even ignoring the million-year lifespan of the Siberian Trap LIP's activity—that would have seen a number of phases to it (so spreading out the impact)—we might very roughly begin to get an idea of its magnitude from the huge (the aforesaid 1–4 million km$^3$) volume of lava involved. So it could be that only a fraction of this was needed, and involved, in a single extinction pulse, even if there were other related but smaller events over the LIP's active lifespan; a lot can happen with up to 4 million km$^3$ of magma over a million years.

To give you an idea of the likely impact, compare this volume with the largest eruption in human historic memory. This occurred in Iceland, 1783–1784, with the eruption of Laki volcano. This lasted eight months and involved just somewhere around a 14 km$^3$ volume of lava, but it still lowered global temperatures by about 1°C. The gasses released also killed almost all of Iceland's crops that year (*8*).

More recently, some clues suggest a single principal cause for much of the terrestrial component to the end-Permian extinction comes from two 2011 papers. The first was in *Nature Geoscience* and by Stephen Grasby, Hamed Sanei, and Benoit Beauchamp. They discovered that much end-Permian strata contained char that looked remarkably like fly ash from modern coal-fired power stations. The rocks were from the Canadian High Arctic and dated to a time close to the mass extinction. They proposed that the char was derived from the combustion of Siberian coal and organic-rich sediments by erupting flood basalts, which was then dispersed globally (*9*).

The second paper was by the previously mentioned (at the end of the preceding subsection) international European team led by Stephan Sobolev and Alexander Sobolev. They came up with a mechanism; a model and field evidence that suggested that recycled oceanic crust rich in volatiles caused a massive gaseous eruption (particularly a massive degassing of carbon dioxide and hydrochloric acid) with the volatiles being released in just a short period even if the overall period of LIP volcanic activity lasted a long while. This short, but considerable, release of volatiles could have been enough to cause the short, sharp extinction we see in the geological record (*10*).

If you like, you can consider it this way in the form of the aftermath of a Hollywood car chase. There is one of the cars involved having crashed upside down and it is in part on fire. The occupants scramble to get out and away. Then the flames reach the fuel tank and boom! Similarly, a LIP of molten lava exists but if it reaches an area of organic rich strata—a coal bed or oil or gas reservoir—then it is like the petrol tank of the burning car exploding.

Taking these two papers together, along with the earlier one by Peter Ward and colleagues (*7*), it seems likely that the extinction was possibly caused by a very short period of intense explosive activity involving organic rich geological strata, in and around the Siberian Traps, that added considerable greenhouse gases to the atmosphere causing major abrupt climate change (substantive warming) in addition to a possible nuclear-winter-like time due to the dust from a major eruption and additional, short-lived cooling from sulphate gases (initial substantial cooling). Subsequently, further degassing of either methane (which would end up as carbon dioxide) and/or carbon dioxide raised ocean pH caused a further marine extinction not to mention for many decades if not centuries, a period of global warming. So the event saw periods of both extreme cooling and warming.

Just after the end-Permian extinction, there is also some evidence of a stromatolite (algal mat) resurgence. This is thought to be possibly due to a decline in marine animals that would otherwise have fed off the algal mats.

Isotopic evidence, using $^{13}C$, suggests disrupted global carbon-cycle activity following the extinction event for some 3 million years. Although the exact details remain unclear, it suggests a relationship between biodiversity and Earth system (biosphere) function. Almost certainly, part of the change in $^{13}C/^{12}C$ was due to the decay of vegetation (and dead animals) releasing carbon rich in $^{12}C$, but some may possibly be due to the release of organic material in geological strata due to volcanic action of the Siberian Traps (*11*) as well as marine methane hydrate (clathrate) disruption (more of which, associated with a different event, shortly).

Again, it is worth emphasising that this end-Permian event was not causally connected to the overall march of life's evolution per se, and so was a wild card event, but it did affect life and evolution at that time. With much of the larger, multicelled life forms wiped out on both land and in the sea with an extinction of the calcareous (shelled) marine creatures, there was a certain re-setting of the evolutionary clock.

The era following the Palaeozoic, the Mesozoic—consisting of the Triassic, Jurassic, and Cretaceous periods—was to see the rise of reptiles and be popularly known as the age of the dinosaurs. That too would end with a wildcard event, but a very different one to that of the end-Permian: it would be the Cretaceous-Tertiary extinction that wiped out the dinosaurs.

## The Cretaceous-Tertiary extinction (66 mya)

The Mesozoic era lasted 186 million years and ended with the Cretaceous-Tertiary extinction 66 million years ago (mya). This had been called by geologists the 'K-T' event but is now known as the 'Cretaceous-Paleogene' or 'K/Pg' extinction, since the term 'Tertiary' period is less favoured (see Appendix 3). Again, I mention this change of nomenclature in case some of you (and especially students) are searching the internet for papers on this topic: searches need to be done using both terms in order to capture more of the academic literature over the years.

Prior to the extinction, much of the Cretaceous—from 66–145 mya—was warm but it appears to have cooled towards its end. For example, the Arctic seems to have gradually cooled by 10°C over 20 million years (*12*).

The K/Pg extinction is probably the most famous mass extinction event of all, it being the one that saw the demise of the dinosaurs. The event saw between 75% and 80% of terrestrial species become extinct (possibly somewhat fewer at higher latitudes).

It is now well established, due to an iridium layer in the geological record, that this was event was caused by an asteroidal impact.

Unlike carbon isotope excursions (CIEs) associated with large igneous provinces (LIPs), which could be said to be part and parcel of the hazards of living on an Earthlike planet (and—we will return to this in Chapters 7 and 9—the Earth is the most Earthlike planet there is by definition), asteroid impacts are an external factor: hence a wild card. Having said that, the risk of asteroid impacts could be considered to be an inevitable risk factor of living in a planetary system: after all, asteroids are simply part of the leftover material from the formation of planets.

A heavy element, iridium layer is a reliable indicator of such an impact due to the way the Earth's crust formed.

When, in the beginning the Earth was completely molten, heavy metals like iridium sank to its core, leaving the Earth's mantle depleted of such metals (see Chapter 1). Conversely, in asteroids with little gravity and no core, there is no separation of heavy and light elements.

In the late 1970s, the Alveraz duo, of father and son and their team in the United States, detected significant levels of iridium in the Cretaceous/Tertiary (Cretaceous/Palaeogene) boundary layer and argued that this pointed to an asteroid impact (*13*). It was this discovery that spawned investigations into possible extraterrestrial causes of other mass extinctions. (Other impacts certainly have taken place, but it now seems that other than this, the Cretaceous-Paleogene (K/Pg) extinction, most of the other major mass extinctions of the past half a billion years were associated with large igneous provinces.)

The K/Pg impact is thought by many to have taken place in Central America at Chicxulub, Yucatan, Mexico, and estimated to have been caused by a six-mile-wide (around 10 km across) asteroid. Whether or not this was the actual impact point that caused the extinction (there is a technical debate that perhaps more than one impact point exists), there was an impact in Central

America somewhere around the extinction time and this caused ejecta that reached up well into the United States. Ejecta dust undoubtedly entered the stratosphere and that would have diminished photosynthesis and caused cooling on a global scale for several years (not to mention provided spectacular sunsets). As with the end-Permian extinction, a fungal and algal spike has been identified (*14*). This fungi-rich short episode was followed by a recovery led by pteridophytes and gymnosperms (ferns and seed-bearing/woody plants). The K/Pg and end-Permian vegetation recoveries represent similar responses to terrestrial ecosystem collapse, although the end-Permian biodiversity crisis was much more prolonged and it was a more severe extinction event.

The impact would have released volatiles including carbon dioxide: this last due to the Chicxulub impact area featuring carbonate geology. The dead biota would also have rotted, releasing methane and again carbon dioxide. The former greenhouse gas would have turned into the latter, which lasted longer in the atmosphere than the cooling dust and so (as with the end-Permian extinction), once the planet's severe cold snap was over, there would have been a period of warming.

In 2018, a small team of researchers looked at the ratio of oxygen isotopes in fish bones: the isotope ratio is indicative of the temperature of the water in which the fish lived. The fish bones dated from the time of the impact. They suggest that there was a period of warming of roughly 5°C warming after the extinction event and that this lasted for somewhere the order of 100,000 years (*15*).

This post-impact warming contributed little to the extinction event itself, as by then the biological damage had been done (even if this warming is of interest to those studying past climate change). How much subsequent warming there was depended on how much carbon dioxide was released. It was the impact—its regional shockwave, hemispheric rain of ejecta, global acid rain, and a dust-shrouded Earth—that was the main cause of the dinosaur extinction worldwide. Subsequently there would have been the longer-term climate change due to the carbon released into the atmosphere from both the impact heating carbonates and the death of a significant proportion of terrestrial life.

Here endeth the summary story. . . . Well, not quite.

There are still some who say that an asteroid did not wipe out the dinosaurs. As a point of pedantry this view, surprisingly, has some merit. To begin with, only some genera of dinosaurs became truly extinct, other genera only suffered a pseudo-extinction: they did not die out as such, but new species evolving at the time survived. For example, birds.

Second, the dinosaurs were already in decline prior to the asteroid impact. The 60-million-year-long Cretaceous period was the third period (the Triassic and Jurassic being the other two) in the Mesozoic era of the dinosaurs, beginning 230 mya. The Mesozoic era saw considerable evolution in the dinosaurs and some dinosaur species that were around in the Triassic period (at the beginning of the Mesozoic) were not around in the Cretaceous period. Conversely, some dinosaurs did not evolve until the Cretaceous. Indeed, the *Tyrannosaurus rex*, star of the book (and film) *Jurassic Park*, is associated with the Cretaceous, not the preceding Jurassic period. So the 'age of the dinosaurs', covering the Triassic, Jurassic, and Cretaceous periods of the Mesozoic era, saw the evolution and decline of many dinosaur species. The bottom line is this: many dinosaur genera were in decline well before the end of the Cretaceous and its asteroid impact, and so the asteroid cannot have been solely responsible for the disappearance of all the dinosaur species we do not see after the end of the Cretaceous: some had died out before the end of the Cretaceous.

The possible culprit behind the—if you will—pre-asteroid dinosaur decline may well be the considerable volcanic activity associated with the Deccan Traps LIP in north-west India. The Traps represent the remains of a continental flood basalt event (like the end-Permian Siberian Traps) which is thought to have been active (albeit to varying degrees) for 8 or 9 million years somewhere around 60–65 mya, although there is some associated geology that is earlier (at 70 mya) than this. The continental flood basalt event would have released cooling dust and ash as well as warming carbon dioxide and methane. Could the cooling dust and ash overwhelm any greenhouse warming, hence explain the 10°C Arctic cooling over 20 million years mentioned earlier?

There is some considerable evidence as to what happened: too much to go into in full here. Though let us look at one illustrative avenue of research. This concerns the element mercury (Hg). Hg is normally found in low background levels but can be released by volcanic activity. With an extensive LIP we might expect significant Hg release. Here Hg has been found in fossils that date from 1 myr to 250 kyr before the asteroid impact 66 mya. This mercury is attributed to volcanism from the Deccan Traps which was the most active volcanic activity of that time. The researchers got around the possible 'it rained on Mongo' mistake (see Chapter 3) by obtaining their fossil samples from various sites on different continents. This showed that whatever they were looking at was not local but global in effect (*16*).

As such, this combination of LIP and a *coup de grâce* asteroidal impact explains both the pre-extinction decline of the dinosaurs and the actual extinction event itself: what may have happened is that the Deccan Traps' flood basalt event, that lasted many millions of years, altered the biosphere sufficiently to stress dinosaurs and then the end-Cretaceous asteroid impact provided the *coup de grâce*.

Though the actual the *coup de grâce* extinction event was due to an asteroid impact, there are some similarities between this event and the end-Permian mass extinction. These include the fungal spike in the geological record similar to that found associated with the end-Permian extinction (*8*). As alluded to earlier, fungi would have flourished in a world of much dead and rotting biomass.

Then again, there was the impact of the event on evolution. Just as the end-Permian event cleared part of the slate for new species to evolve, so the end-Cretaceous extinction allowed some groups of species to increase. The dinosaurs were not the only life on Earth at that time: indeed, there were other newcomers. The late Cretaceous and the early Tertiary were periods of vegetation change due to the appearance, and then increasing dominance, of flowering plants (angiosperms). These greatly increased in number and diversity over the millions of years following the extinction event. Similarly, there are also examples of animals thriving following the age of the reptiles. With the dinosaurs gone, very small mammals diversified and increased in numbers.

The recovery from the extinction event, the critical transition to an asteroid impacted Earth, took time. The fossil record from a site in Colorado, USA, reveals that within about a 100 thousand years (kyr) post-extinction mammalian species richness doubled, and their maximum body mass ranges increased to near extinction levels. By 300 kyr following the extinction event there was a threefold increase in maximum mammalian body mass and dietary specialisation. This last reflected an increase in floral species richness. By 700 kyr the first appearance of the Leguminosae (the pea, bean, and lentil family) and the appearance of additional large mammals (*17*).

Meanwhile in the oceans, a 13 million year (myr) geological record of nanoplankton has revealed that following the asteroid impact there was 1.8 myr of significant plankton species disruption beginning with 90% of species of calcifying plankton going extinct. The return of species richness to pre-extinction levels took 8–10 myr (*18*). So, as you might have already guessed, with this extinction being associated with the dinosaurs' demise, the biological recovery time demonstrates how globally significant an extinction event it was.

Fortunately, for us (human mammals) strikes from large asteroids are rare, and we have not had one as big in the intervening 66 million years since the end-Cretaceous event. Yet there have been other major extinction events which, while not having nearly as big an impact on life as this one, have been significant and global. As with the end-Permian extinction these have been LIP related but, again fortunately, not so extreme, albeit among the more major extinction events in the geological record of the past half a billion years.

Another of these comparatively minor LIP-related events—minor compared to the end Permian event—was experienced by the dinosaurs some 183 million years ago in the Jurassic period at the start of the Toarcian age (*8*). This event was also a carbon isotope excursion (CIE) event and also LIP-coincident with the Karoo-Ferrar LIP being active (see Figure 6.1). Yet another CIE that has attracted a fair amount of study was far more recent (hence there has been more evidence to glean from the geological record). This less-extreme than the end-Permian, but nonetheless global event and also LIP related, took place after the end-Cretaceous dinosaur extinction and at the start of the Eocene some 56 mya. Yet, this event was not so minor, as it is still sufficiently pronounced in the geological record that it defines the beginning of the Eocene epoch. You could say that this CIE initiated the Eocene.

## The initial Eocene CIE (~56 mya)

While this particular carbon isotope excursion (CIE) defines the start of the Eocene epoch, it lasted the order of many thousands of years (over a hundred thousand years), though the Eocene epoch itself lasted 22 million years: the initial Eocene CIE was therefore a geological-scale blip within the much longer Eocene epoch.

The Eocene epoch itself spans the time from 33.9 mya to 56 mya. It is the second of the five epochs in the Tertiary Period (2.58–66 mya). Having said that, these days the geological nomenclature for strata and dating has changed: the term 'Tertiary', though once common, is now rarely used: the 'Paleogene' (23.03–66 mya) is now considered the period within which the Eocene sits (see Appendix 3).

Originally, the term 'Eocene' was defined (along with the aforementioned 'Tertiary') by one of geology's grandfathers, Charles Lyell, back in the 19th century. The name comes from the Greek 'eos' meaning 'dawn', and 'koinos' (or 'ceno') meaning 'recent', as Lyell found the fossils of many present-day related species within strata laid down at that time. So the Eocene represents the time of the dawn of recent species. These include hoofed animals such as early horses, bats, and whales, among many others.

The epoch as a whole saw the Earth at its warmest since the dinosaur extinction, with Eocene carbon dioxide levels almost certainly over 1,000 parts per million (ppm) and at times possibly over 3,000 ppm; there were no large polar ice caps. The Earth's continental outlines, though just about recognisable to us familiar with their present-day configuration, were noticeably differently arranged and part-flooded by the oceans that were very roughly around 100 metres higher than today. Indeed, the Eocene's slightly different continental configuration might have helped keep the planet warmer than today. North and South America had yet to join, so providing greater freedom of movement of heat-carrying water from the equator to the Poles. Similarly, India had not joined with Asia (so no Himalaya with its white-capped ranges reflecting heat), and the Tethys Sea separated Africa and Eurasia. Also, Australia was closer to Antarctica, so no present-day strong circumpolar current climatically partially isolating Antarctica: Antarctica was back then largely ice free (except for its tallest mountains).

Yet while the Eocene epoch's commencement is now defined by this CIE, the nature of this event has only been elucidated in recent decades; the Eocene used to be defined by a sequence of sea-level change, limestone formation, and clay deposition within which the ancestors of many modern animals were found. (Indeed the Eocene's CIEs—of which there were a number, but the first was the greatest—were times when the sea level rose due to thermal expansion to levels higher than the epoch's average.)

The key climatic events are roughly summarised in Figure 6.2.

It is because our understanding of this CIE event only began to mature in recent decades that this event's name has changed a number of times. Knowing that the terminology has evolved is critical for students, as keyword searching using only the recent nomenclature will fail to pick out many previous science papers. So it is important to note that the event has been called the

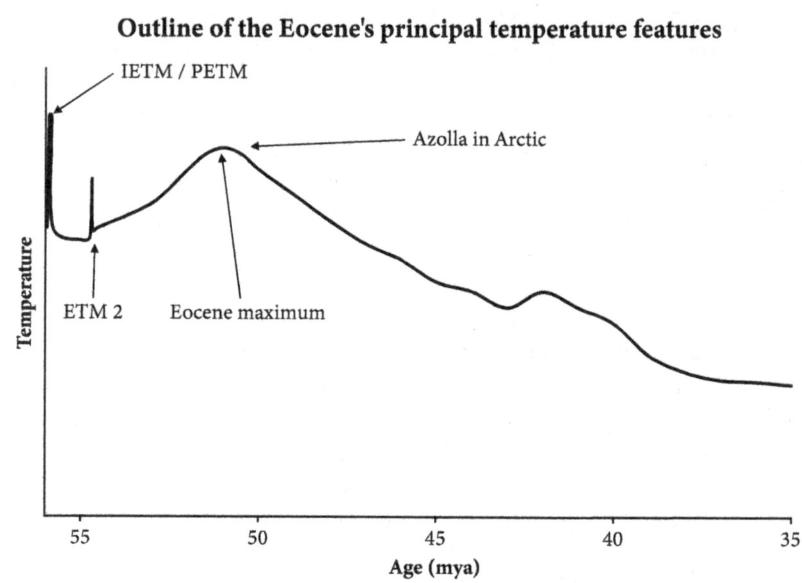


**Figure 6.2** Rough outline sketch of the average global temperature change associated with the principal (only) climatic events found across the Eocene.

*Note*: Here the IETM/PETM is depicted as having a peak temperature a little warmer than the Eocene maximum, but some researchers might debate that it was possibly just a little cooler. (There were a couple of other very minor CIEs not shown.)

'Eocene thermal maximum 1' (as there is an ETM2 and others) but for many years it was known as the 'Initial Eocene Thermal Maximum' (IETM). Today it is commonly referred to as the 'Palaeocene-Eocene Thermal Maximum' (PETM).

Now, this name change is of interest, not just because students need to know it if they are to search the academic literature of a few decades but because it is another example (as was the K/T or K/Pg synonym nomenclature earlier) of how science is not the orderly presentation of knowledge that many suppose: scientific understanding alters as it evolves, and science is done by humans with all too human foibles that do result in nomenclature/terminology changes.

You see the reason that this event was known as the 'Initial Eocene Thermal Maximum' was because it defines the start of the Eocene as had been determined by the International Commission on Stratigraphy: this global warming event took place at the Eocene's 'initiation'. Conversely, the term 'Palaeocene-Eocene Thermal Maximum', that is currently in common vogue, reflects that this event straddles the boundary between the Palaeocene and Eocene epochs.

If you wish, you can consider it a bit like the border between North and South Korea. Where does North Korea end and South Korea begin? The answer is not actually a single specific place: there is a no-man's border zone between the two.

It is therefore somewhat ironic that the term 'Initial Eocene Thermal Maximum' is much less used in the past few years (given that the event by internationally agreed definition marks the beginning of the Eocene) and has been replaced by 'Palaeocene-Eocene Thermal Maximum' (that suggests that the event straddles the border, which the international definition clearly states is not the case). Science can at times be very human. And so as a clear reminder for those wishing to more thoroughly search the academic literature and to highlight this quirk, I will from here on refer to the event as the 'IETM/PETM'.

The other thing to note is that, while the International Commission on Stratigraphy today cites the beginning of this event as taking place 56 mya (*19*), though some research papers give a slightly different date by a few hundred thousand years. Indeed a number of papers, and the International Commission on Stratigraphy, prior to 2010 place this event closer to 55 mya.

So what caused this CIE?

The summary answer is that there was an injection of carbon-12 rich greenhouse gas into the atmosphere and, it appeared at first, this warmed the planet overall by some 5°C. Having said that, regionally temperatures were ~5°C–9°C higher at mid- to high-latitudes than before, with polar temperatures rising by as much as 8°C or 9°C. This warming—which has been confirmed many times—was in some places (high northern latitudes) even greater during the winter compared to winters there before and after the event. This global temperature-rising phase lasted for many thousands of years, but probably peaked for a period of less than 100,000 years as an extreme event (although less significant warming would have lasted longer). This is evidenced by a considerable body of work including that ranging from the then Arctic temperatures by James Kennett and Lowell Stott, both then at the University of California in Santa Barbara (*20*), to Tripati and Elderfield's evidence of the then tropical and subtropical ocean bottom waters (*21*). It being a warm event—distinctly warmer by a few degrees Celsius than before or after—it is known as a 'hyperthermal' (see Appendix 2).

Aside from the temperature rise, the evidence for a CIE from marine sediments is compelling. Wherever in the world we find marine sediments that date from the early Eocene, and we analyse the carbon in their carbonate particles, we see both a spike in the ratio of carbon-13 to carbon-12 isotopes as well as an acidification of the ocean (see Figure 6.3). One of the early significant, and often cited, pieces of work demonstrating that this event was global was by Jim Zachos at the University of California, Santa Cruz, and colleagues from various institutions across the world (*22*). They demonstrated that this CIE was readily detected in sediments in different oceanic basins from the Southern Ocean, the Central Pacific, and the South Atlantic: hence the CIE was truly global.

That is the simple answer, but the CIE's cause is more complex than the injection of carbon-12 rich carbon into the atmosphere. Given that there was an injection of $^{12}C$, what in turn was the injection's cause?

Here there no escaping that there is some debate and rival hypotheses (enough for a book in its own right). However, I will largely confine myself to the ones that have so far garnered a substantive body of evidence, hence arguably have the most traction. Hopefully, any geoscientist readers subscribing to any of the rival scenarios will forgive me; the topic is simply too large, but I will provide key references citing the science involved here. Readers can then use these papers and search-engine out other relevant work to decide whether my judgement is sound or to develop their own thoughts on the matter. (I will, though, make passing reference to some of the other ideas to give you an indication of the debate.)

In 1995 Gerry Dickens and colleagues from the University of Michigan proposed that only geological methane gas had enough light carbon ($^{12}C$) to produce the early Eocene CIE. They proposed that a belch of methane escaped from methane hydrates (methane clathrates) in continental slope sediments as the Earth warmed during the late Palaeocene (*23*). This early idea has some merit in that the mechanism is plausible: methane clathrates do exist in a layer in sea sediments, and frequently on the continental margins, as they are formed by methane release from decaying organic matter and trapped as clathrates under appropriate pressures and temperatures.

What happened, Gerry Dickens and colleagues hypothesised, is this: dead plankton and other organic matter rains down from the sea surface and at the

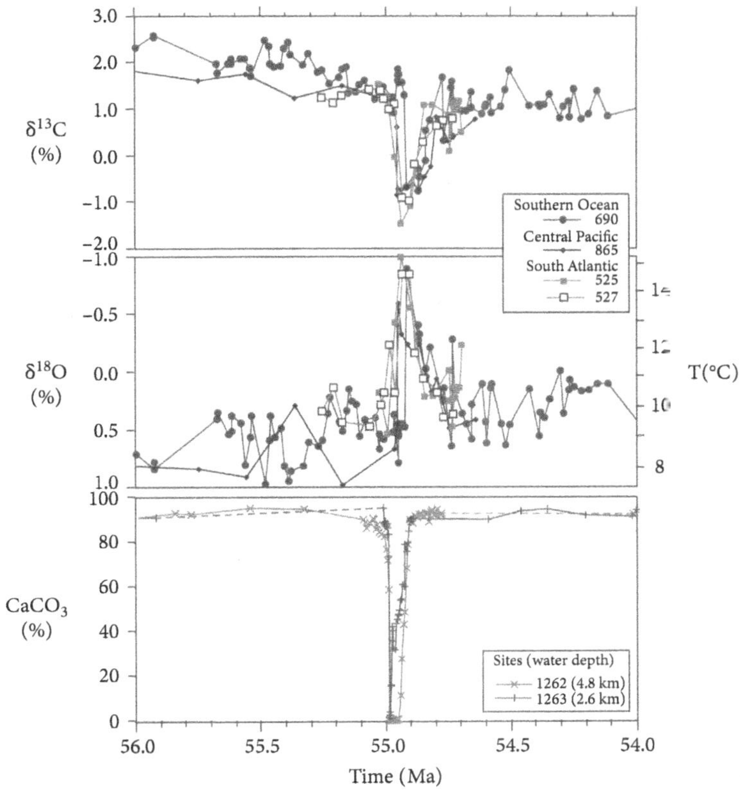


**Figure 6.3** Top panel: The IETM/PETM Carbon Isotope Excursion as revealed by the analysis of benthic (bottom dwelling) foraminifera expressed as VPDB (see Appendices 1 and 2 for an explanation of VPDB and Zachos *et al.*, 2005 [ref. *22*] for details). Middle panel: $^{18}O$ climate proxy temperature analysis of IETM/PETM strata. Bottom panel: carbonate ($CaCO_3$) content of sediments in two cores from the South Atlantic. All three panels reproduced with permission from the Intergovernmental Panel on Climate Change. It appears as **Figure 6.2** in Jansen, E., Overpeck, J., Briffa, K. R., Duplessy, J.-C., Joos, F., Masson-Delmotte, V., Olago, D., Otto-Bliesner, B., Peltier, W. R., Rahmstorf, S., Ramesh, R., Raynaud, D., Rind, D., Solomina, O., Villalba, R., & Zhang, D., Palaeoclimate. In Solomon, S., Qin, D., Manning, M., Chen, Z., Marquis, M., Averyt, K. B., Tignor M., & Miller, H. L., eds., *Climate Change 2007: The Physical Science Basis. Contribution of Working Group I to the Fourth Assessment Report of the Intergovernmental Panel on Climate Change.* Cambridge University Press: Cambridge, United Kingdom and New York, NY, USA (2007). Note, the dating in some subsequent work, one and a half decades on, is different due to improved techniques and understanding of the geoscience. Data based on ref. *22*.

bottom where it settles it is covered by other organic matter, there it undergoes microbial anaerobic digestion and this releases methane. If the water pressure (hence depth) is less than a certain limit and the temperature above a specific point for that depth, then the methane will simply bubble up to the surface. Otherwise what happens, at higher pressures and/or cooler temperatures, is that the methane loosely reacts with water to form a semi-stable hydrate complex (a clathrate). This happens at a specific depth (pressure) for a given ocean abyssal temperature. Clathrate dissociates back to methane and water with rising temperatures to release the strong greenhouse gas methane into the atmosphere.

There are two problems with this hypothesis. The key one is this: something is still needed to destabilise the clathrate. So while methane clathrate dissociation may contribute to warming once ocean warming has commenced, it does not explain what caused the initial warming that triggered the methane release in the first place at the start of the Eocene? The second problem is that it is unlikely that there was nearly enough clathrate globally to cause the CIE by itself (though it could certainly contribute to it) (*24*). Clathrates therefore do not explain the cause of the injection of $^{12}$C-rich greenhouse gas, though they could be involved as a feedback mechanism further enhancing the warming.

What is needed is a specific trigger or, failing that, evidence that there was some sort of trigger either at the start, or just before, the CIE itself.

This last came in 2010, when Ross Secord and colleagues, primarily from the University of Michigan, had their remarkable study of the changing biome in the Bighorn Basin, Wyoming, USA, published in the journal *Nature* (*25*). Using a number of different climate proxies, they pieced together key events of the IETM/PETM.

These proxies included oxygen isotopes ($^{18}$O/$^{16}$O) in mammalian tooth enamel from various remains that date across the CIE and from both phosphate and carbonate. This isotope ratio relates to rainwater's ($H_2O$) oxygen atom: the mammals drink the water and the oxygen atoms in the water end up in the tooth enamel. Given that all the samples were taken from the same place—the same distance inland from the ocean—any heavy oxygen isotope fractionation due to distance from the ocean would be constant (see Chapter 5, 'The second flip and another Snowball' section). This in turn means that any change in the $^{18}$O/$^{16}$O ratio would be likely due to temperature: more of the heavier $^{18}$O in $H_2O$ evaporates in warmer oceanic waters than cooler ones. They also looked at $^{18}$O/$^{16}$O ratio in fossil leaves as well as soil carbonate as other indicators of temperature.

Another proxy from the mammals' tooth enamel was that of $^{13}$C/$^{12}$C carbon isotope ratios. Up to now we have considered $^{13}$C/$^{12}$C in carbonates and the shells of small foraminifera as a proxy for atmospheric $^{13}$C/$^{12}$C of the time the carbonate formed. However, animals that eat plants that have a different $^{13}$C/$^{12}$C ratio reflecting that of the isotope ratio in the plants. But in this instance the $^{13}C/^{12}C$ is determined by something other than carbon drawdown (as discussed in Chapters 1 and 2)!

Here, the thing is that there are broadly two different types of photosynthesis in plants with one type being used by grasses and more tropical species, and the other by species common outside of the tropics. There is no need to go into the detail here, though biology students will know of $C_3$ and $C_4$ plants that respectively use $C_3$ and $C_4$ photosynthesis (see Appendix 2) and that while both types of photosynthesis preferentially take up $^{12}C$ one type does it more than the other. The key point is that $C_4$ photosynthesis can only take place where there is plenty of sunlight and that usually means the tropics.

This means that a change in tooth enamel $^{13}C/^{12}C$ indicates a change in the animals' diet, and here Ross Secord and colleagues detected such a change that indicated that the mammals had been eating temperate plants and then, it is implied with warming, tropical plants. This warming is different to the warming that was separately confirmed by the $^{18}O/^{16}O$ ratio in fossil leaves as well as the isotope ratio change (again $^{18}O/^{16}O$) in mammalian tooth enamel which took place at a different time to the tooth enamel $^{13}C/^{12}C$ change.

Finally, Ross Secord's team looked at $^{13}C/^{12}C$ ratios in soil carbonate. Now, this does relate to the amount of *biogenic* carbon-12 drawdown from, or release to, the atmosphere. (Note the nuance, it does not represent changes in atmospheric carbon dioxide levels per se, just biogenic $^{12}C$ drawdown or release to the atmosphere. You can have a lot or a little carbon dioxide in the atmosphere and yet both have the same $^{13}C/^{12}C$ ratio.)

Together these various oxygen and carbon isotopic signatures, from soil carbonate and mammals, represent various environmental parameters: type of vegetation (tropical or not), temperature, and atmospheric carbon that came from organic matter.

Together they painted the following picture of the early Eocene that had two key components.

First, the temperature increased and this changed the vegetation at the onset of the Eocene nearly 56 mya to that of a more tropical kind (evidenced by a change in mammalian diet). This warming was evidence of some sort of trigger before the CIE.

Secondly, and a little later (a few tens of thousands of years, which is next to nothing in geological terms), the carbon isotope excursion (CIE) set in, indicating that the atmosphere had a subsequent influx of organic carbon rich in $^{12}C$, be it in the form of either carbon dioxide or methane; either way there was further and more sustained warming. (The warming being evidenced by oxygen isotopes and the CIE by carbon isotopes.)

What seems to have happened was that the warming was not caused by the CIE itself but by an inorganic carbon trigger; presumably lava-induced (inorganic) carbon dioxide (with more of the trace $^{13}C$ in the mix) venting into the atmosphere. This in turn subsequently destabilised some reservoirs of organic carbon and it was this last that we detect in Zachos and colleagues' CIE found in marine sediments in different ocean basins and also detected by Secord and his team in the soil carbonate (but not the tooth enamel which reflected dietary change).

This idea of a non-biological trigger was later supported by work of a small international team led by Marcus Gutjahr of the National Oceanography Centre at the University of Southampton, using observations of the geological record, and which included Andy Ridgwell of Bristol University, who undertook computer modelling of the data gathered. They looked at forams from a Deep Sea Drilling Project Site in the northeast Atlantic Ocean and the isotopes therein. These were: $^{18}O/^{16}O$ (indicating temperature), $^{13}C/^{12}C$ (indicating organic carbon release/drawdown into/from the atmosphere), and boron (indicating marine acidity, which in turn implies the concentration of carbonic acid in the seas, which reflects carbon dioxide in the atmosphere irrespective of a biogenic or inorganic source). In this way the research could not only determine temperature change with time back in the early Eocene but also tease apart the amount of both organic and inorganic carbon dioxide in the atmosphere: the boron was indicative of the total carbon dioxide content and the $^{13}C/^{12}C$ the organic carbon component. Relating all these was done by a paired carbon isotope record in an Earth system model created by Andy Ridgwell in order to reconstruct the unfolding carbon-cycle dynamics during the event (*26*).

As with the Ross Secord team's work, the Marcus Gutjahr–led researchers detected an inorganic-carbon, warming trigger that was subsequently amplified by a subsequent release of organic carbon into the atmosphere.

Fair enough, there was an inorganic carbon trigger, but what exactly was this trigger?

As noted earlier this chapter in the section 'Large igneous provinces (LIPs)', we have today the remnants of a formerly fully active LIP in the form of the Iceland hotspot. This today coincidentally sits at the northern end of the mid-Atlantic ridge between separating plates. But back in the early Eocene this hot plume spot was beneath Greenland with hot material spreading beneath, and in some places through, the crust to as far south as the northern coast of Scotland and the north-east Atlantic off of Norway; back then the plates had not spread so far and so Greenland was closer to north-western Europe. This LIP is called the North Atlantic Igneous Province (see Figure 6.1).

Back in 2004, Henrik Svensen and fellow Norwegian colleagues announced they had found evidence of the remains of literally thousands of hydrothermal vents in a geological seabed complex off the More and Voring basins in the

Norwegian Sea. These date from the Palaeocene–Eocene boundary. Could the intrusion of mantle-derived melts in the carbon-rich sedimentary strata in the north-east Atlantic have caused an explosive release of carbon a little like Svensen contemplated (*27*)?

A similar structure on land is associated with the Siberian Traps that resulted in the end-Permian extinction approximately 250 mya (mentioned earlier this chapter). Yet, the complex of the More and Voring basins is part of the greater North Atlantic volcanic province.

Henrik Svensen related these hydrothermal vents to a release of organic carbon as the CIE trigger, yet the subsequent work of the Ross Secord and Marcus Gutjahr groups rule out an organic carbon source as the trigger for the CIE causing the *initial* warming. So while a lava intrusion from the North Atlantic Igneous Province into in organic carbonate sediments would have released a massive injection of carbon-12, conversely an intrusion into inorganic carbonate would release inorganic carbon (as carbon dioxide with more of the minority carbon-13 in the mix) into the atmosphere. If this happened, it too would result in an initial warming, but warming with more carbon-13 in the mix. This then subsequently triggered further release of organic carbon (with more carbon-12), that we see as a subsequent carbon isotope excursion (CIE).

So the warming of the early Eocene was down to an increase in carbon greenhouse gas in the atmosphere, but an increase that had two phases. First an injection of inorganic carbon and then a second injection of organic carbon. Only this last saw carbon with more of the majority $^{12}C$ (and less of the minority $^{13}C$) that is discernible as a CIE in the geological record.

That there was massive flood basalt volcanism for a million years in the runup to the CIE was elucidated a couple of years later (after the Henrik Svensen paper) in 2007 by Michael Storey and colleagues. They used $^{40}Ar/^{39}Ar$ age determinations of Danish ash strata. Their work supports the view that the IETM/PETM was triggered by greenhouse gas release during an episode of magma interaction with carbon-rich sedimentary rocks close to the plate boundary between Greenland and Europe (*28*).

Further evidence that the North Atlantic Igneous Province saw hot mantle plume activity prior to the IETM/PETM CIE comes from mapping strata now buried below the sea bed in an area north of Scotland by researchers from Cambridge University and British Petroleum: BP had an interest in geological mapping of the area, which lies almost adjacent to oil and gas field regions. The area mapped was approximately 10,000 km$^2$ on the north-west continental shelf of Europe. They found a landscape of buried hills and valleys with the remains of drainage channels (rivers). This meant that, when this landscape, now buried beneath marine sediment, formed, it must have been above sea level!

The scenario they conjecture is this. For roughly 1.5 million years, and ending before about 55 million years ago, there was a period of geological uplift in the region. The rate of uplift was small and mostly less than 3 mm $yr^{-1}$, but, because it lasted so long, the total uplift was almost a kilometre in height. This took the landscape above sea level, where, for hundreds of thousands of years, rain eroded the landscape forming the valleys and river channels seen in the survey. Before 55 mya the land subsided (*29*).

Could it be that part of this uplift was caused by a build-up of gas from the melt intrusions from the hot spot into inorganic carbonate sediments? If so, at some point, the strata fractured and the gas was released in a pulse that in turn caused the first surge of early Eocene warming. This trigger warming then caused other biosphere reservoirs of organic carbon to destabilise, so resulting in additional warming by organic carbon-12, and this caused the CIE detected in the geological record.

Of course, the question that remains is, which were the biosphere reservoirs of organic carbon that were destabilised? As noted, one of the early hypotheses was that the $^{12}C$ in the CIE came from marine methane clathrates, but this has to be discounted as the sole source of the organic carbon as there is not enough clathrate in ocean sediments to produce such a marked CIE ($^{13}C/^{12}C$ swing) as seen in the geological record.

Alternatively, the organic carbon could come from, say, the combustion of terrestrial plant material on a more or less global scale, but that would necessitate an extinction level not seen in the geological record of the time. It must be something else.

One significant biosphere reservoir of organic carbon is carbon in soils (see Chapter 2). It is estimated that today some 18.8 million square kilometres of northern hemisphere soils hold about 1,700 billion tonnes of organic carbon (which is equivalent to the same in gigatonnes or petagrams)—'organic carbon' being the remains of plants and animals that have been accumulating in the soil over thousands of years. That is equivalent to about four times more than all the carbon emitted by human activity since the industrial revolution and roughly twice as much as is present in the early 21st-century atmosphere (*30*). Here, some 185 ± 70 billion tonnes of carbon that is today trapped in 1.7 ± 0.5 million $km^2$ in northern hemisphere permafrost soil is particularly sensitive to release with warming (*31*). However, at the end of the Palaeocene (before the Eocene) the Earth was warmer than today and there were less extensive permafrosts in the northern hemisphere, but there were some in the southern hemisphere. Antarctica at the time had no extensive ice caps and it was likely even warm enough for there to be woodland in places. But where it was colder, there would have been permafrosts.

In 2011 Robert DeConto of the University of Massachusetts and colleagues contemplated that the Antarctic permafrosts could have been destabilised by the initial warming. (Though they put this trigger down to Earth orbit forcing, which the works, discussed earlier, of Ross Secord [25] and Marcus Gutjahr [26] disprove). DeConto and colleagues modelled the Earth of the late Palaeocene with permafrost stores of carbon including a substantive one in Antarctica. Their results show that the permafrost carbon released with a little warming could account for much of the IETM/PETM CIE (*32*).

While they contemplated an orbital-induced warming trigger for the event, a North Atlantic Igneous Province injection of carbon would better serve, as we know that an inorganic carbon warming trigger was involved. Where the DeConto and colleagues conjecture does score is that it also predicts a number of smaller follow-up warming events to the IETM/PETM that are found in the geological record and which do follow orbital timings.

The reason the IETM/PETM warming event is of particular interest today is that the amount of organic carbon involved is similar in order of magnitude to the amount of organic carbon humanity is likely to emit this (21st) century (*33*). It could be that today we are heading towards a hyperthermal of our own making. Yet, while this makes understanding the IETM/PETM important, in terms of the life-induced flips of the Earth state it is irrelevant. Instead, determining the unique characteristics of life-induced flips of the Earth state, we can then differentiate between these largely-life-evolving-induced events and the wild card joker flips of state examined in this chapter, even if there are some similarities: such as, all involve perturbations of the carbon cycle and all involve global climate change.

## The characteristics of life-driven flips of Earth state

The key characteristics of the flips of state so far discussed are summarised in Table 6.1.

Here it can be seen that all the events saw a perturbation of the carbon cycle and all had associated climate change. But there are differences. The end-Permian and end-Cretaceous extinctions as well as the Initial Eocene Thermal Maximum/Palaeocene-Eocene Thermal Maximum did not have *key* evolution developments (such as oxygenic photosynthesis or multicellularity) associated with them or enhanced information processing, even if there were species extinctions and/or speciation. Nor did they see the large-scale use by life of a new environment. (This last is a bit of a broad-brush, inexact statement, as some cyanobacteria probably lived at the sea surface before eukaryotes evolved.)

**Table 6.1** Characteristics of the flips of state so far discussed.

| Event | Carbon cycle perturbation (CIE) | Climate change | Evolution development | New environment colonised | Information processing enhanced |
|---|---|---|---|---|---|
| Life begins | ✓ | ? (Likely) | ✓ (Life) | ✓ (Hot springs/vents) | ✓ (First nucleic acid) |
| Snowball Earth I | ✓ | ✓ | ✓s (Eukaryotic OP) | ✓ (Sea surface) | ✓ (Incorporation of prokaryotic DNA) |
| Snowball Earth II | ✓ | ✓ | ✓ (Multicellularity) | ✓ (Land) | ✓ (development of animal brains) |
| End-Permian extinction | ✓ | ✓ | X | X | X |
| End Cretaceous extinction | ✓ | ✓ | X | X | X |
| IETM/PETM | ✓ | ✓ | X | X | X |

Note that only three involved a major evolutionary development and information processing. The other three (End-Permian End-Cretaceous and the IETM/PETM) are wild card joker flips of state that saw no major evolutionary development or information processing. The 'new environment' column is inexact as almost certainly some cyanobacteria used the sea surface before eukaryotes even if sea surface photosynthesis likely took off with the evolution of eukaryotes.

The important thing is that the key flips of state at the heart of this book all relate to key evolutionary developments.

There are other differences too. While all were critical transitions to a new state, the last three—the jokers and wild cards—were reversible, with the global environment eventually returning to (near) its previous state. However, with both the Snowball Earth events the planet flipped to a new, long-lasting state: first from anaerobic to ~2% oxygen, and then from that to hovering around 21% oxygen.

In short, it is patently possible to distinguish between a wildcard or joker event and those associated with key evolutionary developments.

There is another characteristic that (currently) is less clear-cut but nonetheless there. This is that, with the biologically driven flips of state, there is the beginning of periods that see the emergence of new mineral species.

Now, the term 'species' is typically associated with biology but it is used in geology too. Quite simply, a 'mineral species' is one that has a unique chemical formula together with a unique structure. So, for example, while both diamond and graphite have the same chemical formula, in that they both consist of just carbon, they have different atomic structures. Graphite consists of layers of hexagonal rings of carbon, while diamond sees its carbon in a tetrahedral structure: graphite and diamond are therefore two distinct mineral species.

In 2019, Robert Hazen of the Deep Carbon Observatory and colleagues from a number of US university geoscience departments, together with a UK- and also a Russian-based contributor, published a review of data-driven mineralogy that included an analysis of the age of the oldest sample found of a mineral species. This analysis is summarised in Figure 6.4 (*34*).

First, it is important to recognise that this is a work in progress and that as other samples are found, it is likely that some of these new samples will be older than the current oldest. (Remember the Signor–Lipps effect discussed in Chapter 4.) So expect the bar graph in Figure 6.4 to move a little to the left.

Robert Hazen and colleagues' interpretation is that generally the rise of new mineral species is associated with the Wilson cycle of formation and

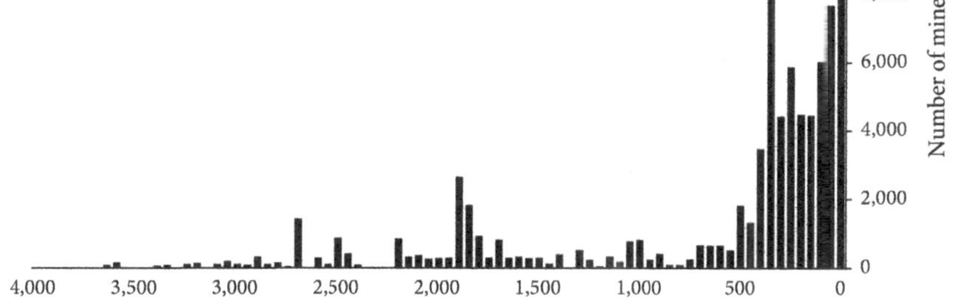


**Figure 6.4** Number of oldest mineral species so far found by 2019.

Adapted/data re-plotted from Hazen, R. M., Downs, R. T., Eleish, A., *et al.*, Data-driven discovery in mineralogy: Recent advances in data resources, analysis, and visualization. *Engineering*, **5**, 397–405 (2019).

fragmentation of supercontinents. The formation of supercontinents sees the beginning of a period of new mineral species. This period lasts a few hundreds of millions of years.

If you like, as plate tectonics sees material carried to greater depth and pressure, so new mineral species are produced. It is as if the Earth's crust is generating new information in the form of new mineral species: more information is needed to define the Earth 's crust if there are more mineral species. (Hence the final column in Table 6.1—we will return to information processing in more detail in Chapters 8 and 9.)

As to these new species, Robert Hazen and colleagues note that the younger new mineral species tend to have a higher oxidation state. This last comes about because—as noted earlier—the Earth has gone through flips of state from being anaerobic, to a ~2% of atmosphere of oxygen, before a final flip to near present-day levels.

But let's unpack the rise-of-new-mineral-species-with-the-Wilson-cycle thing. The first mineral species at around 3.6 billion years ago (bya) represent the closest we have to primordial crust. There are some new mineral species around 3.6 bya, and remember it was around this time that plate tectonics likely got going. We see a pulse at around 2.5 bya, and this could be associated with the supercontinent Kenorland whose breakup coincided with Snowball Earth I (Chapter 3). We also see a pulse of new mineral species at around 1.9 bya, and the time before and after this coincides with formation and breakup of the supercontinent Columbia. This was during the boring billion. Could it be that oceanic sediments and sediments formed from land erosion under the new, post-Snowball I environment of ~2% atmospheric oxygen, were for the first time worked in the Wilson cycle? The coming together, uplift, and volcanic working of these new partially oxidised strata would provide new mineral species.

Though supercontinent formation—or rather their breakup, two of which are associated with a flip to a new Earth state—is comparatively transient, geologically speaking, lasting several of tens of millions of years, the flip of state lasts longer with the period of critical transition (Chapter 2) lasting the order of over a hundred million years (Chapter 3)—largely the Cryogenian epoch (see Appendix 3). Yet, each period of new mineral species formation lasts even longer still, and so Figure 6.4 does not have the clearer grouping around the two flips of state associated with jumps in oxygen levels as reflected by changes in the carbon cycle revealed by carbon-13 ($^{13}C$) over the broad sweep of the Earth's history in Figure 2.1.

As to why the period of new mineral species arising lasts longer, it is because while a flip of Earth state is associated with a *major* evolutionary development (the rise of oxygenic photosynthesis and then multicellularity) finally making its mark on the state of the Earth system, biological evolution nonetheless continues as an ongoing process and this too contributes to the generation of new mineral species.

For example, true multicellular plants rose long after the proto-lichen colonisation of land associated with Snowball Earth II (Chapter 5). These plants had better roots to penetrate rock, releasing minerals some of which were transformed by the penetrating root microenvironment. 'True plants' continued to evolve, and one development was that of the hard, structural organic compound lignin found in trees as we commonly know them today. One of the first such trees was *Archaepteris*, in the mid-Devonian somewhere around 390 mya. Plants featuring lignin are harder to decompose, and this is one reason why we do not see extensive coal beds (themselves composed of mineral species) until the Carboniferous (359–299 mya) even though Snowball Earth II took place 720–635 mya.

But we are getting ahead of ourselves, though you will get the idea that biology can continue to give rise to new mineral species long after either of the two major flips of Earth state and their associated major evolutionary leaps.

Following 1.9 bya and the supercontinent Columbia, the next small, but significant, pulse of new mineral species comes at 750 to 600 mya and is associated with the supercontinent Rodinia. Its breakup saw Snowball Earth II and the rise in atmospheric oxygen from ~2% to near modern-day levels. As noted, it saw the likely colonisation of land by proto-lichens. Yet, evolution was to continue.

In 2018 Jennifer Morris, Mark N. Puttick, James Clark, and colleagues conducted a molecular clock analysis from extant species but calibrated it with thirty-seven fossils to dated geological strata. Their study estimates land plants to have emerged in a middle Cambrian to Early Ordovician interval (roughly somewhere between 520 and 470 mya), and the more advanced vascular plants to have emerged in the Late Ordovician—Silurian (or roughly somewhere 450–430 mya) (35). These might account for the slight rise in new mineral species seen at that time.

The real take-off of new mineral species began somewhere around 400 mya. The Silurian geological period (444–419 mya) saw the further, extensive rise of larger, multicellular land plants, though these at first had no true roots as we would recognise them today, nor tree trunks. However, this extensive biota when decaying and buried would generate new mineral species. The Devonian period (419–359 mya) saw the diversification of fish species as well as the diversification of land plants. The Silurian and the Devonian also saw multicellular animal species leave the marine environment and colonise the land. Feeding off terrestrial biota, animals, as well as the diversified more advanced plants, meant that new mineral species from their remains and waste products were possible and so these times saw a rise in new minerals. New mineral species continued to rise with a final pulse the last few million years. The supercontinent Pangaea formed around 335 mya and broke up about 175 mya. This too under the Robert Hazen and colleagues' view likely generated new mineral species.

The modern distribution of continents is a result of the dispersal of Pangaean fragments. However it is the final surge in new mineral species that is of note because it cannot be associated with supercontinent rather, as Robert Hazen himself notes (and previously Jan Zalasiewicz), instead it is due to humans and the waste products of technology (*36, 37*).

All this raises an intriguing question: Was each critical flip of the Earth's state essentially driven by (or at least had it as the major factor) biology (evolution), and did each represent a change in the way carbon cycles, albeit mediated by geology (nutrients from supercontinents), that heralded the beginning of a period of the formation of new mineral species? If this is so, then what of the new mineral species humanity is creating and discarding to enter the geological record today? Does it mean that we now are in the process of a critical transition in the Earth's state? Indeed, what are the implications of this narrative for life elsewhere?

The latter question will begin to be addressed in the next chapter, after which we will return to the question, Are we today undergoing a critical flip of planetary state?

## References


1. Galway-Witham, J., & Sringer, C., How did *Homo sapiens* evolve? *Science*, **360**, 1296–1298 (2018).
2. Wignall, P. B., Lethal volcanism. *Nature*, **477**, 285–286 (2011).
3. Sobolev, S. V., Sobolev, A. V., Kuzmin, D. V., *et al.*, Linking mantle plumes, large igneous provinces and environmental catastrophes. *Nature*, **477**, 312–316 (2011).
4. Clarkson, M. O., Kasemann, S. A., Wood, R. A., *et al.*, Ocean acidification and the Permo-Triassic mass extinction. *Science*, **348**, 229–232 (2015).
5. Kump, L., Drivers of the end Permian hyperthermal: system failure? In Foster, G., Zachos, J., Lunt, D., Hull, P., & Broecker, W., eds., Abstracts for the symposium *Hyperthermals: Rapid and Extreme Global Warming in Our Geological Past*, 25th–26th September, Royal Society, London (2017).
6. Kump, L. R., Prolonged Late Permian–Early Triassic hyperthermal: Failure of climate regulation. *Philosophical Transactions (A) of the Royal Society*, **376**, 20170078 (2018).
7. Ward, P. D., Montgomery, D. R., & Smith, R., Altered river morphology in South Africa related to the Permian-Triassic extinction. *Science*, **289**, 1740–1743. (2000).
8. Cowie, J. *Climate Change: Biological and Human Aspects*, 2nd edition. Cambridge: Cambridge University Press, Great Britain (2013).
9. Grasby, S., Sanei, H., & Beauchamp, B., Catastrophic dispersion of coal fly ash into oceans during the latest Permian extinction. *Nature Geoscience*, **4**, 104–107 (2011).
10. Sobolev, S. V., Sobolev, A. V., Kuzmin, D. V., *et al.*, Linking mantle plumes, large igneous provinces and environmental catastrophes. *Nature*, **477**, 312–316. (2011).

11. Payne, J. L., Lehrmann, D. J., Wei, J., *et al.*, Large perturbations of the carbon cycle during recovery from the end-Permian extinction. *Science*, **305**, 506–509 (2004).
12. Jenkyns, H. C., Forster, A., Schouten, S., & Damsté, S., High temperatures in the late Cretaceous Arctic ocean. *Nature*, **432**, 888–892 (2004).
13. Alveraz, L. W., Alveraz, W., Asaro, F., & Michel, H. V., Extraterrestrial cause for the Cretaceous-Tertiary extinction—Experimental results and theoretical implications. *Science*, **208**, 1095–1108 (1980).
14. Vajda, V., & McLoughlin, S., Fungal proliferation at the Cretaceous–Tertiary boundary. *Science*, **303**, 1489 (2004).
15. MacLeod, K. G., Quinton, P. C., Sepulveda, J., & Negra, M. H., Postimpact earliest Paleogene warming shown by fish debris oxygen isotopes (El Kef, Tunisia) *Science*, **360**, 1467–1469 (2018).
16. Kyle W. Meyer, K. W., Petersen, S. V., Lohmann, K. C., *et al.*, Biogenic carbonate mercury and marine temperature records reveal global influence of Late Cretaceous Deccan Traps. *Nature Communications*, **10**, 5356 (2019).
17. Lyson, T. R., Miller, I. M., Bercovici, A. D., *et al.*, Exceptional continental record of biotic recovery after the Cretaceous–Paleogene mass extinction. *Science*, **366**, 977–983 (2019).
18. Alvarez, S. A., Samantha J. Gibbs, S. J., Bown, P. R., Kim, H., Sheward, R. M., & Ridgwell, A., Diversity decoupled from ecosystem function and resilience during mass extinction recovery. *Nature*, **574**, 242–245 (2019).
19. International Commission on Stratigraphy, *International Chronostratigraphic Chart v 2020/01*. International Union of Geological Sciences: Beijing, China (2020).
20. Kennett, J. P., & Stott, L. D., Abrupt deep sea warming: Palaeoceanographic changes and benthic extinctions at the end of the Palaeocene. *Nature*, **353**, 225–229 (1991).
21. Tripati, A., & Elderfield, H., Deep-sea temperature and circulation changes at the Paleocene-Eocene Thermal Maximum. *Science*, **308**, 1894–1898 (2005).
22. Zachos, J. C., Röhl, U., & Schellenberg, S. A., *et al.*, Rapid acidification of the ocean during the Paleocene-Eocene Thermal Maximum. *Science*, **308**, 1611–1615 (2005).
23. Dickens, G. R., O'Neil, J. R., Rea, D. K., & Owen, R. M., Dissociation of methane hydrate as a cause of isotope excursion at the end of the Paleocene. *Paleoceanography*, **10**, 965–971 (1995).
24. Pagani, M., Caldeira, K., Archer, D., & Zachos, J. C., An ancient carbon mystery. Science, 314, 1556–1557 (2006).
25. Secord, R., Gingerich, P. D., Lohmann, K. C., & MacLeod, K. G., Continental warming preceding the Palaeocene–Eocene thermal maximum. *Nature*, **467**, 955–958 (2010).
26. Gutjahr, M., Ridgwell, A., Sexton, P. F., *et al.*, Very large release of mostly volcanic carbon during the Palaeocene–Eocene Thermal Maximum. *Nature*, **548**, 573–577 (2017).
27. Svensen H., Planke, S., Malthe-Sørenssen, A., *et al.*, Release of methane from a volcanic basin as a mechanism for initial Eocene global warming. *Nature*, **429**, 542–545 (2004).
28. Storey, M., Duncan, R. A., & Swisher, C. C., III, Paleocene-Eocene Thermal Maximum and the opening of the Northeast Atlantic. *Science*, **316**, 587–589 (2007).
29. Hartley, R. A., Roberts, G. G., White, N., & Richardson, C., (2011) Transient convective uplift of an ancient buried landscape. *Nature Geoscientist*, **4** (8), 562–565.
30. Schuur, E. A. G., Abbott, B., & the Permafrost Carbon Network. High risk of permafrost thaw. *Nature*, **480**, 32–33 (2011).
31. Hugelius, G., Loisel, J., Chadburn, S., *et al.*, Large stocks of peatland carbon and nitrogen are vulnerable to permafrost thaw. *Proceedings of the National Academies of Sciences USA*, **117** (34), 20438–20446 (2020).
32. DeConto, R. M., Simone Galeotti, S., Pagani, M., *et al.*, Past extreme warming events linked to massive carbon release from thawing permafrost. *Nature*, **484**, 87–91 (2011).
33. Intergovernmental Panel on Climate Change, *Climate Change 2007: The Physical Science Basis. Working Group I Contribution to the Fourth Assessment of the IPCC*. Cambridge: Cambridge University Press, Great Britain (2007).

34. Hazen, R. M., Robert T. Downs, R. T., Eleish, A., *et al.*, Data-driven discovery in mineralogy: Recent advances in data resources, analysis, and visualization. *Engineering*, **5**, 397–405 (2019).
35. Jennifer L. Morris, J. L., Puttick, M. N., Clark, J. W., *et al.*, The timescale of early land plant evolution. *Proceedings of the National Academies of Sciences USA*, 115 (10), e2214–e2274 (2018).
36. Zalasiewicz, J., Kryza, R., & Williams, M. The mineral signature of the Anthropocene in its deep-time context. *Geological Society of London Special Publ.*, **395**(1), 109–117 (2014).
37. Hazen, R. M., Grew E. S., Origlier, M. J., & Downs R. T., On the mineralogy of the 'Anthropocene Epoch'. *American Mineralogy*, **102** (3), 595–611 (2017).

# 7
# Exobiological Implications

The narrative I have been developing has two principal strands. First, a geological one: the evolution of the planet from its molten beginnings to being a world with an ocean, land, and then plate tectonics with growing continents with the Wilson cycle of supercontinent formation and fragmentation and pulses of mineral species formation. And second, a biological one: the rise of life, biological innovation with oxygenic photosynthesis and eukaryotes, and finally true multicellular species, with each of these stages seeing an increase in information processing (which we will return to in detail in Chapters 8 and 9). Through all this, the Earth has gone from a state of having an atmosphere unperturbed by life to a new one with it, a chemically reducing atmosphere all but devoid of oxygen before flipping state to having ~2% oxygen atmosphere, and then a second great flip to one where oxygen hovers around its present-day values of 21%, with both flips seeing a disruption of the global carbon cycle manifesting climate change in these instances with two Snowball Earths.

Both these strands might be brought together in the concept of the co-evolution of life and planet, with its interaction between biology and geology in the biosphere.

In presenting this narrative, I have occasionally mentioned that it is relevant to the question of the possibility of life on other worlds. Also, at the end of the last chapter, there was the question, Are we now close to, or even in, another change of planetary state? I will return to this last in Chapter 9, but meanwhile there are implications arising from the notion of co-evolution of life and planet Earth with that of possible life elsewhere. We turn to this now.

## Implications for possible life elsewhere in the Solar system

Other than on Earth, there are just a few places in the Solar system capable of harbouring carbon-based life, assuming that water is the biological solvent (the liquid in which life's organic molecules reside). There are good reasons for these assumptions, but alas not the space to go into detail here. While it is possible to postulate non-carbon- and non-water-based life, this really is more the stuff of science fiction. In his novel *Iceworld* (1953) The chemistry teacher and science fiction author Hal Clement explored the concept of such a life form—in this case

*Earth's Flips of State*. Jonathan Cowie, Oxford University Press. © Oxford University Press (2026)
DOI: 10.1093/9780197825402.003.0007

an alien living at high temperatures who made the planet Mercury his temporary base while exploring the strange third planet in the Solar System (Earth) which was so cold that sulphur was a solid and not a gas (*1*). The value of this novel is not so much in the imaginary biology of the alien explorer but rather in the reader considering our human environment as being normal. In fact, life can exist in environments that we might consider extreme. Though in this chapter only possible life will be considered that is carbon-based and relies on liquid water. I am therefore mainly what the astrophysicist Carl Sagan would describe as a 'carbon chauvinist' (*2*).

Nonetheless, putative life forms elsewhere in the Solar System are likely to be what we consider to be extremophiles. However, as for Mercury itself, being the planet closest to the Sun, its environment is too extreme for even extremophiles—even those of a biologist's wildest imagination—with Mercurial equatorial temperatures ranging from −173°C at night to 427°C during the day; and having almost no atmosphere. But there are at least four other places in the Solar System that might (contentiously) support life, if not today then in the past. Their and Earth's respective key features of mass, gravity, surface atmospheric pressure, and median surface temperature are outlined in Table 7.1. This table lists three planets and three moons of gas giants. Of the planets, in addition to Earth, two likely had abundant liquid water on their surfaces in the distant past: billions of years ago when the Solar System was young. Of the moons, one has an atmospheric chemistry that might well reflect aspects of the primordial Earth's, while the other two have an abundance of liquid water. Here, it should be noted that there are other moons in the Solar System that may have some liquid water (they certainly have ice). It has even been hypothesised that there is a subsurface ocean on the largest of the dwarf planets, Pluto (*3*). But the chance of such places having all the necessary conditions for life to arise, let alone thrive, is not thought to be nearly as great. So, for time's sake, only the bodies other than Earth in Table 7.1 will be considered.

## The possibility of life on Venus

Venus, the next planet out from the Sun after Mercury, also has surface conditions too extreme for life. Under an atmosphere whose pressure is ninety-two times that of Earth, or roughly that found 900 metres undersea on Earth, its waterless surface has an average temperature that exceeds 462°C: even hotter than equatorial Mercury during the day. The atmosphere itself mostly consists of around 96.5% carbon dioxide and approximately 3.5% nitrogen. There is also sulphur dioxide and argon at concentration levels of parts *per* million (ppm), and even trace water vapour at around 20 ppm. These trace gases make for

**Table 7.1** Principal features of Solar System bodies (in order of distance from the Sun) that have, or had, conditions likely to support life. Their mass, gravity and atmospheric pressure is given as a proportion in relation to that on Earth.

| Body | Mass | Gravity | Atmospheric pressure | Average temperature |
|---|---|---|---|---|
| Venus | 0.815 | 0.904 | 92 | 467°C |
| Earth | 1.0 | 1.0 | 1.0 | 15°C |
| Mars | 0.11 | 0.38 | 0.6 | -54°C |
| Europa | 0.008 | 0.134 | 10–12 | −170°C |
| Titan | 0.0225 | 0.14 | 1.45 | -179°C |
| Enceladus | 1.8 × 10–5 | 0.0113 | trace | −198°C |

an important part of Venus's atmospheric chemistry, for the carbon dioxide, water, and sulphur dioxide react to create clouds of sulphuric acid that dominate the atmosphere at around 50–75 km above the surface, beneath which is a band of sulphuric acid haze down to roughly 30 km above the surface. It even rains sulphuric acid, but this evaporates long before it reaches the Venusian surface.

Yet Venus, away from its surface, is not entirely inhospitable. As on Earth, its atmospheric temperature decreases with altitude before rising again to a temperature near that of sunlight in space. Consequently, there is an altitude—between 40 and 50 km above the surface—at which the temperature is just a few degrees Celsius. At that altitude, the atmospheric pressure is only a little above that of Earth's at sea level. It has even been contemplated that this band of atmosphere could be conducive to life (4, 5)!

Of course, if there were say, single-cell, perhaps prokaryotic analogue, life in this Venusian atmospheric level, then it would have to contend with being continually flushed with atmospheric circulation into inhabitable levels of the atmosphere. So the possibility of extant life today high in the Venusian atmosphere, it would seem, is unlikely.

Equally obvious is that if there were life in Venus's upper atmosphere (assuming it survived constant flushing over the eons through some protection mode) then it must have got there somehow. If life began, as it is thought to have on Earth, in an aquatic environment, then there must have been a time when Venus had surface water. In actuality, it is possible that at one time Venus did have oceans, though this is debated.

Just as on Earth water ($H_2O$) vapour with the minority heavier oxygen-18 isotope ($^{18}O$) preferentially precipitates so that rain falling far inland in large continents has much less of the $^{18}O$ than that closer to the coast (see Chapters 5 and 6), so the heavier deuterium hydrogen isotope (with double the atomic weight of the majority hydrogen isotope) is also less likely to escape Venus's atmosphere. Now, Venus presumably formed at the same time as the Earth 4.6 billion years ago (bya) with a similar Late Heavy Bombardment (LHB) delivering a 'late veneer' of volatiles including water 4.1–3.8 bya (see Chapter 1), then Venus might at this time sustained an ocean to nurture putative life. Given this, and knowing that Venus is now effectively waterless (aside from trace amounts in the atmosphere), then there must have been considerable water loss in the past (*6*). Here, as the normal hydrogen would preferentially escape leaving behind more of the heavier deuterium isotope, this would mean that Venus would today have a higher proportion of deuterium to the normal hydrogen ratio (its D/H ratio) than the Earth, which retained most of its water. Indeed, observations of Venus have shown that it truly does have a higher D/H ratio: 150 ± 30 times that of Earth's (*7*). So it is likely that Venus did once have oceans or, if not that then, considerable atmospheric water vapour.

The question that then arises is, whether an early Venus could have retained significant oceans (assuming it had them) for the time that it took life to get going? Here, there are a number of competing factors.

First, the young Sun at the time after planetary formation had less of an output energy (see Chapter 3). It was a faint Sun that has since been steadily increasing—as do all main sequence stars—in level (by nearly 30%) to today's output. This fainter young Sun would have meant that Venus would have had less chance for a greenhouse runaway effect. Second, Venus has a less protective magnetosphere than the Earth and this facilitates Solar wind stripping atoms and molecules from the top of its upper atmosphere. This is also enhanced compared to the Earth, as Venus, being closer to the Sun, experiences a little stronger Solar wind. Lighter atom and molecule loss from the top of the upper atmosphere is also enhanced with Venus compared to the Earth, as Venus is a little lighter in mass than the Earth and so has a lower (~10% lower) gravity, that also facilitates loss from the upper atmosphere. Conversely, a younger Venus is likely to have a more active core than today and so a more pronounced protective magnetosphere than is seen today. Third, Venus is closer to the Sun and so exposed to more charged particles and radiation than Earth. Hypothetically speaking, this could have enhanced the upper-atmosphere tholin haze that would exist with a reducing atmosphere mix of carbon dioxide and methane. This enhanced tholin haze, if it existed, would have conferred a greater cooling of the lower atmosphere and surface. Finally, for the longevity of any primordial Venusian oceans, much depends on the size of these oceans: the more ocean to start off with, the longer it takes to see water loss to the trace levels seen today.

There is even the possibility that a young Venus with a more active core than today would have a more active mantle and possibly a primordial version of plate tectonics (even if a more developed plate tectonic system—as seen on Earth today—had not got going) especially if water was available as a lubricant. We do know that today Venus has no plate tectonics and has been volcanically resurfaced. An estimate as to when this (last) happened is in the order of hundreds of millions of years ago (*8*): presumably Venus must have lost its putative oceans before then and most likely a lot longer ago.

In 2016, a small group of researchers led by Michael Way and Anthony Del Genio number-crunched a few of these factors to calculate the likely longevity of hypothesised, early Venusian oceans, hence how long Venus might have been habitable before the greenhouse runaway effect turned it into the inhospitable world we see today. To do this, they employed a modified general circulation climate model used in climate change research, but one especially adapted by NASA for considering other planets. Their simulations all used Venus's radius, gravity, modern obliquity (angle of tilt), and modern orbital eccentricity (elliptical deviation from a true circle), and an estimated Solar energy flux 2.9 bya (~77% of present-day Solar luminosity) as well as separately 0.715 bya (~94% of present-day Solar luminosity). In terms of the energy flux Venus receives compared to Earth, the value is higher, as Venus is closer to the Sun. 2.9 bya is after the LHB, not to mention the time when it is thought a more developed plate tectonic system began on Earth, where life had already risen. Conversely, 0.715 bya is the time the researchers considered for purposes of their model when volcanic resurfacing of Venus took place and so unlikely for there to have been any significant ocean present and hence, just prior to this, the latest possible time for liquid water on its surface.

Their simulations show that a slow-rotation Venus, with land/ocean fraction of ~40/60% (today Earth's is ~30/70%) could have seen moderate temperatures with an average not too dissimilar to Earth's today despite an incident solar flux 46%–70% higher than Earth received. Such conditions could, they suggest, have continued up to 0.715 bya provided there was a significant ocean. Their results also suggest that such a Venus-like a planet with a modest amount of surface liquid water is more conducive to habitability over a wide range of stellar fluxes than a planet largely or completely covered by water. (Remember, land reflects more sunlight as well as retaining less heat than oceans and so having more land has a cooling effect.) (*9*)

The researchers subsequently refined their calculations that supported their original conclusions. They further speculated that large igneous provinces and the global resurfacing hundreds of millions of years ago played key roles in ending the clement period in Venus's history and presenting the planet we see today. Their results have implications for what astronomers term our planetary system's 'habitable zone', and if Venus-like exoplanets exist with clement conditions akin to modern Earth they proposed to place them in what they term the 'optimistic Venus zone' (*10*).

Of course, Venus's surface could have been volcanically reworked earlier than 0.715 bya: all we know is that Venus was reworked roughly then, it could still have experienced previous episodes of reworking. Much depends on the initial amount of water present: too little and the ocean would quickly go, too much and the oceanic albedo absorbing more Solar energy would see the planet heat up towards greenhouse runaway. In short, the simulations suggest that an ocean was possible, but not how long it lasted.

Turning now towards our co-evolution of life and the planet Earth narrative. Because life on Earth seems to have established itself quickly, it is perfectly feasible that an early Venus with oceans and land with thermal vent ponds might allow for the ultraviolet-mediated chemical reactions leading to nucleic acid precursors (see Chapters 1 and 2). Additionally, a very early Venus would likely have less carbon dioxide, as it would not have had the time to accumulate from volcanic outgassing. Further, the ocean would initially be a modest carbon sink. The Michael Way and Anthony Del Genio team simulations assumed a largely nitrogen early Venusian atmosphere at roughly today's Earth pressure.

Critical to any putative Venusian ocean's longevity would be these carbon sinks. If there were some plate subduction, even if full-blown planetary plate tectonics had yet to get going, then this would have provided a carbon sink. And if there was an ocean to lubricate subduction zones, then this sink might have lasted as long as the ocean did. However, once Venus's core dynamo, hence magnetosphere, declined, so water loss from upper atmosphere ablation would have increased. Once the ocean was gone then any volcanic resurfacing would have transformed any sink carbonate to carbon dioxide, and any volcanic emission of carbon dioxide would remain in the atmosphere. Hence the Venusian atmosphere we see today.

In terms of our co-evolution of life and the planet Earth narrative, optimistically, assuming the 0.715 bya was the first complete volcanic reworking of Venus's surface, assuming there were carbon sinks in an oceanic early Venus, and assuming a similar timeline to Earth's narrative of a great oxidation 2.4–2.2 bya, oxygenic photosynthesis might have evolved on Venus.

In terms of our co-evolution of life and the planet Earth narrative, aside from a relatively early change of state from one of purely chemical equilibrium to one perturbed by life, the Earth's roughly 4.5-billion-year history saw it go through two major critical transitions, flips of state, each lasting a couple of hundred million years. Assuming (and it is an assumption) a similar-length critical transition for Venus flipping from a planet with significant liquid water to one with none, then we might possibly have seen this significant water begin to disappear somewhere around 0.9 bya assuming that 0.715 bya saw the first volcanic reworking of Venus.

This, though, is the optimistic view. Having said that, assuming just a short-lived early Venus with an ocean to a little over 3 bya, it would still not be completely unreasonable to assume that there were Venusian analogues of prokaryotes. In short, there might have been a period when there was life on both Earth and Venus at the same time. The issue regarding Venus therefore becomes one of whether this life could have become airborne?

Today, with no oceans and a hostile surface, there is no possibility of marine- or land-based life on Venus. Yet, if some sort of Venusian analogue to Earth's prokaryotes had evolved in some long-gone ocean or, more likely, terrestrial hot spring, could it have survived today? This is rather a fanciful question, but as you will recall, Venus, away from its surface, is not entirely inhospitable. It has been contemplated that life might exist higher in the atmosphere. Such musings have not only been recent but as long ago as the 1960s, including by the respected astrophysicist Carl Sagan (*11*). Yet this musing remained purely fanciful until 2020 and the detection of phosphine ($PH_3$) in Venus's atmosphere, by a team of British astronomers led by Jane Greaves, Anita Richards, and William Bains, using two different radio telescopes.

Phosphine is known to be produced by life, yet no mechanism is known for its purely chemical (non-biological) production given Venus's atmospheric constituents. That there may be microbial life in Venus's atmosphere is a tantalising prospect, but as yet is unproven (*12*). At the time of writing, this detection is being challenged: it is important in science for there to be independent confirmation of any experiment or observation. Some consider that they misinterpreted their data, which instead should suggest that there is no significant quantity of phosphine in Venus's upper atmosphere (*13*). Others, that one of the telescopes used could not detect the gas (*14*). Finally, some are of the view that what was detected was not phosphine but sulphur dioxide ($SO_2$) (*15*).

Another blow to the notion of extant life on Venus, high in its clouds, is the lack of sufficient water. One measure of water availability for life is the 'water activity' (see Appendix 2) of a solution in which life lives. Extremophile species of fungi can live in an environment with a water activity as low as 0.585. However, the water activity in Venus's atmosphere is less than 0.004, two orders of magnitude below the 0.585 limit. It therefore seems that Venus today is simply too dry for life as we know it, or can currently conceive it (*16*).

Intriguing as it is for there to be the possibility for a chemical biomarker in Venus's upper atmosphere (a question that will eventually be answered one way or another), the possibility for there to have been Venusian life on the early primordial planet is real.

Yet before getting too excited, do note the word 'possibility' and especially that part of the above case for early Venusian oceans (hence life) is based on a computer model. Here a line from *Monty Python and the Holy Grail*, when the castle of Camelot comes into view, is most apt:

> King Arthur (boldly): 'Ahhh, Camelot!'
> Courtier (quietly): 'Shhh, it's only a model'. (*17*)

The thing is that models—including computer models—are not reality even if they are an attempt to represent it: they have flaws and omissions as well as guesses and estimates. The aforementioned Michael Way and Anthony Del Genio team's modified general circulation climate model was just that: a model. One-dimensional numerical climate models cannot account for the effects of atmospheric circulation and clouds.

In 2021, Martin Turbet and his small team, primarily from the University of Geneva, published results from a more sophisticated model. First, this model included the effect of clouds (not just incident Solar radiation and atmospheric greenhouse gases). Secondly, it recognised that an oceanic world was not Venus's (or Earth's) initial condition. When the planet was forming, during a period of gas and dust accretion, heat would be released, just as when you clap or rub your hands heat is released. Initially, the entire world was a molten ball with volatiles such as water being gaseous. Only as the planet cooled sufficiently could water condense out on the surface as it plainly did on Earth. However, the Martin Turbet team's model showed that the Venusian surface never cooled enough for water to form. Instead, what happened was that only on Venus' night side and over the poles would steam begin to condense and form clouds. These clouds trapped in the daytime heat on Venus's night side just as on Earth cloudy nights are warmer than clear nights. Meanwhile, Venus' day side was still receiving heat. The energy balance, the Martin Turbet team's results suggest, was such that Venus was always too hot for liquid water to have ever existed on its surface. This was so in spite of the faint young Sun 4.5 billion years ago (see earlier this chapter and Chapter 3). Indeed, a faint young Sun might have been necessary for Earth to get its oceans (*18*, *19*).

Having said that, remember *Monty Python*, the Turbet team also based their case on a computer model. Could this model also lack a critical feature? Here, I venture, tholins caused by the action of sunlight on methane and carbon dioxide (see Chapter 3) would provide a cooling effect on Venus's day side. Could this be sufficient?

Clearly, as is so often the case in science, more research is needed. But our contemplation of Venus, nonetheless, brings us on to the question of possible life on Earth's other planetary neighbour: Mars.

## The possibility of life on Mars

The possibility of life on other worlds has been entertained for centuries, with much popular currency though to the present day, accruing traction in both science and more recently its arts genre counterpart, science fiction.

One of the most widespread episodes of scientific interest, specifically in the idea of life on Mars, took place at the end of the 19th century and into the early 20th. This period of attention was arguably galvanised by the Italian astronomer Giovanni Schiaparelli, the foremost Mars cartographer of the 19th century. He thought he saw channels (*canali*) that, he contemplated, were the work of intelligent beings. This notion was then famously elaborated on by the American astronomer Percival Lovell, who in 1894, being independently wealthy, built an observatory specifically to study Mars. Percival quickly formulated a theory (which he subsequently spent years elaborating) that Mars was a dying world whose intelligent inhabitants built canals to transport water from the poles to arid areas closer to the Mars equator (*20*).

Such fascination in these ideas stemming from the astronomical community fuelled fiction. Perhaps the most famous science fiction novel of the time, and still in print today, is H. G. Wells's *The War of the Worlds* (1898). Having said that, while its founding premise was that there was intelligent life on a dying Mars, the motivation behind the novel—as with much notable science fiction—was social comment: the Martian invasion was an observation on British atrocities, especially the Tasmanian genocide of the 1830s, to which he returned to explicitly in some of his other works. To overcome the militarily strong Britain, he needed a powerful foe. Hence, he devised the powerful, invading Martians (*21*). Wells himself was a scientist with a First Class Honours BSc from the Normal School of Science (now part of Imperial College London) in biology (he was also a Fellow of the Zoological Society): so science of the day permeates his fiction. With regards to *The War of the Worlds* itself, biology was central to the novel's denouement, with the Martians succumbing to Earth's native microbes to which they had no resistance. (We will shortly return to this.)

Much science fiction, relating to both possible life on Mars and exobiological hypotheticals generally, has subsequently been written, and much has been based on the science (the scientific understanding) of the respective days they were written. Some of these now seem terribly dated, such has been the advancement in our understanding of the planet Mars the past half-century. Yet such has been the fascination in the prospect of life on that planet, and elsewhere away from Earth, that it is now an established genre trope together with that of space travel. Indeed, without such popular interest, one wonders whether there would have been the somewhat considerable investment given, over the latter half of the 20th century through to today, on interplanetary space missions?

One notable science fiction example of the human exploration of Mars—again based on the science of the day—is Arthur Clarke's 1951 novel *The Sands of Mars* (*22*): Like Wells, Clarke had an academic background in science, though his was physics, and an interest in astronomy. That novel portrayed Mars as having a thin, but almost breathable atmosphere.

Back in 1950, one of the few ways that the Martian atmosphere's density was estimated was to see, using Earth-bound telescopes, how the brightness of bright patches on the surface varied between when viewed at the equator (looking through the least thickness of atmosphere) and when the same patch appeared at the edge of the Martian disc looking through more atmosphere. This uses the effect similar to that of the Sun appearing dimmer and more red at sunrise or sunset, as more yellow and blue end of the spectrum is filtered out by the greater thickness of atmosphere light has to traverse at that low angle.

Such early estimates, between 1940 and 1948, gave a range of results of Mars's atmospheric pressure at its surface of between 80 and 120 millibars (8,000 and 12,000 Pa) (*23*). (For comparison, the air pressure at sea level on Earth is just a little over 1,000 millibars or 100,000 Pa.) So a mid-range 1940s estimate for the Martian atmosphere, of say 100 millibars (10,000 Pa), is a pressure roughly 10% that of Earth's at sea level and roughly the Earth's atmospheric pressure 11 miles up (very roughly twice the height of Everest). This is the pressure Clarke seems to be using in his novel: indeed, he himself refers to it being equivalent to the pressure of Earth's a few miles above that of Everest. At such a pressure, all a human would need to survive would be a pressure suit of the sort high-altitude, jet fighter pilots use and an airtight helmet with an oxygen supply. In Clarke's *The Sands of Mars*, all people need on the surface of Mars is a light pressure suit and an oxygen mask.

This is important because Clarke and the science of the day, in 1951, envisioned a Mars with an atmospheric pressure that allowed liquid water on its surface.

Today, post the 1976 Viking missions we now know that the Martian atmosphere's density around its lowlands is only around 6 millibars (600 Pa) or 0.6% that of Earth's. But don't blame Clarke for not knowing that over two-thirds of a century ago. This pressure is very close to the triple point of water: the pressure at which water at a temperature of 0.16°C can exist in all three phases of solid (ice), liquid and gas (vapour). Consequently, by convention, the height of land on Mars where the atmospheric pressure is on average 6.105 mbar (the exact water triple point pressure), or 610.5 Pa, is deemed the zero elevation (in light of the fact that Mars today has no sea, hence no sea level). This is equivalent to Earth's atmospheric pressure at an altitude of 22 miles (35 km) Indeed the two 1976 Viking probes themselves reported Martian atmospheric pressures of around 7 mbar (700 Pa) (*23*). So, as on Earth, Mars's atmospheric pressure depending on surface elevation varies from a low of 0.3 mbar (30 Pa) at Mars's highest point, *Olympus Mons* (which is over three times the height of Everest's 5.5 miles or 8.8 km), to over 11.55 mbar (1,155 Pa) in *Hellas Planitia* at Mars's lowest elevation. The latter is still only roughly 1% of Earth's atmospheric pressure and 10% of that Clarke, and that of the late 1940s science, considered Mars' atmospheric pressure to be. Yet, these higher Martian pressures below the aforementioned zero elevation, can see water exist on the surface in liquid form and, if the water is salty, in liquid form over a range of several degrees centigrade.

There are two reasons for Mars having a thinner atmosphere than the Earth. First, Mars is smaller than Earth and is very roughly only half its diameter. Because of the cube-square law (by which volume of a sphere is proportional to the cube of its diameter but its surface area is proportional to only the square of its diameter), and because the planet is less dense, it only has 11% the mass of the Earth. This in turn means it has low gravity, equivalent to about 38% that of Earth's. Mars's low gravity means that it is easier for the planet's atmosphere to escape into space. Second, Mars no longer has a magnetosphere (magnetic field) protecting its upper atmosphere from Solar wind ablation. Both factors serve to see Mars having gradually lost much of its atmosphere and water over time.

With a thinner atmosphere (less greenhouse warming) and Mars being further from the Sun than the Earth (it receives just 43% of the amount of Solar energy), the Martian surface temperatures vary from lows of about −143°C (at the winter polar caps) to highs of up to 35°C. So Mars's size means that for the most part there is little liquid water, hence little prospect for life. As the author Harry Turtledove noted in the introductory page of his science fiction novel *A World of Difference*, 'Mars is boring. Turns out it's too darn small. But what if it weren't . . .' (*24*).

If Mars were significantly larger—at the very least over half Earth's mass instead of 11%—then there would be a number of consequences. First, with greater gravity, the loss of atmosphere would be reduced. Second, it would have retained a core dynamo longer (hence also a magnetosphere) and so possibly, at least for a while, progressed into developing proper plate tectonics. As it is, the beginnings of Martian plate tectonic movement did take place at least 4 billion years ago (bya). This is evidenced in areas of ancient Martian crust by magnetic reversal lines (similar to those by the mid-Atlantic spreading plate zone on Earth) detected by the Mars Global Surveyor probe's magnetometer experiment in 1999 (*25*). With a longer-lasting, active core dynamo Mars would have had a magnetosphere protecting its upper atmosphere from Solar wind ablation for more of its history than the possibly several hundred million years it did have (*26*). This too would have helped Mars conserve its atmosphere. Both these factors would mean that back then (~4 bya) it would have had a thicker atmosphere with reduced day-night temperature variations and a warmer average temperature. This in turn would mean a greater proportion of Mars's surface area seeing liquid water. And, of course, if atmospheric loss was reduced, then so would be the loss of water.

Today water is limited on Mars, but it is present: it exists as ice (mixed with carbon dioxide) at its poles and also as a liquid beneath the said poles' ice as well as ice in permafrosts and subsurface liquid brine in some areas (*27*, *28*). Briny water has also been observed on Mars escaping permafrost on some slopes (*29*). Further, its geology, as revealed by lander probes, shows that some considerable amount of water is bound to minerals (*30*). It has been calculated that in excess of 9% by volume of the Martian mantle may contain hydrous mineral species (*31*). A recent analysis by Californian researchers suggest that both atmospheric loss of water as well as the incorporation of water into minerals best explains the deuterium-to-hydrogen ratios detected by various space craft, rovers, and meteorites. Mineral incorporation, they say, could account for between 30% and 99% of ancient liquid water loss: a wide range admittedly, but opening the door to mineral incorporation possibly being the dominant mechanism (*32*).

Given there is water on Mars today, albeit much of it locked up, what of the distant past before water escape, along with atmosphere loss, took its toll, and back when Mars had some protection with a magnetosphere from Solar wind atmospheric ablation? As it turns out, there is considerable evidence that the primordial Mars did have an ocean, including firm evidence of ancient shorelines and river deltas (*33*). It is thought that there could have been an ocean covering much of Mars's northern, with a small sea in its southern, hemisphere. These were likely formed before 4 bya and possibly lasted to after 3.5 bya but not as long as to 3 bya (*34*).

Further, some areas of Mars in its southern hemisphere, where there is ancient surface dating from 3.7 bya, show branching river drainage channels. The question then arises, Did these form due to groundwater leaching out (as is seen today) or from actual rainfall? Fortunately, we can see on Earth, river channels formed mainly from groundwater (such as in the lower Amazon basin), as opposed to from rainwater falling on an otherwise dry surface (such as in eastern Algeria), have different, distinctive branching patterns. It turns out that the pattern of channels seen on the ancient Martian southern hemisphere have a branching pattern that is very similar to the rainwater type on Earth. Given that Mars's early ocean dominated its northern hemisphere and land its southern, it suggests that Mars at the time had a fully functioning water cycle with ocean evaporation and rainfall over land (*35*).

Fitting all this into our co-evolution of life and planet narrative, we need to assume that very roughly the timescales for the key evolutionary events on Earth similarly took place on Mars. Of course, we could contemplate interplanetary biological cross-contamination whereby a meteor strike on one planet threw up debris into space that ultimately ended up on another planet. Here it is easier for putative Martian life to transfer material to contaminate the Earth than the other way around: Earth has a deeper gravitational well to overcome as well as any debris would have to have sufficient energy to overcome the difference between the Earth's potential in the Sun's gravitational well to that of Mars. So cross-contamination from Earth to Mars is much less likely.

Ignoring the remote possibility of Earth being contaminated by early Martian life—that is assuming life arose on Earth as discussed in Chapters 1 and 2—it is not unreasonable to assume that if life arose in an aquatic environment early on Earth, that it did so too on Mars.

Here, assuming a similar co-evolution of life and planet timeline as on Earth, then life on Mars quite likely was established by about 3.8 bya, but possibly (though less likely) could have arisen as early as around 4.28 bya (see Chapter 1). This was shortly (in the geological sense) after the planet's formation ~4.5 bya, and if so any early life would have to survive the Late Heavy Bombardment of 4.1–3.8 bya (again, see Chapter 1). If there was an ocean on Mars before 4 bya, then there would have been a habitat for life. In short, if a similar pattern of co-evolution of life and planet narrative took place on Mars as it had on Earth, there could well have been life on the very early Mars in the form of something analogous to prokaryotic life on Earth, and quite likely something analogous to chemolithoautotrophs on our planet before later other types of prokaryotic analogues.

The next question is, could oxygenic photosynthesis have arisen on Mars? The answer to this is less clear, but we can at least see whether it is a theoretical possibility or not.

The Earth 3.4 bya likely had a warm ocean, roughly around 70°C having possibly cooled roughly 10°C the previous 500 million years, and which subsequently further cooled to around 20°C about 800 mya later (see Chapter 3). This warm ocean had significant geothermal heat input from a thinner crust and possibly insulated with a thicker carbon dioxide atmosphere that also conferred greater greenhouse warming than today. Prokaryotes can survive such high temperatures whereas eukaryotes are more sensitive and cannot, so could only evolve much later. Having said that, Mars, being much smaller, has a higher surface area to volume ratio than the Earth and so cooled faster. That plate tectonics did not properly get going on Mars testifies to that. Mars's ocean would have cooled earlier than Earth's primordial oceans. So in terms of temperature something analogous to Earth's eukaryotes could have evolved. But in terms of time—assuming the timing of events on Earth is relevant—there was not enough before much of the large Martian ocean was lost. And with the early shutting down of Mars's core dynamo, Solar wind ablation of its upper atmosphere would have ramped up. That a significant Martian ocean did not survive until sometime before 3 bya meant that it was a race against time. Given that the rise of eukaryotes on Earth took place around 2.2 bya (possibly even just a little earlier—see Chapter 4) this was long after the loss of a significant Martian ocean. With the ocean gone after 3 bya, water vapour (itself a greenhouse gas) in the now thinner (conferring even less greenhouse warming) Martian atmosphere would see a surface less conducive to life. Indeed, as one of the ways water loss took place was the photolysis of water (we will come to this shortly), chlorates and hypochlorites were created which would make the surface even more inhospitable. The likelihood of the evolution on Mars of something analogous to a eukaryote is therefore remote. So in turn, true multicellular life rising on Mars is simply not in the cards: there just was not the time and nor were the conditions conducive.

However, early Martian life did have two other things going for it in addition to earlier ocean cooling. The first was its ocean was large and shallow together with a large continent: remember, at that time the early Earth was largely water with proto-continents (before plate tectonics got fully going) smaller with little land above sea level (see Chapter 3). Yet Mars already had a large area of land and so some rain-induced erosion of nutrients into the ocean was possible (even if it was not the erosion rate that would have been had there been multicellular life). This would have delivered phosphorous to the ocean in a way not possible on the very early Earth with its far smaller continents (not to mention a larger ocean volume diluting nutrients). Indeed the smaller southern hemisphere sea on Mars would have been surrounded by, in effect, a supercontinent size of land. Nutrient concentrations in this sea would have been higher than that on the early Earth.

Another factor delivering nutrients to the Martian ocean and sea is erosion caused by lake breaching. Mars is peppered by craters, and these formed lakes. But when it rained, and as these filled from any streams entering them, they overspilled at the craters' lowest points and where the surrounding land descended. There are some 265 of these palaeolakes, of which 234 have water channels leaving them. US geoscientists Timothy Goudge and Caleb Fassett with colleagues have considered the erosion potential from these lake spillovers. They found that lake breach flooding was responsible for eroding at least 24% of the volume of incised valleys on early Mars, despite representing only approximately 3% of Mars' total, water-incised valley length (*36*).

This is an important point regarding the commencement of life. You will recall (Chapters 1 and 2) that the synthesis of nucleic acids from much simpler molecules requires periods of desiccation as well as ultraviolet light and that this suggests that life first arose in land-based pools (not around deep sea vents even if those provided a good habitat once life became established). If much of the primordial Martian water cycle involved crater palaeolakes then some of these (presumably those with hot springs) would have provided a crucible for life.

At this point, step up Jonathan Toner and David Catling of the University of Washington in Seattle. They consider carbonate rich lakes tend to accumulate phosphate (you will recall the importance of phosphate from Chapter 5). This is because the carbonate sequesters calcium, which would otherwise react with phosphate to form the mineral apatite, so removing phosphorous from being available to life. On the prebiotic Earth, and presumably Mars, carbonate-rich lakes were likely abundant (Martian carbonate exists today) and phosphate-rich due to the enhanced chemical weathering of phosphate minerals under carbon-dioxide-rich atmospheres. This phosphate would then be available to any primordial life. Jonathan Toner and David Catling say that this points to an origin of life in carbonate-rich lakes (*37*). It also suggests that primordial Mars, with its large number of crater palaeolakes, would have been a better place than Earth for life to get going even if conditions later deteriorated.

In short, the Martian landscape lent itself towards nutrient delivery to its northern hemisphere ocean and southern hemisphere sea more so than on the Earth, which at the time had much less land and (with a thicker atmosphere) less craters.

Secondly, the ocean and sea having extensive area of shallows would provide a greater opportunity for photosynthesis to evolve that used visible light rather than just lower-grade (in energy terms) infrared. This potentially would have given Martian life a head start on the road to oxygenic photosynthesis.

Even so, while anaerobic photosynthesis could well have evolved, assuming an evolutionary trajectory similar to the co-evolution of life and planet Earth narrative, it would have been touch and go for the evolution of oxygenic photosynthesis in some sort of analogue to Earth's prokaryotes: something analogous to cyanobacteria (again, see Chapter 3). But 'touch and go' does not mean 'impossible'.

For now, let's assume for argument's sake that some sort of Martian analogue to prokaryotic life had evolved, did it die out with the ancient Martian ocean, or could it have survived to today?

As it happens, there are some tantalising teasers from the present-day Martian atmosphere and also from some similar to present-day Martian environments here on Earth.

At first glance, a perfunctory analysis of the Martian atmosphere does not suggest the presence of life, contrary to a basic analysis of Earth's. (See Table 7.2.)

Life, where it exists, perturbs a planetary atmosphere's chemical balance. For example, as previously noted (see Chapter 3), oxygen is highly reactive. In a lifeless world, there should only be trace amounts of oxygen from the breakdown of water vapour by high-energy UV light (photo-dissociation or photolysis—see Chapter 3) and this oxygen would be absorbed by minerals. Further, with life there would be a disequilibrium if there was both persistent oxygen and

**Table 7.2** Principal atmospheric gases by proportion of the present-day Earth and Mars.

| Gas/Planet | Earth | Mars |
|---|---|---|
| Carbon dioxide | 0.04% | 95% |
| Nitrogen | 79% | 2.7% |
| Oxygen | 21% | 0.13% |
| Argon | 1.6% | 1% |

Mars data from Mahaffy, P. R., Webster, C. R., Atreya, S. K., *et al.*, Abundance and isotopic composition of gases in the Martian atmosphere from the *Curiosity* rover. *Science*, **341**, 263–266 (2013).

methane in the atmosphere; on a lifeless world the oxygen and methane would react to form water and carbon dioxide. On the Earth methane does exist albeit in trace amounts, on Mars methane does not exist in such quantities. This line of reasoning is one that James Lovelock (whom you may recall from Chapter 2) used in his book, *Gaia: A New Look at Life on Earth*, to argue that Mars is a lifeless world (*38*). James Lovelock enlarged on this in his follow-up work, *The Ages of Gaia* (*39*). In it, he recounts how he was recruited by NASA to help with devising ways the Viking Lander might detect life. He called for a detailed atmospheric analysis but NASA decided not to go down that route, rather they used two experiments employing radioactive carbon ($^{14}C$) and one biochemical experiment to see if metabolic reactions were taking place. In 1976 when the two Viking missions to Mars took place, apart from one experiment on one of the two landers, the experiments failed to show any indication of such reactions that might indicate life. As only one experiment on one lander showed any sign of something that might be a metabolic process, it was decided to consider this experiment as a chance outlier (caused by instrument malfunction) and so not evidence that there was life on Mars (*40*). Indeed, a chemical reaction with hypochlorite could be an explanation for Viking's anomalous experimental result.

Nearly three decades later, in the 21st century, the ability to detect lower levels of gases in Mars's atmosphere revealed that there actually is methane at a concentration of around 7 parts *per* billion by volume (ppbv) or $0.007 \times 10^{-6}$%. For comparison, this concentration is two orders of magnitude (or one hundredth) lower than on our living planet Earth at 1,800 ppbv (or $1.8 \times 10^{-6}$%). A year later, in 2004, the Mars Express Orbiter and ground-based observations suggested the presence of methane in the atmosphere at a level of about 10 ppbv ($0.01 \times 10^{-6}$%) (*41*). In 2018 the *Curiosity* rover showed that Martian methane pulsed with the seasons with release peaking in the summer (*42*).

The source of this methane is currently a mystery. While on Earth there are methane-producing prokaryotes (methanogens), and while—according to the co-evolution of life and planet narrative presented in this book—it is in theory possible that some sort of life analogous to prokaryotes arose on Mars, this methane is not proof that life evolved on Mars or, even if it did, that it survived through to today: methane can be produced in a few non-biological ways. Of these, one of the most likely, is a process involving water, carbon dioxide, and the mineral quite common on Mars, olivine. Here hydrogen is released that then reacts with carbon dioxide to form methane and water. A problem is that for this to happen, a higher temperature and pressure is required than is found on the Martian surface. Yet such higher temperature and pressure conditions for non-biological methane release from olivine may exist deeper within the crust (*43*). The methane produced might then migrate towards the surface where it cools and possibly forms clathrates in a subsurface briny liquid rich layer and/or in a permafrost layer. Summer temperatures then see the clathrate dissociate, releasing the methane. There are also other non-biological hypotheses for the seasonal release from near the surface.

Nonetheless, some chemolithoautotrophs can also live off olivine giving off methane and these do so at lower temperatures (see Chapter 1). Therefore, something analogous to such life could, in theory, exist on Mars. Which means there are both explanations involving life, as well as those that are purely geological, to account for the seasonal pulsing of Martian methane.

So, if seasonal methane on Mars is not conclusive of life, is there any other evidence?

Here, the discovery of another seasonal variation in the Martian atmosphere, that of oxygen, is very interesting.

In 2019, a team of primarily US-based researchers led by Melissa Trainer, Michael Wong, and Timothy McConnochie reported on the 3-Martian-year long (2012 to late 2017) monitoring by the *Curiosity* rover of Mars's atmosphere in the Gale Crater. (Remember, Mars orbits further out from the Sun than Earth and so has a longer year, roughly equalling 1.9 Earth years. So 4 Earth years roughly approximates 2 Martian ones.) They detected seasonal variations of 13% in trace levels of oxygen ($O_2$) concentration around 0.1% of the atmosphere. Could this oxygen be produced by seasonal oxygenic photosynthesis: in other words life?

An announcement of the detection of a sign of life on Mars would be sensational news; equally, a premature announcement that turned out to be a false alarm would be as discrediting for those making it. This is no doubt why Melissa Trainer, Michael Wong, and Timothy McConnochie did not attribute the presence of seasonal oxygen to life. Instead, they looked at various possible abiotic causes (including atmospheric chemistry photolysis mechanisms) but dismissed each of these in turn as each would have other consequences that should have been detectable but were not. So this work does not conclude the detection of life on Mars, but instead teasingly said: 'Oxygen has been observed to show significant seasonal and year-to-year variability, suggesting an unknown atmospheric or surface process at work' (*44*). Having said that, if oxygenic photosynthetic (OP) life were in the future discovered on Mars then it is likely that Melissa Trainer, Michael Wong, and Timothy McConnochie's work would be cited by the discoverers, as those researchers know what was meant by Wong and McConnochie's paper's teasing conclusion. Certainly, OP life could easily explain the variations in Martian gases *Curiosity* detected.

The final question we might ask (given the knowledge we have so far) is, could any species on Earth live in the Martian environment today? If they could then it would bode well for the possibility of early Martian life surviving through to today.

The surface of Mars is not hospitable to most of the life on Earth today, including microbial life. This is the result of one of the ways early Mars lost its ocean—photolysis of water vapour ($H_2O$) generated hydrogen and oxygen radicals that in turn reacted with water and minerals to form perchlorates. (As we will shortly come to, we do find perchlorates on Earth.) Alas, in turn the ultraviolet (UV) photolysis of perchlorates results in chemicals (including hypochlorite, which is the active constituent of household bleach) that are toxic to many forms of life. So it could be that the surface of Mars would be lethal to most prokaryotes or Martian analogues of prokaryotes (*45*). However, we have seen on Earth that prokaryotes can evolve to live in extremely hostile environments (*e.g.* antibiotic resistant bacteria and extremophiles). Such evolution happens quite easily and quickly: antibiotic resistance arises well within the scale of human lifetime. So adapting to the slower change from an ocean Mars to a drier one over many millions of years, is not beyond the realms of possibility. But, even so, could life survive such conditions?

The Atacama Desert in Chile is one of the driest places on Earth. There, the annual average rainfall is less than 20 mm. This is so low that it reduces weathering rates and leaching losses to below the accumulation rates of atmospheric delivered salts and dust. This means that the amount and concentration of dust and salts increases, so resulting in an extremely dry and high-salt-concentration conditions. Its altitude also means that it has a high UV irradiation which, though not as great as that on Mars due to Earth's thicker atmosphere, is still considerable, as the Earth is closer to the Sun. Consequently, as on Mars, UV light does enable photolysis of chlorates to perchlorates. In this sense, the Atacama Desert is analogous to Mars. Having said all that, rainfall does occur roughly once a decade and also occasionally water condenses directly onto surfaces (as garden dew does). Here, this water could be considered as analogous to the subsurface water and condensed water vapour on Mars.

Yet life in the driest and saltiest parts of the Atacama does exist, as an international collaboration, led by Dirk Schulze-Makuch, Dirk Wagner, and Samuel Kounaves, reported in 2018 (*46*). They found that prokaryotes survived even in the driest areas. Species found included those that belonged to the *Geodermatophilaceae*. They are known to colonise hyperarid habitats and tolerate high levels of oxidative stress, desiccation, salts, and metals. They are also somewhat UV resistant. Additionally, traces of some fungi were found belonging to the *Ascomycota* and *Basidiomycota* but mostly at a depth 20–30 cm below the surface. The Dirk Schulze-Makuch collaboration's results suggest that any ancient, putative Martian life could have possibly survived the transition from Mars's early aquatic stage, through increasing drier cycles, to its present arid state, and perhaps they may even found a briny subsurface niche beneath today's severely dry surface.

Today on Earth, we do find cyanobacteria beneath thin, translucent stones. Morning dew condensing on the stones, though quickly evaporating, sees enough water to moisten the ground beneath them. Meanwhile, sufficient light penetrates the stones for them to photosynthesise. Could there be some Martian analogue to cyanobacteria on Mars?

While the ability for some forms of Earth life to exist under Martian-like conditions is interesting, it does not in itself prove that there is life on Mars. And even if there was life in the early Martian ocean (again, as yet unproven), that conjecture does not establish that there is life on Mars today. Nor does the presence of present-day seasonal methane (which may be produced abiotically) or oxygen (even though its seasonal presence on Mars is mysterious without life). Nothing short of actually seeing life on Mars today will be hard proof that there is native biology on the planet. However, employing the co-evolution of life and planet narrative added to the 'circumstantial' present-day evidence is tantalisingly suggestive as to the genuine possibility of life on Mars.

Before leaving the topic of possible life on Mars, it is worth briefly turning back to Martian life in literature. In the previously mentioned H. G. Wells's seminal novel, *The War of the Worlds* (1898), the invading Martians were finally laid low by Earth's native microbes to which they had no resistance (*21*). Today, NASA's *Curiosity* rover is steering clear of areas most likely to harbour Martian life lest it contaminate the site with microbes it carried from Earth: though the lander and rover were assembled in clean rooms, leaving Earth they were not sterile. We are going to have to go to greater lengths to protect from contamination sites that are potential homes to putative Martian life. Not only does this mean that we need to think carefully about a human-crewed mission to Mars, but also the symmetry here with Wells's novel is notable (*47*).

## The possibilities for life elsewhere in the Solar System

Beyond the Solar System's inner, rocky planets there are the gas giants of Jupiter, Saturn, Uranus and Neptune. The pressures and conditions there make them unlikely homes for life as we know it. However, the possibility of primitive life on some of these gas giants' moons is intriguing, though remote: certainly organic chemicals have been detected.

Of these moons, the largest of Saturn, Titan, has been likened in some respects to a primordial Earth. Like Earth, Titan has an atmosphere dominated by nitrogen, but Titan has higher percentage of nitrogen (98%). Like the early Earth, before its second (or first major) flip of state, Titan's atmosphere lacks oxygen, but it does have some methane (1.4%). So it is possible that some aspects of Titan's atmospheric chemistry had similarities with that of the early Earth's, in particular with regard to the action ultraviolet light (see Chapter 3). Even though Titan and Saturn are over nine times further out from the Sun than Earth, and so only receive 1% of the Solar energy, the ultraviolet (UV) light still causes photolysis of Titan's methane and nitrogen, so generating various organics (such as ethane, diacetylene, methylacetylene, acetylene, and propane, as well as nitrogen organics such as cyanoacetylene and cyanogen), including large organic molecules (*48*), and upper atmosphere tholins (*49*). Because nitrogen and methane molecules photo-dissociate (undergo photolysis) under UV light, so they absorb the UV and in this way protect Titan's surface from UV much as Earth's ozone does today. Indeed, tholins likely played such a protective role on the primordial Earth (see Chapter 3).

However, there the similarities end. With a mass just a couple of per cent of Earth's, gravity less than a fifth, no atmospheric water vapour, and a median surface temperature of $-198\,^{\circ}C$, its surface is inhospitable to life as we know it.

Now, you may be wondering, how come Titan has such a thick atmosphere given that Mars lost its atmosphere despite having a stronger gravity (see the previous subsection)? There are two reasons. First, the temperature of Titan's atmosphere is much lower than Mars's, and so its molecules do not move so fast and so do not have the energy to escape the moon. Second, Titan's orbit is largely within Saturn's magnetosphere and so Titan's atmosphere is part protected from Solar wind ablation. (Also remember, Titan is so much further out from the Sun than Mars that the Solar wind has less strength.)

Yet there is something of a mystery in Titan's atmosphere. Calculations suggest that the rate of atmospheric methane destruction is such that all the atmospheric methane would be eliminated within several million years. This is much less time than the age of the Solar System, so something must be replenishing the atmospheric methane in addition to the methane lakes detected on Titan's surface (*50*). Some have speculated a subsurface ocean of methane. Others that there is some cycling of methane through the upper atmosphere's organics and tholins as UV light would break these down. One possibility that has been considered is that life is replenishing the methane.

This notion is not entirely in the realms of fantasy. It is possible that, irrespective of a subsurface liquid methane layer, there is a watery ocean (possibly containing ammonia) and that this might harbour life. This ocean would be kept liquid by both the presence of ammonia (which acts as an antifreeze) and the pressure of ice above it. It is thought it might have an average temperature of around temperature of 176°K (−97°C) (*51, 52*).

Applying our co-evolution of life and planet narrative, it is unlikely that oxygenic photosynthesis evolved in a subsurface ocean, as the ocean would be cut off from sunlight, which in any case would be less than the 1% of solar energy received by Titan which would be further diluted by the atmospheric methane and tholin haze. Further, any subsurface oceanic life would not see ultraviolet light that is thought was possibly required for the formation of the precursors to both DNA and RNA on Earth, likely in terrestrial thermal springs (see Chapters 1 and 2) (*53, 54*).

Yet, as the rise of life on Earth seems to have been an 'easy' evolutionary step (see Chapter 1), this does not preclude a different, 'harder' route to life. Though carbon-based, this would not be life exactly as we know it.

So, a different type of life might possibly exist on Titan's surface. This life, it has been mused, might use hydrogen as an electron donor in a variation to the reaction discussed in Chapter 2, and respire using hydrogen in place of oxygen, reacting it with acetylene instead of sugars, and exhale methane instead of carbon dioxide. But at such cold temperatures, it would have to use methane or ethane as a solvent and not water, which life uses on Earth (*55*). Other than Earth's early life likely using hydrogen, the only thing otherwise possibly analogous would be that putative Titan's life might be a chemolithoautotroph. But given Titan's layered structure of core, ocean, possible methane layer, and a water-ice surface with liquid methane/ethane lakes, it is difficult to see how such surface life would access nutrients from minerals from its core. Also, while the respiration reactions would in theory, energetically work, they would have to operate within a different, functioning biochemical system, and we have no idea what that might look like!

So, once more, with regards to Titan's surface and applying our co-evolution of life and planet narrative, it is unlikely that life, recognisably as we know it on Earth, exists on its surface. It seems that the best Titan might be is in providing a laboratory for some of the atmospheric chemistry that took place on the primordial Earth.

One place that might harbour life is the smallest of the moons of Jupiter Galileo Galilei discovered in 1610 (*56*): Europa. Interest in Europa peaked with the two *Voyager* flybys in 1979; these roused considerable public interest especially when astronomers considered the possibility of it being a home for life. This was further fuelled in 1995 to 2003 when the *Galileo* probe explored Jupiter space (*57, 58*), and from 2016 when *Juno* reached Jupiter. Indeed, the author Arthur C. Clarke, who you will recall has a degree in physics and an interest in astronomy, employed the notion of extraterrestrial intelligence protecting and nurturing life on Europa in his novel *2010: Odyssey Two* (a novel that was, Hugo Award winningly, cinematically adapted in 1984). Humanity is told of the Solar System 'All these worlds are yours—except Europa. Attempt no landings there' (*59*). Here, if you recall from earlier the concern over contaminating putative Martian life, so seriously is Europa considered a possible home for life that at the end of its mission the *Galileo* orbiter was deliberately crashed into Jupiter (on 21st September 2003) to avoid any possibility of it biologically contaminating the moon.

This brings us to the question as to why Europa is considered sufficiently hospitable for life. Indeed, at first glance, it seems an unlikely place to sustain any biological forms. With a diameter of close to 1,900 miles (3,110 km) it is slightly smaller than the Moon and has a surface gravity to match. Like the Moon, Europa has a negligible atmosphere, near to vacuum, but what it has is mainly oxygen. This oxygen, unlike that of Earth's which is of biological oxygenic photosynthesis, comes from photolysis. (We will return to this shortly.)

With regard to the moon's surface water, unlike the Moon (other than in the permanent shade of polar craters), Europa has retained its water and is covered by an ice shield. This retention is because Europa's surface water froze: its equatorial surface temperature averages −160°C (it is even colder at its poles). At this temperature, water ice is so cold that it remains as ice with physical properties more like those of granite. The water ice also reflects sunlight so making the surface cooler than that of Titan which also benefits from some greenhouse gas warming.

Yet while Europa's surface is distinctly inhospitable, beneath its ice shield conditions are very different, it is warmer and the water exists as liquid, not ice. This is because Europa orbits Jupiter in resonance with fellow Jovian moons, Ganymede and Io. The changing gravitational tug from these other moons on Europa as it orbits Jupiter, results in tidal stress, hence internal warming, and this sustains a warmer subsurface ocean of water (*60*). It could be that the ice shield is several miles thick covering an ocean 60 or so miles deep (*61*). If this is so, then the volume of Europa's subsurface ocean would be three times that of all of the Earth's.

However thick the ice shield, what is known from the Hubble space telescope observations and the *Galileo* probe, is that there are places where it is fractured so allowing the subsurface ocean waters to erupt in geysers that rise in a plumes to ~125 miles (~200 km) above the surface (*62*). These geysers originate from tidal-stress cracks in Europa's ice shield and are geologically short-lived with water welling up from the subsurface ocean freezing.

Furthermore, it might possibly be that Europa's subsurface ocean is oxygenated. Though Europa receives less sunlight than Mars, its negligible atmosphere means that UV and energetic particles can split Europa's surface water ice molecules. (Here, the energetic particles come from Jupiter's magnetosphere, as Europa's position within the magnetosphere is such that instead of protecting the moon from Solar particles, it concentrates them.) $H_2O$ dissociates, through photolysis, into hydrogen (which escapes) and hydroxyl radicals that in turn forms oxygen molecules ($O_2$) and hydrogen peroxide ($H_2O_2$) (*63*). (This is effectively the same process discussed in Chapter 5.) Oxygen on the surface of the ice then escapes and forms the very thin, ionised oxygen atmosphere, but oxygen within the surface ice remains trapped. This oxygen may well then be slowly transported to the ocean by two mechanisms. First, as water vapour geysers replenish the surface with fresh ice, so the surface becomes buried. Over time ice moves from the surface towards the base of the ice shield where it melts on contact with the subsurface ocean and any oxygen transported this way is released. Second, the surface ice shield is completely separated, detached, from Europa's solid crust by the ocean. This has been deduced from the cracks in Europa's ice shield that should orientate with respect to the plane of the Galilean moons' orbits about Jupiter. That the cracks show a number of different orientations is thought to be due to the ice shield moving independently over the subsurface ocean with respect to Europa's solid surface beneath. It also allows for the possibility that ice shield movement, somewhat roughly akin to plate tectonics on Earth, may take place with areas of subduction. This subduction would also carry surface ice down to the ocean.

The process of oxygen transportation to the ocean would be slow. It has been estimated that to oxygenate the ocean to the level of Earth's oceans would take many millions of years (*64*). This would not be fast enough a replenishment rate to sustain anything but an extremely small population of oxygen-respiring species.

Leaving the question of an oxygenated Europan subsurface ocean aside for a moment, a non-oxygenated subsurface ocean could still provide a home to sustain life.

In terms of the co-evolution of life and planet narrative, if Europa's tidal flexing provides enough energy for there to be abyssal thermal vents into its subsurface ocean, then hydrogen and other volatiles might sustain putative microbes analogous to prokaryotes and there might be chemolithoautotrophs.

There is, though, a problem with the rise of life in the first place. Much depends on whether the UV-mediated nucleic acid precursor synthesis, that has been presented as the route to life, is the sole, likely synthesis mechanism. If it is, and as there is no UV light in the subsurface ocean, then life would not rise on Europa even if its ocean could support it. Conversely, if there are other nucleic acid precursor synthesis mechanisms that did not require UV mediation, then the rise of life on Europa would be possible. We simply do not know. Yet, given that the reactions that have been theorised also give rise to the handed amino acids life on Earth uses, alternate mechanisms would likely give rise to alternate amino acids and handedness. Europan life, if it exists, could be biochemically very different from life on Earth or even putative Martian life.

What we can say is that, with no light reaching the subsurface ocean, there would not be the possibility for oxygenic photosynthesis (OP) to evolve: you will recall that OP needs the higher-energy, visible light photons in addition to the bolting together of two photosystems (see Chapter 3). However, if there are thermal vents then photosynthesis using lower-energy infrared would be possible. Though such photosynthesis would not sustain as large a biosphere (one with more net primary production—see Appendix 2).

The question of the rise of something analogous to eukaryotes is harder. On Earth, the formation of land enabled acidic erosion and the release of nutrients, and the formation of supercontinents and their breakup allowed for nutrients to build up in the narrow inland seas and so spur evolution. On Europa there is no land (other than the ice shield) as such, so no acid-facilitated erosion as on Earth. The question is, therefore, are there other nutrient delivery mechanisms and cycling taking place on Europa? We simply do not know, but if there is, then something akin to eukaryotes are in theory a possibility especially that given on Earth we know it happened a number of times and so was an 'easy' evolutionary step (Chapter 4).

According to the co-evolution of life and planet narrative presented earlier, the rise of true multicellularity was also an 'easy' evolutionary step (Chapter 4) but it did require oxygen even if only at the level of around 4% (see Chapter 5). Without OP the Europan subsurface ocean would have to rely on oxygen produced by photolysis from the surface being delivered. As noted, such delivery is a slow process, so even if after many millions of years the ocean became oxygenated to the level of Earth's today, as soon as any Europan life evolved oxygen respiration to deliver the energy true multicellularity needed, then the subsurface ocean oxygen level would be depleted to the point where consumption equalled the slow delivery rate. This makes true multicellularity on Europa unlikely.

For all these reasons, life on—or in—Europa seems rather unlikely.

Yet, if life got going on Europa—note the 'if'—something more or less akin to prokaryotes on Europa is a distinct possibility. Nor could we totally rule out the eventual evolution of something analogous to eukaryotes. However, there is a limit to stretching the wildest of credulity: true multicellularity is most unlikely.

The final Solar System body in Table 7.1 is Enceladus. Like Titan, Enceladus is a moon of Saturn, only Enceladus is one of Saturn's inner moons. Like Europa, it has a negligible atmosphere, hence has no benefit from greenhouse warming. Also like Europa, it is covered by ice that gives it a high albedo (great reflectivity). Both these factors make its surface cold with a mean surface temperature of −198°C (75°K).

Being an inner moon (close to Saturn) and being in a 2:1 orbital resonance with Dione (another moon) causes repeated, varied tidal stress that serves to warm Enceladus's interior. While the degree of Enceladus's interior warming is currently something of a mystery—calculations of the various warming factors simply don't add up—the moon seems to have a southern hemisphere sea beneath an ice shield. Fractures, caused by the changing orbital tidal stress, allow water to escape through geysers (65).

In 2015, NASA's *Cassini* probe flew through Enceladus's geysers detecting methane ($CH_4$), carbon dioxide ($CO_2$), ammonia ($NH_3$), molecular nitrogen ($N_2$), and molecular hydrogen ($H_2$). In addition, carbon monoxide (CO) and ethene ($C_2H_4$) were found among other substances including organic compounds with atomic weights exceeding 200 (66). The presence of hydrogen and methane is intriguing for biologists as, you will recall, some chemolithoautotrophs can metabolise hydrogen for energy (see Chapter 2). So if both methane and hydrogen are present in Enceladus's geyser plumes could this signal life? Well, as noted earlier this chapter when discussing Mars, there are both biological and non-biological means of releasing methane from certain minerals. This means that we will need more evidence from future probes before we can conclusively ascertain whether there is life in the Enceladus subsurface ocean.

Disappointing as for now this may be, there is still some circumstantial evidence that, though not conclusive, piques interest: could life as we know it on Earth live under the conditions thought to exist in Enceladus's subsurface ocean? This was the question a collaboration of Austrian and Germans sought to answer in a paper published in *Nature Communications* in 2018 (67). Given that Enceladus has a substantive ice shield, it is thought that the subsurface ocean must have an average temperature below 100°C (if it were above that then the ocean would melt the ice shield).

Olivine hydrolysis at low temperatures is a key process for sustaining chemolithoautotrophic life on Earth, and if hydrogen is produced in significant amounts on Enceladus, then it could also serve as a substrate for biological $CH_4$ production. Methanogenic Archaea are the only known microorganisms that are capable of performing biological methane production in the absence of oxygen. On Earth, methanogens are found in a wide range of the acid-alkali spectrum (pH 4.5–10.2), temperatures (from below 0°C to 122°C), and pressures (0.005–759 bar). So what the Austrian and German researchers did was to grow different species of Archaea in vats under conditions that might exist in the moon's subsurface ocean and in the presence of chemicals in various concentrations around those found in the plume through which *Cassini* flew. They found that two species, called *Methanothermococcus okinawensis* and *M. villosus*, could survive under all the conditions tried. *M. okinawensis* was the best in the most reproducible way when the experiments were repeated.

The researchers also devised a geological model of Enceladus and found that if there were life there that was somewhat analogous to *M. okinawensis* and with its biological methane production, then it would always compete with any abiotic methane generation processes: so any methane production on Enceladus would be from a mix of biological and geological processes. Regardless, their bottom-line conclusion was that life analogous to some of Earth's chemolithoautotrophs is, in theory, possible.

Of course, all this depends on how any putative life on Enceladus arose in the first place. Even if a subsurface ocean on Enceladus (or for that matter Europa) could sustain Earth's prokaryotes should they ever be transported there ('translocated', as ecologists say), for there to be native biota it would need to have first evolved from scratch. If, as has been hypothesised (see Chapters 1 and 2), ultraviolet light is needed for the non-biological reactions forming the precursors to both RNA and DNA then, as the ice shield would prevent this light reaching the vents where life likely originated, life simply would not arise. Furthermore, the absence of periodic drying (as is hypothesised took place on the edges of terrestrial hot springs, and which is also necessary for nucleic acid formation) is another impedance for life arising in gas giant moons' subsurface oceans. Yet we cannot discount other alternate routes to life that may not require UV or indeed the cycling of a primordial soup *via* geysers—so exposing organics to UV as well as freeze-drying—and back to the subsurface water much as happens on Europa. This, though, seems a rather remote possibility.

Yet, even if indigenous life in gas giant moons' subsurface oceans seems unlikely, these places may still inform us as to primordial organic chemistry (the chemistry of largely carbon and hydrogen, and carbon and hydrogen and oxygen, compounds). This might provide further insights as to how life on Earth (and possibly Mars) arose in the first place.

Yet, if life were ever found on Enceladus, or even Europa, then it would be a momentous discovery. If it was, we could be confident that microbes at least would be extremely common across the Galaxy (not to mention other galaxies) and found in many planetary star systems, even if oxygenated photosynthesis and multicellularity were not. Even so, Galaxy-wide more complex, multicellular life, as seen on Earth, is likely to be—by comparison to prokaryotic analogues—a rarity.

## Life beyond the Solar System

As just discussed, any signs of life found on Mars—should it have existed or exist—is likely to be analogous to a prokaryote irrespective of whether Martian oxygenic photosynthesis (OP) arose as it did in Earth's prokaryotic cyanobacteria. There is also a lesser, but (theoretically given our current knowledge) not a zero, probability that hypothetically something akin to prokaryotes might reside in a gas giant's moon's subsurface ocean or sea, but even if that is the case, OP there is unlikely. Therefore, should the signs of life in the Solar System—be it past or present—be clearly detected somewhere else other than Earth, then we might reasonably conclude that prokaryotic analogues are fairly common, if not almost ubiquitous, in stars' planetary systems that have worlds in habitable zones or subsurface seas. Consequently, discussion in this section on the possibility of life in other planetary systems will be confined to the more difficult question of true, multicellular life beyond the Solar System.

Before we go there, we might consider our bias as humans living on the planet Earth.

Given multicellular life now abounds on our home planet, the Earth, discussions of the possibility of life elsewhere in the Galaxy often include the term 'Earthlike planet' as if the Earth somehow represented the pinnacle of probability for life evolving. Though it is natural to bat for the home team, we do need to be wary of this bias. Taking a somewhat Copernican view that there is nothing special about the Earth, and with regards to this book's narrative, that there is nothing special about the Earth as a home to multicellular life, then we might reasonably speculate that of all such planetary homes the Earth is average. That is to say, compared to the Earth, that some multicellular-life-bearing planets are *less* conducive to nurturing multicellular life and some *more* conducive. If the latter, then there must be some worlds out there whose conditions enable multicellular life to better flourish (greater biodiversity), flourish more (greater biomass), and flourish for longer (greater biosphere durability).

That there are worlds less conducive to life is easy to imagine: in our own system Mercury is one; it is decidedly not an Earthlike world. Venus might have been an Earthlike world but it was in the wrong place just that bit too close to the Sun. Mars, on the other hand, is more Earthlike than Venus and was even more Earthlike in its primordial time: as discussed, it might even have given rise to life that possibly might even be extant today. Location (in the case of Venus too close to the Sun) and size (in the case of Mars unable to retain an atmosphere and maintain an ocean) are just two of the variables that determine a world's long-term habitability.

On the other hand, could it be that there are worlds that could better provide a home for life? Such a world might be able to sport more living biomass, and with a greater biodiversity, than the Earth. In 2014 René Heller and John Armstrong suggested that one should not consider worlds to be either habitable or non-habitable: it is not a binary situation whether one is or is not. Instead, they considered degrees of habitability and that even that the Earth might not be the most 'habitable'. They proposed that there might be 'superhabitable worlds' (*68*).

One could imagine a planet the right distance from its star that was perhaps a little bigger than the Earth. A bigger Earth, that gave rise to life as Earth did and which had a similar proportion the same age as our Earth is of land to sea, would have a bigger surface area and so would see more life flourish. However, for a variety of reasons, it could not be too big. For instance, it has been argued that a reasonably larger-than-Earth planet would have an extended hydrogen atmosphere that would protect a nitrogen rich atmosphere closer to its surface from Solar wind ablation (*69*). However, according to the co-evolution of life and planet narrative, it could not be this large, with a stronger gravity, so as to prevent hydrogen escape. We saw, in Chapter 3, how hydrogen escape was necessary during, and for, the critical transition from Earth's nearly completely anaerobic atmosphere to one of just a couple or so per cent oxygen.

Then again, just as the world's location about its star and its size are factors in its habitability, so is the duration of its habitability. Mars, if it ever was a home for prokaryotes, could only have been properly so back in its earliest days when it had an ocean. It is certainly less habitable now even if we can't preclude the possibility (or even likelihood) of subsurface life today: there was a comparatively short time, compared with the Earth, when Martian life might have thrived. But could it be that there are superhabitable worlds where life had a longer-lasting window of opportunity than on Earth?

Before we can answer this, we need to ascertain how long can we expect life to be sustainable on Earth? Here the key determinant may well be the stability of the Earth system discussed in Chapter 2. You will recall from that chapter, and also Chapter 3, the brief mention that throughout the Earth's history, the Sun has increased its output (grown warmer): it has not remained the same! This slow increase in output is common in all main sequence stars (see Appendix 2). It is thought that around 2.5 to 3 billion years ago the Sun's luminosity was between 80% and 85% of its current value and that when the Sun was born it was only about 70% as bright as it is today. With regards to the Earth, it is likely that there were higher concentrations of both the greenhouse gases carbon dioxide and methane, and also possibly more nitrogen in the atmosphere (which slowly became included into minerals), and that this larger atmosphere with more greenhouse gases countered the paradox of the 'faint Sun' (*70*). The two Snowball Earths excepted, homeostatic feedbacks (discussed in Chapter 2) kept conditions on Earth comparatively stable—at least stable enough for life to flourish. But as we move into the future, the Sun will continue to warm. In about 600 million years it will give off roughly 5% more energy than today.

There are a number of feedbacks that help stabilise the global climate. One of the key ones is based on silicate weathering (see Chapter 5). As the Earth warms, so there is more evaporation from the ocean which washes carbon dioxide out of the air (as carbonic acid) to react with silicate rocks that traps the carbon as carbonate. A cooler Earth sees less rain, hence less silicate weathering, hence less carbon drawdown. This reaction is countered by life. A cold world in a glacial sees less photosynthesis hence less atmospheric carbon drawdown, and a warmer world more drawdown. As the Sun slowly gets brighter both these reactions result in more carbon dioxide drawdown, a smaller greenhouse effect and so a cooling counteracting the warmer Sun.

All well and good, but there are limits. Over geological time, as the level of the Earth's atmospheric carbon dioxide gets lower, there will come a time when the concentration will not be great enough to sustain photosynthesis. At that time we will start to see plants dying. The first to go will be the $C_3$ plants ($C_3$ and $C_4$ plants were touched upon in Chapter 6). $C_4$ plants (maize, sugar cane, and other tropical grasses) can continue to photosynthesise with lower levels of carbon dioxide, but as $C_3$ plants include trees and 95% of all plants, the biosphere will be radically different.

It is thought that the crunch time for $C_3$ plants will be in roughly half a billion years and $C_4$ plants a couple of hundred million years or so later than that. (*71*) After this, as the Earth continues to warm, more water vapour enters the atmosphere due to evaporation from the oceans' surface. Water is itself a greenhouse gas, and so, as matters progress, there will come a point at which there is a runaway warming and the very thermal structure of the atmosphere changes. At the moment, as you go higher so the air becomes cooler. Leaving the lower level of the atmosphere and entering the stratosphere, nearly all the water freezes out of the air. This traps water in the lower atmosphere, below the ozone layer and the worse effects of ultraviolet light, which would otherwise photo-dissociate water. Water molecules undergoing such photolysis would see hydrogen escape into space (as in Chapter 3 for the brief time the Earth had neither an ozone layer nor protective tholins). In a hot Earth, there is no such freezing water trap: water splits into hydrogen and oxygen with the hydrogen escaping and there being a net water loss (as happened on Mars).

All this has implications for worlds with large, complex multicellular species. These require oxygen levels close to that of the present-day Earth, and the Earth has only had that since around not quite 700 mya (see Chapter 5). Consequently,

Earthlike worlds, positioned so that they last a long time in their star system's habitable zone as it migrates outwards as its main sequence star warms, will see a biosphere only sport large, multicellular species in its late middle-age onwards.

Now we can return to the question of superhabitability. A superhabitable world would be one in which the duration of its star's time on the main sequence was longer (hence its sun's warming over geological time slower) than our sun, the Sun.

Stars come in a range of sizes, and size is critical to a star's lifetime: large stars 'burn' brightly lasting a shorter time and smaller stars give of less energy, even for their small size, but last a long time on the main sequence. Because small stars last so long there are more of them. Small stars called red dwarfs (or M-type stars) have lifetimes of trillions of years; even the larger red dwarfs have a lifetime of a trillion years. Conversely, our Sun is a mid-range, G-type star with main sequence lifetime of around 10 billion years. Consequently, red dwarfs spend over a hundred times more time on the main sequence than the Sun.

Given all this, you might think that a superhabitable world would be one orbiting a red dwarf. Unfortunately, this is not so for two reasons. First, red dwarfs are so cool that a world in its habitable zone would orbit closely to it and quickly become tidally locked. One side of such a planet would continually face the star, baking, and the other side in freezing, permanent night.

At least that was the view commonly held. In the Solar System, its second planet, Venus, is almost tidally locked with the Sun but isn't quite: its rotation is longer than its year and a day on Venus is equivalent to 243 Earth days. (The Earth, being further out from the than Venus, escapes being almost tidally locked.) The reason for Venus not being absolutely tidally locked is due to its atmosphere. It has a non-instantaneous thermal response to mid-day: there is a delay between noon and when the atmosphere becomes the hottest time of day. This in turn causes a bulge in the atmosphere that is offset to one side of the line between Venus's core and the centre of the Sun. This bulge generates a tidal drag that knocks the planet out of being exactly tidally locked.

In 2015, the possible implications for all this on an Earthlike planet orbiting a red dwarf in its habitable zone, were calculated by a small team of astrophysicists from the University of Toronto led by Jérémy Leconte and Hanbo Wu (*72*). They showed, in theory at least, that an Earthlike planet did not need to have as thick an atmosphere as Venus's to prevent becoming fully tidally locked: an atmosphere only as dense as Earth's would suffice. This in turn would mean that there being no permanent night side, the planet's water would not end up trapped as ice on the night side leaving the day side desiccated. In short, in terms of the narrative discussed in the preceding chapters of co-evolution of life and planet, that prokaryotic life might indeed rise on an Earth-sized planet in a red dwarf's habitable zone.

Next comes the question, could life, let alone oxygenic photosynthesis (OP), evolve on such a world?

Alas, here there is a problem. M class, red dwarf stars are red. This means that they give off very little, barely any of the ultraviolet light it has been hypothecated as necessary to mediate the chemical reactions necessary for the formation of the precursors to both RNA and DNA (see Chapters 1 and 2). Yet, even if there is another way to generate these chemicals required for life as we know it, there is another problem with red dwarfs being red.

Red dwarfs being red fundamentally alters the way photosynthesis can evolve from being anaerobic to being OP compared to Earth bathed in more white light. As you will recall (Chapter 2), the energy contained in photons of light depends on their frequency (colour): frequency being inversely proportional to wavelength, so high-frequency photons have shorter wavelength and *vice versa*. Infrared light contains less energy *per* photon than red light, which in turn contains less energy than blue light. Blue light of a wavelength of 500 nanometres has an energy of 2.48 eV *per* photon, whereas red light with a wavelength of 700 nm has only 1.77 eV/photon (*73*). Yet, as discussed (Chapter 3) OP requires considerable energy and so uses visible light, including blue light, as well as the bolting together of two photosystems. The trouble is that red dwarf stars do not give off the higher-energy blue light needed for nearly all OP on Earth.

There is one hypothetical way around this, and that is just as on Earth cyanobacteria bolted together two prokaryotic photosystems, so an OP organism on a world around a red dwarf might bolt three photosystems together. If this was possible, it would take more time to evolve than OP did on Earth: presumably life would have to bolt two photosystems together first before adding a third later. Fortunately, as said, red dwarf stars are longer-lived compared to our Sun (which is a brighter G-type star).

This raises the question, why did a three photosystem photosynthesis never evolve on Earth? Most likely this is because a two photosystem photosynthesis works fine with light of our star's (the Sun's) spectrum and once established could not be biologically out-competed: three photosystem photosynthesis would require a greater proportion of the energy it harvests for maintenance if only to ensure that the connection between the first two systems was not interfered with by the connection between the second and third. Therefore three photosystem oxygenic photosynthesis—assuming it is possible (and here the discussion is highly speculative)—could only work on planets such as those around red dwarfs, where two photosystem photosynthesis could not provide the energy to split water.

If OP did evolve on a planet orbiting a red dwarf—and that is a big 'if'—then something analogous to eukaryotic life would be a reasonable likelihood. As was noted (see Chapter 5), symbiogenesis has taken place a number of times and is an easy evolutionary step, so *if* OP did arise then something analogous to eukaryotes would be likely in a red dwarf system with a suitable planet.

There is, though, another problem—in addition to extremely long days and nights in a near-tidally locked world—with life on a terrestrial planet around a red dwarf: red dwarfs are highly variable stars. Their output can vary by as much as 40% for months at a time. These factors together serve as to work against the likely evolution of large, complex, multicellular, land-based animals.

This discussion is all very hypothetical, but we might in a few years' time get an answer as to life around a red dwarf star. The European Space Agency (ESA) has launched its CHEOPS observatory. CHEOPS stands for 'CHaracterising ExO-Planet Satellite'. Its mission will take spectra of starlight as it passes through the atmosphere of an intervening exoplanet and so enable elucidation of the planet's atmosphere's chemical composition. If oxygen (ideally with traces of methane) is detected, then OP will be a distinct possibility. (The presence of methane along with greater proportion of oxygen, or *vice versa*, would be compelling as one would react with the other: both would indicate something was replacing these molecules and life would be the likely candidate.)

A type of star that is better suited than are red dwarfs to be the home to a superhabitable planet is a K-type main sequence star. These are warmer than M-type red dwarfs hence have a shorter time on their main sequence. Being warmer they are orange with more yellow and some green and a little blue light in their spectra and so should support OP species with just two photosystems bolted together. Also being warmer, exoEarths in the habitable zone around a K-type star are not so close to their star and so may escape the fate of (near) tidal locking. They are cooler than G-type stars like the Sun, hence are longer lived: their main sequence lifetimes are 20 to 70 billion years, compared to 10 billion for the Sun. Having said that, there may be a problem in synthesising the precursor molecules for life as they produce less UV than G-type, Sun-like stars. K-type stars are, nonetheless, of considerable interest to those involved with exobiology (*68*).

There are stars hotter than the Sun, but these have an even shorter life. Consequently, while prokaryotic life might abound on some planets around these stars, it would be touch and go whether their lifetime was long enough for oxygenic photosynthesis to evolve, or evolve and transform a reducing biosphere to an oxidising one. Similarly, complex multicellular life would be unlikely.

Taking a pessimistic—if not a Copernican—perspective, it could be that Earthlike planets orbiting a G-type stars are the only ones capable of generating OP hence multicellular life. It might be that G-type stars are in a stellar sweet spot: they are bright enough, with a spectrum to match, to have OP and yet just long-lived enough on the main sequence to allow for the rise of complex multicellular life.

Optimistically, we might find complex, multicellular life around K-type stars.

Alas, either way it seems that the majority of stars are unsuitable for the evolution of complex life.

Looking at our galaxy (The Milky Way) we can see in our spiral arm that red dwarf stars are the most populous. G-type stars are very much in the minority and even adding in K-type does not even allow them to rival the number of red dwarfs.

Figure 7.1 is a histogram of the stars within a little over 30 light years (10 parsecs) of the Earth. It reflects a broader picture that the Sun is brighter than 85% of the stars in the Galaxy. About 1 in 13 of the main sequence stars in the Solar neighbourhood are G-type stars and roughly 1 in 8 K-type stars. Nonetheless, the Galaxy is big. Back in the 1970s it was thought that the Galaxy was at most nearly 100,000 light years in diameter with a hundred billion stars (*76*). Since then estimates of its size have increased due to better telescopes and a new, outer spiral arm has been discovered which puts the diameter of the Galaxy in excess

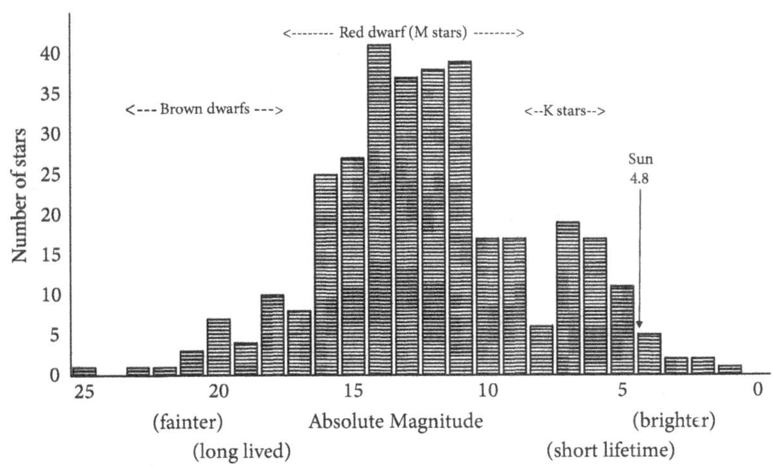


**Figure 7.1** Star types in our part of the Galaxy, within 10 parsecs (32.6 light years) of the Sun.

Adapted and modified from Lenton, T., & Watson, A., *Revolutions That Made the Earth*. Oxford University Press: Oxford, Great Britain (2011). Data from RECONS (the Research Consortium on Nearby Stars).

of 160,000 light years and twice as many stars: some suggest four times as many (*77*, *78*). There could be as many as 200 million main sequence K- and G-type stars combined.

Having said this, not all of the Galaxy is habitable. Too close to its centre and there are too many supernovae in the densely packed star population for there to be enough stability for long enough for complex, multicellular life to arise. Conversely, too far from the Galactic centre and there has not been enough stellar cycling for sufficient heavy elements (heavier than hydrogen) for Earth-like planets and life. Nonetheless, it could be that there are over 100 million stars in our Galaxy that could—given the presence of an Earth-like planet in their habitable zone—harbour life. Galactic habitable zone theory suggests that roughly three-quarters are older than the Sun and on average these would be about a billion years older (*79*). As we shall shortly see, these worlds would be too old to any longer harbour complex life: their G-type suns would be so far through their main sequence that once-habitable Earthlike planets would now be too hot for life. So even if the over-100 million stars that-could-harbour-life were all G-type (some would be longer-lived K-type) and three-quarters were too old (and this is an average—not all would be) that would still leave us with somewhere around over 25 million suitable stars. Even if 99% of these did not give rise to complex life with intelligence and eventually technology, that would still leave, very conservatively, 250,000 that potentially could.

It is sobering to think from the perspective of our arm's-length journey through the Earth's existence (Chapter 1.) that our post-industrial revolution has been going for a couple of centuries yet only represents a length smaller than that of a bacterium on the tip of the finger (1/500 billionths the length of the arm). On that basis, even with a couple of hundred million K- and G-type stars, there may not even be a handful of technological civilisations in the Galaxy. Even if technological civilisations lasted for hundreds of millions of years (and were uniformly distributed) then the nearest would be over a thousand light years away (assuming it had not migrated anywhere). Much depends on the longevity of technological civilisations. (This last we will return to shortly and will also be briefly discussed in the final chapter.)

Now, I do not want to get into questions of galactic ecology here (far too speculative), but if the co-evolution of life narrative has any substance, then in the Galaxy simple species of life analogous to prokaryotes should be extremely common, analogue eukaryotes less so, complex multicellular life being something that is comparatively rare, and extant technological civilisations extremely rare (*80*).

All this might explain the Fermi Paradox—the likelihood, yet apparent absence, of technological civilisations other than on Earth—in two ways: (1) the afore's rarity (which assumes they had not undertaken a diaspora) simply because extant technological civilisations are extremely rare; (2) they could predict—through a more advanced understanding of the co-evolution of life narrative—worlds likely to give rise to complex multicellular life and eventually sentience. In doing so, they may give these worlds a wide berth on biological conservation and ethical grounds.

First, a word on the Fermi Paradox. It is named after the Nobel laureate Enrico Fermi. Around 1950, he and three colleagues (Emil Konoponski, Edward Teller, and Herbert York) who had all worked on the Manhattan Project developing the United States' nuclear bomb, were discussing a cartoon of an alien and a flying saucer. This led to Fermi posing the question, where is everybody? (Meaning, where are the aliens?) Apparently, they were discussing whether faster-than-light travel was possible and concluded it was not, because if it was we would have firm evidence of aliens today or alternatively have definitive evidence of past visitations (a fossilised *Star Trek* type tricorder or something). They concluded that either: (1) aliens did not want contact; (2) no faster-than-light travel is possible and they are too far away if going slower than light; or, (3) there are no advanced civilisations interested or capable of interstellar travel either in person or by probe (*81*). Some other discussion on the Fermi Paradox also concludes that technological civilisations are capable of destroying themselves (such as with nuclear weapons) and so do not last long. However given our galaxy's size and age, even travelling at 1% or 2% the speed of light, it would be possible to traverse the Galaxy scores of times in its history to date and so colonise it to such an extent that any aliens would be bound to be in our galactic neighbourhood even if that—on the astronomical scale of things—meant they were just a couple of hundred light years away.

In addressing the Fermi paradox, much depends on the actual longevity of technological civilisations. Yet, assuming they were more advanced than us and their civilisation had longevity, as alluded to above, they would have a more developed understanding of exobiology and a more refined understanding of the co-evolution of life and planet narrative. They would be able to predict which planets were capable of a biosphere that might spawn complex multicellular species that might, in time, engender a technological civilisation. Such worlds might be off limits, just as today we set aside nature reserves. Similarly, they would be able to identify worlds that while unsuitable to bring forth their own technological civilisation, might be able to have their biosphere tweaked to allow them to be colonised.

Such discussion is, of course, highly speculative (though it will be briefly returned to in Chapter 9). What, though, is clear from the co-evolution of life and planet narrative—with biospheres crossing critical transitions, flipping through various states as they develop—is that while life analogous to prokaryotes may be reasonably common, the more advanced (more Earthlike or superhabitable) the biosphere type the less common they are and (leaving aside the thorny question of persistence), so that technological civilisations rarely arise.

In the context of the Fermi Paradox, this rarity, it is argued, is due to filters (Fermi Filters) or even a 'great filter' a term coined by Robin Hanson in 1998. (*82*) These 'filters' are factors that prevent, or hinder, the rise of extraterrestrial life capable of communicating with us. Over the years there has been considerable discussion and debate regarding the Fermi Paradox and the consideration in identifying possible Fermi filters; too much to go into detail here. However, should you wish to check out the range of the debate then for a sober assessment you might see some the relevant 'alien life' episodes of New York's City University's, astrophysicist Matt O'Dowd's *PBS Space Time* (*83, 84*).

Other than an astronomical catastrophe (such as a nearby supernova) in general, Fermi filters can be assigned to one of two categories: 'early' or 'late'. Examples of early filters include the difficulty or improbability of: life arising, photosynthesis, and the evolution of eukaryotes as well as multicellular life. However, according to the co-evolution of life and planet narrative, considering each of these individually as significant Fermi filters does not make sense. Life got going very early in the Earth's history, shortly after if not hypothetically before, the Late Heavy Bombardment (Chapter 1), so the rise of life, shift from geochemistry to biochemistry, if you will, seems to be an 'easy' evolutionary step. Photosynthesis arose a number of times (Chapter 2), only oxygenic photosynthesis (OP) evolved once (Chapter 3). So photosynthesis per se is also an easy evolutionary step, only OP may be a difficult step: though maybe it was such a successful innovation that once it arose it out-competed any embryonic rivals. (Indeed having evolved once, OP has been refined many times: there are over 60 lineages of $C_4$ plants (*85*). Similarly, both endosymbiosis for eukaryotes (Chapter 4) and true multicellularity (Chapters 4 and 5) occurred a number of times. The only genuine argument for there being a biological early Fermi filter is that each and every one of these evolutionary steps has to occur and also they must take place in strict sequential order. It is taking them all together that takes the time and so possibly makes them a difficult collective step to make within a certain time frame. This, for the Earth, has been of the order of 5 billion years.

Even so, ignoring a technological-wielding species, with the Earth we have had multicellular life on both land and in the oceans for over 600 million years and are likely to have it for another 700 million years (*70*). Combined, this is over 1.3 billion years of potential complex multicellular lifetime or over 17% of the Sun's main sequence lifetime. (Note: This discussion assumes a uniform star age distribution, which is not the case: the Galaxy has been around for a while. Also, not all of the 75% of suitable stars being older than the Earth would be too old for complex life. Both these factors serve to underestimate potential worlds with complex life.) For longer-lived K-type stars, assuming co-evolution of life and planet timing broadly similar to that of Earth's, a greater proportion of its main sequence time would see complex multicellular life.

Taking our earlier conservative (and again remember we ignored the possibility of longer-lived, more K-type stars) estimate of 250,000 stars in the Galaxy having complex life, and then in turn 17% of those having sufficient time to see the rise of a technology-wielding species, then that gives us the possibility of 42,500 such species extant in our Galaxy today. (Remember from earlier, this assumes a 99% fail rate: if the fail rate was 98%, or even 97%, then this rough, broad-brush figure could be two or three times greater. Conversely, if the fail rate was higher at 99.5%, then the guesstimate figure of 42,500 planets with a technology wielding would be halved.)

Do please treat this 42,500 guesstimate figure with a fairly large pinch of salt. While it is based on a pessimistic estimate on the number of worlds on which a technology-wielding species arises, it is also based on an extremely optimistic assumption as to the civilisation's longevity lasting hundreds of millions of years. Finally, if having a large moon (see Chapter 1) was a prerequisite for the stability needed for a technological species to arise and if the chances of that are very slim—pessimistically say at one in ten thousand—then there would only be a single-figure number of technological species extant in the Galaxy today. At the moment, we simply do not have the data to make even a vaguely reasonable estimate.

Yet even if there were hundreds (let alone thousands) of extant, technology-wielding species in the Galaxy today, such is the size of our galaxy, that we may not detect them. Assuming they had a technology at least as advanced as ours, then their most powerful transmitter would be a twin of the former 1,000 ft antenna, used as a planetary radar, at Arecibo, Puerto Rico. It could detect a similar device at a distance of 30 light years, or 300 light years if both were targeted at each other (*86*). So, even if there were hundreds of alien technologies in our galaxy communicating in the same way as ourselves, then they would be too far away: we may well be blind to them, and they to us (*80*). This might be an explanation for the Fermi paradox if biological early Fermi filters are unlikely. (There are other possible explanations, two of which I have already touched upon—rarity and distance—and will again in the final chapter along with a new explanation for Fermi's paradox.)

The true hurdle that makes for an early Fermi filter would seem to be the rise of a technological civilisation itself. As noted, ours has so far been going for roughly 1/500 billionths of the time our Sun has been on its main sequence to date. So the question arises as to how long a species with technological capability can exist? If such technology-wielding species can exist for tens of million, or even hundreds of millions of years then as we have seen, on a purely random, statistical chance basis, that would mean that there would likely be many extant in our galaxy today (our conservative 42,500 guestimate *only* relates to such long-lived technological civilisations: being so long-lived could be too good to be true).

This brings us on to late Fermi filters: if humanity and life on Earth has not yet seen a Fermi filter, and if humanity is the only technological species in the Galaxy (with something curtailing the longevity of others), then the late filter is ahead of us, for us to encounter in the future. Could it be that, as mentioned earlier, technology-wielding civilisations wipe themselves out by, for example, nuclear warfare?

Back in Enrico Fermi's time when his paradox was first discussed, we had had nuclear weapons for half a decade, so they were science fact; yet the first humans would not step on the Moon for nearly another two decades and so that was then still science fiction. So, back in 1950, the notion that we might destroy ourselves in a nuclear conflict seemed much more realistic than that we might leave the Earth and explore space and so eventually ensure our species' survival.

Self-destruction aside, much of the discussion of late Fermi filters really borders that of science fiction (*87*). As such, any notions thrown up by that debate are unverifiable and so beyond the scope of this book. Nonetheless, if there is a late Fermi filter inhibiting long-lived civilisations, then our pinch-of-salt 42,500 estimate would be greatly reduced.

All of which brings us back to the present day. On our arm's-length scale journey down the Earth's history, we have arrived at our human civilisation. As noted (see Chapter 1), on this scale, the few thousands of years that our species has had writing amounts to the size of a bacterium on the end of our forefinger, and the time since the dawn of the industrial revolution less so.

Yet, one question remains. Just as the biosphere has gone through past critical transitions and changed state, with each transition marking an evolutionary development, is there a transition to come or, indeed, are we in the midst of such a transition now?

## References

1. Clement, H., *Iceworld*. Del Rey: New York, USA (1953).
2. Sagan, C., *The Cosmic Connection*. Anchor Books, Anchor Press/Doubleday: New York, USA (1973).

3. Hussmann, H., Sohl, F., & Spohn, T., Subsurface oceans and deep interiors of medium-sized outer planet satellites and large trans-Neptunian objects. *Icarus*, **185** (1), 258–273 (2006).
4. Landis, G. A., Astrobiology: The case for Venus. *Journal of the British Interplanetary Society*, **56** (7–8), 250–254 (2003).
5. Dartnell, L. R., Nordheim, T. A., Patel, M. R., *et al.*, Constraints on a potential aerial biosphere on Venus: I. Cosmic rays. *Icarus*, **257**, 396–405 (2015).
6. Kasting, J. F., Pollack J. B., & Ackerman T. P., Response of Earth's atmosphere to increases in Solar flux and implications for loss of water from Venus. *Icarus*, **57**, 335–355 (1984).
7. Donahue, T. M., New analysis of hydrogen and deuterium escape from Venus. *Icarus*, **141**, 226–235 (1999).
8. McKinnon, W. B., Zahnle, K. J.,. Ivanov, B. A., & Melosh, H. J., Cratering on Venus: Modelling and observations, in Bougher, S. W., Hunten, D. M., & Phillips, R. J., eds., *Venus II Geology, Geophysics, Atmosphere and Solar Wind Environment*, University of Arizona Press: Tucson, USA, pages 585–414 (1997).
9. Way, M. J., Del Genio, A. D., Kiang, N. Y., *et al.*, Was Venus the first habitable world of our Solar System? *Geophysical Research Letters*, **43**, 8376–8383 (2016).
10. Way, M. J., & Del Genio, A. D., Venusian habitable climate scenarios: Modelling Venus through time and applications to slowly rotating Venus-like exoplanets. *Journal of Geophysical Research Planets*, **125**, e06276 (2020).
11. Sagan, C., & Morowitz, H., Life in the clouds of Venus? *Nature*, **215**, 1259–1260 (1967).
12. Greaves, J. S., Richards, A. M. S., Bains, W., *et al.*, Phosphine gas in the cloud decks of Venus. *Nature Astronomy*, **5**, 655–664 (2020).
13. Snellen, I. A. G., Guzma Ramirez, L., Hygate, F. P. S., & van der Tak, F. F. S., Re-analysis of the 267 gigahertz ALMA observations of Venus—No statistically significant detection of phosphine. *Astronomy & Astrophysics*, **644**, L2 (2020).
14. Akins, A. B., Lincowski, A. P., Meadows, V. S., & Steffes, P., Complications in the ALMA detection of phosphine at Venus. *Astrophysical Journal Letters*, **907** (2), L27 (2021).
15. Lincowski, A. P., Meadows, V. S., Crisp, D., Akins, A. B., *et al.*, Claimed detection of $PH_3$ in the clouds of Venus is consistent with mesoferic $SO_2$. *Astrophysical Journal Letters*, **908**, L44 (2021).
16. Hallsworth, J. E., Koop, T., Dallas, T. D., *et al.*, Water activity in Venus's uninhabitable clouds and other planetary atmospheres. *Nature Astronomy*, **5**, 665–675 (2021).
17. Gilliam, T., & Jones, T., *Monty Python and the Holy Grail.* Python (Monty) Pictures: London, Great Britain (1975).
18. Turbet, M., Bolmont, E., Chaverot, G., *et al.*, Day–night cloud asymmetry prevents early oceans on Venus but not on Earth. *Nature*, **598**, 276–280 (2021).
19. Kasting, J. F., & Harman, C. E., Venus might never have been habitable. *Nature*, **598**, 259–260 (2021).
20. Plaxo, K. W., & Gross, M., *Astrobiology*, 2nd edition. John Hopkins University Press: Baltimore, USA (2011).
21. Jones, D., 'Introduction' to Wells, H. G., *The War of the Worlds*. Oxford University Press: Oxford, Great Britain (2017 edition), the latter originally published 1898. William Heinemann: London, Great Britain.
22. Clarke, A. C., *The Sands of Mars*. Sidgwick & Jackson: London, Great Britain (1951).
23. Moore, P., *Guide to Mars*. Lutterworth Press: London, Great Britain (1977).
24. Turtledove, H., *A World of Difference*. Hodder Stoughton: London, Great Britain (1998).
25. Connerney, J. E. P., Acuña, M. H., Wasilewski, P. J., *et al.*, Magnetic lineations in the ancient crust of Mars. *Science*, **284**, 794–798 (1999).
26. Stevenson, D. J., Mars core and magnetism. *Nature*, **412**, 214–219 (2001).
27. Orosei, R., Lauro, S. E., Pettinelli, E., *et al.*, Radar evidence of subglacial liquid water on Mars. *Science*, **361**, 490–493 (2018).

28. Usui, T., Alexander, C. M. O. D., Wang, J., Simon, J. I., & Jones, J. H., Meteoritic evidence for a previously unrecognized hydrogen reservoir on Mars. *Earth Planet. Science Letters*, **410**, 140–151 (2015).
29. Ojha, L., Wilhelm, M. B., Murchie, S. L., *et al.*, Spectral evidence for hydrated salts in recurring slope lineae on Mars. *Nature Geoscience*, **8** (11), 829–832 (2015).
30. Carter, J., Poulet, F., Bibring, J. P., Mangold, N., & Murchie, S., Hydrous minerals on Mars as seen by the CRISM and OMEGA imaging spectrometers: updated global view. *Journal of Geophyical Research: Planets*, **118**, 831–858 (2013).
31. Wade, J., Dyck, B., Palin, R. M., Moore J. D. P., & Smye, A. J., The divergent fates of primitive hydrospheric water on Earth and Mars. *Nature*, **552**, 391–394 (2017).
32. Scheller, E. L., Ehlmann, B. L., Hu, R., Adams, D. J., & Yung, Y. L., Long-term drying of Mars by sequestration of ocean-scale volumes of water in the crust. *Science*, **372**, 56–62 (2021).
33. Di Achille, G., & Hynek, B. M., Ancient ocean on Mars supported by global distribution of deltas and valleys. *Nature Geoscience*, **3**, 459–463 (2010).
34. Citron, R. I., Manga, M., & Hemingway, D. J., Timing of oceans on Mars from shoreline deformation. *Nature*, **555**, 643–646 (2018).
35. Seybold, H. J., Kite, E., & Kirchner J. W., Branching geometry of valley networks on Mars and Earth and its implications for early Martian climate. *Science Advances*, **4**, eaar6692 (2018).
36. Goudge, T. A., Morgan, A. M., de Quay, D. S., & Fassett, C. I., The importance of lake breach floods for valley incision on early Mars. *Nature*, **597**, 645–649 (2021).
37. Toner, J. D., & Catling, D. C., A carbonate-rich lake solution to the phosphate problem of the origin of life. *Proc. Natl. Acad. Sci.*, **117** (2), 883–888 (2020).
38. Lovelock, J. E., *Gaia: A New Look at Life on Earth*. Oxford University Press: Oxford, Great Britain (1979).
39. Lovelock, J. E., *The Ages of Gaia: A Biography of Our Living Earth*. Oxford University Press: Oxford, Great Britain (1988).
40. Horowitz, N. H., *To Utopia and Back: The Search for Life in the Solar System*. W. H. Freeman: New York, USA (1986).
41. Formisano, V., Atreya, S., Encrenaz, T., *et al.*, Detection of methane in the atmosphere of Mars. *Science*, **306**, 1758–1761 (2004).
42. Webster, C. R., Mahaffy, P. R., Atreya, S. K., *et al.*, Background levels of methane in Mars' atmosphere show strong seasonal variations. *Science*, **360**, 1093–1096 (2018).
43. Oze, C., & Sharma, M., Have olivine, will gas: Serpentinization and the abiogenic production of methane on Mars. *Geophysics. Research Letters*, **32** (10), L10203 (2005).
44. Trainer, M. G., Wong, M. H., McConnochie, T. H., *et al.*, Seasonal variations in atmospheric composition as measured in Gale Crater, Mars. *Journal of Geophysical Research: Planets*, **124**, 3000–3024 (2019).
45. Wadsworth, J., & Cockell, C. S., Perchlorates on Mars enhance the bacteriocidal effects of UV light. *Scientific Reports*, **7**, 4662 (2017).
46. Schulze-Makuch, D., Wagner, D., Samuel P. Kounaves, S. P., *et al.*, Transitory microbial habitat in the hyperarid Atacama Desert. *Proceedings of the National Academies of Sciences USA*, **15** (1), 2670–2675 (2018).
47. Cowie, J., Correspondence—Martian dance of fact and fiction. *Nature*, **538**, 317 (2016).
48. Niemann, H. B., Atreya, F. K., & Bauer, S. J., *et al.*, The abundances of constituents of Titan's atmosphere from the GCMS instrument on the Huygens probe. *Nature*, **438**, 779–784 (2005).
49. Waite, J. H., Jr., Young, D. T., Cravens, T. E., *et al.*, The process of tholin formation in Titan's upper atmosphere. *Science*, **316**, 870–875 (2007).
50. Stofan, E. R., Elachi, C., Lunine, J. I., *et al.*, The lakes of Titan. *Nature*, **445**, 61–64 (2007).

51. Grasset, O., Sotin, C., & Deschamps, F., On the internal structure and dynamic of Titan. *Planetary and Space Science*, **48** (7–8), 617–636 (2000).
52. Fortes, A. D., Exobiological implications of a possible ammonia-water ocean inside Titan. *Icarus*, **146** (2), 444–452 (2000).
53. Xu, J., Chmela, V., Green, N. J., *et al.*, Selective prebiotic formation of RNA pyrimidine and DNA purine synthesis. *Nature*, **582**, 60–66 (2020).
54. Le Vay, K., & Mutschler, H., A plausible route to the first genetic alphabet. *Nature*, **582**, 33–34 (2020).
55. McKay, C. P., & Smith, H. D., Possibilities for methanogenic life in liquid methane on the surface of Titan. *Icarus*, **178** (1), 274–276 (2005).
56. Robinson, K. S., *Galileo's Dream*. Harper Voyager: London, Great Britain (2009).
57. Kargel, J. S., Kaye, J. Z., Head, J. W., III, *et al.*, Europa's crust and ocean: Origin, composition, and the prospects for life. *Icarus*, **148** (1), 226–265 (2000).
58. Marion, G. M., Fritsen, Christian H., Eicken, H., & Payne, M. C., The search for life on Europa: Limiting environmental factors, potential habitats, and Earth analogues. *Astrobiology*, **3** (4), 785–811 (2003).
59. Clarke, A. C., *2010: Odyssey Two*. Granada: London, Great Britain (1982).
60. Pappalardo, R. T., Belton, M. J. F., Breneman, H. H., *et al.*, Does Europa have a subsurface ocean? Evaluation of the geological evidence. *Journal of Geophysical Research*, **104** (E10), 24015–24055 (1999).
61. Park, R. S., Bills, B., & Buffington, B. B., Improved detection of tides at Europa with radiometric and optical tracking during flybys. *Planetary and Space Science*, **112**, 10–14 (2015).
62. Jia, X., Kivelson, M. G., Khurana, K. K., & Kurth, W. S., Evidence of a plume on Europa from Galileo magnetic and plasma wave signatures. *Nature Astronomy*, **2** (6), 459–464 (2018).
63. Carlson, R. W., Anderson M. S., Johnson, R. E., *et al.*, Hydrogen peroxide on the surface of Europa. *Science*, **283**, 262–264 (1999).
64. Hand, K. P., Carlson, R. W., & Chyba, C. F., Energy, chemical disequilibrium, and geological constraints on Europa. *Astrobiology*, **7** (6), 1,006–1,022 (2007).
65. Porco, C. C., Helfenstein, P., Thomas, P. C., *et al.*, Cassini observes the active South Pole of Enceladus. *Science*, **311**, 1393–1401 (2006).
66. Postberg, F., Khawaja, N., Abel, B., *et al.*, Macromolecular organic compounds from the depths of Enceladus. *Nature*, **558**, 564–567 (2018).
67. Taubner, R-T., Pappenreiter, P., Zwicker, J., *et al.*, Biological methane production under putative Enceladus-like conditions. *Nature Communications*, **9**, 748 (2018)
68. Heller, R., & Armstrong, J., Superhabitable worlds. *Astrobiology*, **14**, 50–66 (2014).
69. Lammer, H., *Origin and Evolution of Planetary Atmospheres: Implications for Habitability*. Springer Briefs in Astronomy. Springer: Berlin, Germany (2013).
70. Lenton, T., & Watson, A., *Revolutions That Made the Earth*. Oxford University Press: Oxford, Great Britain (2011).
71. Caldeira, K., & Kasting, J. F., The life-span of the biosphere revisited. *Nature*, **360**, 721–723 (1992).
72. Leconte, J., Wu, H., Menou, K., & Murray, N., Asynchronous rotation of Earth-mass planets in the habitable zone of lower-mass stars. *Science*, **347**, 632–635 (2015).
73. Hall, D. O., & Rao, K. K., *Photosynthesis—Studies in Biology*, 6th edition. Institute of Biology and Cambridge University Press: London & Cambridge, Great Britain (1999).
74. Nicolson, I., *Astronomy: A Dictionary of Space and the Universe*. Arrow: London, Great Britain (1977).
75. Xu, Y., Newberg, H. J., Carlin, J. L., *et al.*, Rings and radial waves in the disk of the Milky Way. *The Astrophysical Journal*, **801**, 105 (2015).
76. López-Corredoira, M., Prieto, C. A., Garzón, F., *et al.*, Disk stars in the Milky Way detected beyond 25 kpc from its center. *Astronomy & Astrophysics*, **612**, L8 (2018).

77. Lineweaver, C. H., Fenner, Y., & Gibson, B. K., The Galactic habitable zone and the age distribution of complex life in the Milky Way. *Science*, **303**, 59–62 (2004).
78. Cowie, J., The probability of extraterrestrial intelligence. *Biologist*, **38** (1), 9 (1991), text box in Cohen, J., The possibility of life on other planets. *Biologist*, **38** (1), 7–10 (1991).
79. Schrijver, J., *One of Ten Billion Earths: How We Learn about Our Planet's Past and Future from Distant Exoplanets*. Oxford University Press: Oxford, Great Britain (2018).
80. Hanson, R., The great filter—Are we almost past it? http://mason.gmu.edu/~rhanson/greatfilter.html (1998) (accessed October 2021).
81. O'Dowd, M., Why we haven't found alien life. *PBS Space Time*, **1**, 32. PBS Digital Studios. https://www.youtube.com/watch?v=cJONS7sqi0o (2015).
82. O'Dowd, M., Will we ever find alien life? *PBS Space Time*, **4**, 42. PBS Digital Studios. https://www.youtube.com/watch?v=uhJ9lJPt09k (2018).
83. Sage, R. F., Christin, P-A., & Edwards, E. J., The $C_4$ plant lineages of planet Earth. *Journal of Experimental Botany*, **62** (9), 3155–3169 (2011).
84. Regis, E. ed., *Extraterrestrials: Science and Alien Intelligence*. Cambridge University Press: Cambridge, Great Britain (1985).
85. Arthur, I., Fermi Paradox: Exploring the Great Filters–Late Filters. *Science & Futurism with Isaac Arthur*, **5**, 47 (#213) https://www.youtube.com/watch?v=cCzpYxedIe0 (2019).

# 8
# A New Flip?

First, let's discount the obvious. When life ends on Earth there will be another flip of the planet's state. As discussed in the last chapter, as the Sun proceeds though the main sequence stage of its life it will become more luminous and warm the Earth. There will come a time when the carbon feedback mechanisms that regulate the planet's temperature will be overcome: the concentration of atmospheric carbon dioxide will become so low that plant life will be unable to undertake photosynthesis (*1*). Then will begin the process—through a critical transition—possibly akin to that which transformed Venus into the planet we see today. So there is most certainly another flip of planetary state to come.

Discounting this doomsday event, a more interesting question is before then will there be another life-driven flip of planetary state?

Whether or not there will be another, pre-doomsday, flip of the Earth's state in the far future is, alas, unknowable. It is theoretically possible that some evolutionary development may take place that alters the global carbon cycle. Yet as to what that might be would be pure speculation.

Instead, what we can do is look at the present for signs that might portend a near-future change in the planetary state and consider whether it is likely that we are about to encounter, or even be in the midst of, a critical transition to a new state.

This is not an impossible question to address. Knowing the characteristics of past flips of state, we can see whether any, or preferably all, of these characteristics are present today.

Chapter 6 saw the teasing out of the differences between the co-evolution of life and planet flips of state and wild-card or joker events such as the eruption of large igneous provinces (LIPs) and asteroid strikes. Table 6.1 shows that these last did not see a major evolutionary development or innovation even if LIPs and asteroid strikes caused extinction events and subsequent speciation (biological species evolving into other species): there was no new major evolutionary innovation such as oxygenic photosynthesis (OP), possibly eukaryotic OP, or multicellularity arising to influence if not trigger the respective flip of state.

Table 6.1 depicts that these co-evolution of life and planet transitions—assuming that eukaryotic OP was a critical factor involved—also saw some enhancement of information processing. The rise of eukaryotes saw the incorporation of prokaryotic nucleic acid (genetic information) through symbiogenesis

*Earth's Flips of State*. Jonathan Cowie, Oxford University Press. © Oxford University Press (2026).
DOI: 10.1093/9780197825402.003.0008

that ended up as mitochondrial DNA within eukaryotic cells. Subsequently, the rise of true multicellular life also saw enhanced data-processing with intercellular signalling as well as animals having neural networks and brains: both saw life undertaking additional information processing.

Most importantly, these co-evolution of life and planet transitions featured a modification of the carbon cycle on a global scale. Of less significance, but worth noting, it can be argued that these transitions involved the significant colonisation of a new environment. With the first major transition, oxygenic eukaryotes thrived on the sunlight-bathed, sea surface, positively blooming in the inland seas that formed during supercontinental breakup and were fed by nutrient rich run-off from the said supercontinent. (Previously, prokaryotes that could not harvest as much of the electromagnetic spectrum, could not exploit this niche so successfully.) With the second major transition there was the colonisation of land by multicellular life that enhanced nutrient release. (Here, previously life's terrestrial presence was much less: for example, largely confined to terrestrial thermal springs and associated freshwater systems that, being aquatic, were not strictly terrestrial, as well as biofilms on or in proto-soils.) Both transitions (associated with OP and multicellularity) saw life significantly developing in a new environment.

Finally, the discussion in Chapter 6 noted that these critical transitions marked the beginning of a long-term period of new mineral species formation.

Let's look at each of these dimensions and see whether they relate to the present day and/or the foreseeable future.

## The present-day carbon isotope excursion and climate change

That we are currently in a period of global warming is common knowledge. What is less well known is that we are also in the middle of a carbon isotope excursion (CIE). Yet, having seen Chapters 2 and 6, we must now be in a CIE. Current global warming is driven by greenhouse gas emissions primarily from fossil fuel use as well as from farm ruminants and agricultural wetlands (such as rice paddy fields). This carbon, due to its photosynthetic origin, is rich in the carbon-12 ($\delta^{12}C$) isotope and depleted in carbon-13 ($\delta^{13}C$): you will recall, photosynthesis preferentially selects $^{12}C$ and is depleted in the minority $^{13}C$ (see Chapter 1). The consequence is that, just as carbon dioxide ($CO_2$) and methane ($CH_4$) today are building up in the atmosphere, the proportion of that carbon that is $^{13}C$ is decreasing.

The increase in atmospheric carbon dioxide from the late 1950s to 2020 is depicted in Figure 8.1A. The reason this growth is not smooth but wobbles, or cycles annually, is due to the yearly northern hemisphere seasonal cycle of

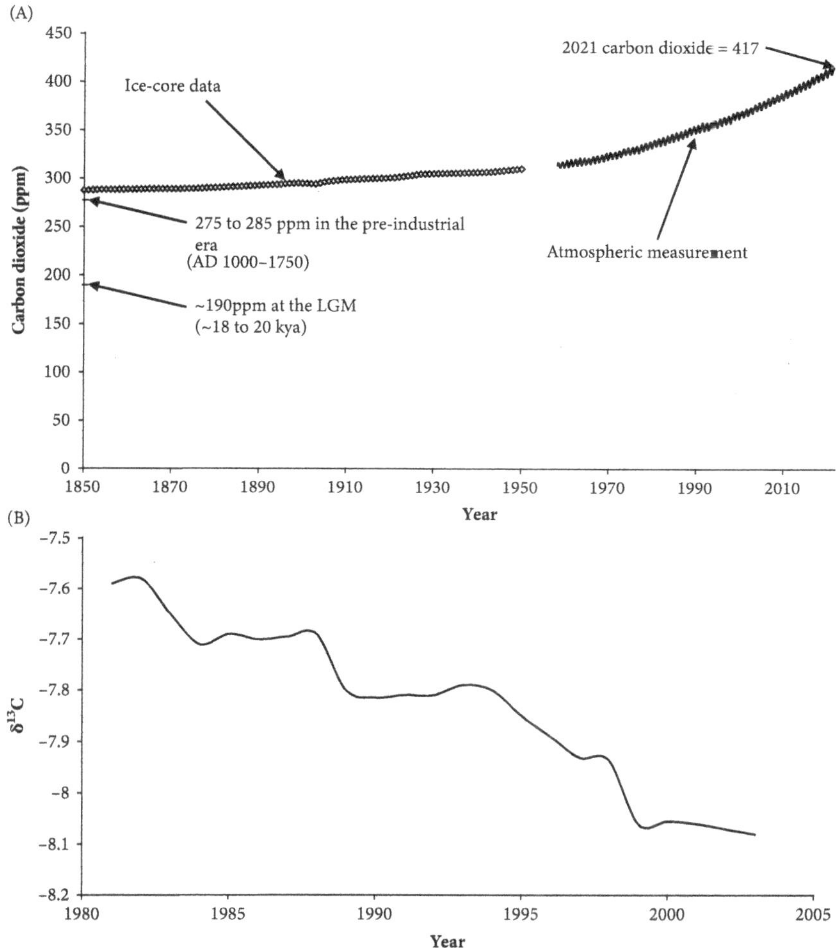


**Figure 8.1** (A) $CO_2$ concentrations (monthly averages) measured in parts per million by volume (ppmv) by continuous analysers over the period 1969 to 2021 from Mauna Loa, Hawai'i. The data points 1850–1950 are taken from atmospheric bubbles in Antarctic ice cores. Also indicated is the concentration at the time of the last glacial maximum (LGM). (Redrawn and updated from ref. *98*.)
(B) Annual averages of the $^{13}C/^{12}C$ ratio measured in atmospheric $CO_2$ at Mauna Loa from 1981 to 2002. The isotope data are expressed as $\delta^{13}C(CO_2)$ ‰ (per mil) deviation from the VPDB calibration standard, see Appendices 1 and 2. Data from http://scrippsco2.ucsd.edu.

plant growth and dormancy: these measurements were taken in Hawai'i in the northern hemisphere. Regardless, the inexorable rise in atmospheric carbon dioxide concentration is plain to see.

That the atmospheric carbon mix of minority $^{13}C$ and majority $^{12}C$ is being diluted by biological carbon that contains even more $^{12}C$ results in a decrease in the proportion of $^{13}C$ in atmospheric carbon dioxide. Figure 8.1B shows this decrease from the 1980s to early 2000s. This change in $^{13}C$ ($\delta^{13}C$) is measured in parts *per* thousand (‰) relative to a standard $^{13}C$ to $^{12}C$ mix (the VDPB standard). That the proportion of atmospheric $^{13}C$ has been declining is also plain to see, even though the rate of change is not constant. This is because the dilution of atmospheric $^{13}C$ by $^{12}C$ is not constant but affected by a number of factors, not least including changes in: fossil fuel carbon emissions; emissions from wetlands and permafrosts, which is related to the climate that year; and the rate of absorption of carbon dioxide by the oceans (which is also climate-related and also affected by changes in ocean circulation). That atmospheric $^{13}C$ is declining puts paid to any argument by climate deniers that the increase in atmospheric carbon dioxide is due to volcanic activity (remember, volcanically emitted carbon is richer in $^{13}C$ than carbon that has been photosynthetically processed).

The bottom line is that this decline in $^{13}C$ clearly shows that we are experiencing a CIE.

All this extra carbon in the atmosphere is having an impact on the global climate, and the theory behind this is very well known, having been so for over one and a half centuries! Despite this, some—including some leading politicians—still debate whether humans are having an impact on the global climate. It is therefore worthwhile summarising the development of understanding of human-induced global warming.

The first person to realise that some gases can trap heat was Jean-Baptiste-Joseph Fourier in the early 19th century. His early career saw him being a civil servant following the French Revolution. It was a perilous period, and Fourier himself was arrested and released a number of times, once barely escaping execution. In 1795, with the worst excesses of the French Revolution over, he managed to get nominated to the École Normale and so establish an academic career.

What interested Fourier was the puzzle, why was the Earth so warm? Why did not incoming heat (sunlight) get reflected back off clouds and out into space, and what kept the Earth warm at night without the Sun? He then sketched out a notion. Perhaps not all the Sun's heat was re-radiated back out into space but some of it was trapped by the atmosphere and radiated back towards the ground. Fourier likened this to there being an invisible dome over the Earth's surface, it being a container as if made of glass but actually of invisible gases and water vapour.

These ideas were presented in an essay which translated to English is titled 'General Remarks on the Temperature of the Terrestrial Globe and Planetary Spaces' in 1824. Then in 1827, almost the same essay appeared as a paper in the journal *Mémoires de l'Académie Royal des Sciences*. These works captured the essence of the concept of the greenhouse effect but without actually using the term 'greenhouse' (though they came close) and so the fundamental science underpinning our modern understanding of our current global warming stems from this time (*2*).

The next significant development in greenhouse gas theory came in 1861 with the Irish polymath John Tyndall. He described the greenhouse role of some gases and succeeded in quantifying their heat-absorbing properties.

Originally, Tyndall was an engineer who worked helping to build England's rail network in the 1840s. Not surprisingly, being an engineer and working on the steam railways, Tyndall had an interest in gases and heat. He set about studying the heat properties of various gases found in the atmosphere, including water vapour, carbon dioxide, and ozone. He published a paper on his conclusions in 1861 (*3*).

But it was the Swedish chemist and Nobel laureate Svante August Arrhenius who, over one-and-a-quarter centuries ago in 1896, first proposed that the human addition of carbon dioxide to the atmosphere would result in global warming, although he himself did not use the benign-sounding term 'greenhouse', but 'hothouse'; an inkling—if one was needed, and it is—of the seriousness of the phenomena (*4*).

At this time (the mid-1890s), the concentration of atmospheric carbon dioxide was around 290 parts *per* million (ppm): remember that it has since steadily risen and today (2021) it is well over 400 ppm. Yet taking the mid-1890s value Arrhenius's calculation was that doubling carbon dioxide concentrations (to near 600 ppm) would increase temperatures by 5°C or 6°C. Similarly, halving it (as happens during the deepest parts of glacials) would lower the temperature by 4°C or 5°C.

Now, this estimate is not that far off (just a degree or two) of the modern Intergovernmental Panel on Climate Change (IPCC) estimates based on state-of-the art computer models (Arrhenius made his lengthy calculations by hand): it is well within the same order of magnitude (power of 10).

Remember that carbon dioxide is not the only greenhouse gas. Methane and CFCs (chlorofluorocarbons) are just two of the others: CFCs are human-made and methane has (like carbon dioxide) had a human-generated contribution to the atmosphere, but Arrhenius did not allow for these (indeed, there were no CFCs in Arrhenius's time). So while carbon dioxide has increased by a little over 30% in the Earth's atmosphere since Arrhenius's time, it only accounts for about half of the degree or so of warming to date that climate scientists attribute to human action. Taking all this into account, we can see that Arrhenius's 1896 estimate was not that bad. Similarly, we can see from the warming that has happened since Arrhenius's time, as well as his calculations, that carbon dioxide will be the largest single greenhouse contributor to the warming through to the end of the 21st century of the few degrees (°C) the UN's IPCC predicts. As such, carbon dioxide we now know is the single most powerful driver of current human-induced warming and Arrhenius was arguably right to focus his attention on it.

Svante Arrhenius deserves recognition for his warming calculations. However, he did not get it all right and—though it seems churlish to criticise him with our privileged position of hindsight—some others of his related predictions were very wrong, as he could not see what was to come in the future.

Using the then-current rates of fossil fuel consumption, he predicted that it would take some 3,000 years for atmospheric concentrations of carbon dioxide to double (*4*). He could not know that, following the 1890s, the global population would soar throughout the next century, especially as it had seemed more stable the previous millennium (see Figure 8.2A). Nor could he know that fossil fuel consumption *per* person would also grow by so much as it did throughout the 20th century to the present. Clearly then he could not realise that 20th century carbon dioxide emissions would increase in such an unprecedented way (see Figure 8.3). Taking these factors together it is easy to see why his carbon dioxide doubling time estimate of 3,000-years was more than a little out.

The other area where Svante Arrhenius's conclusions differ from the modern climate science consensus relates to the environmental effect of global warming. He assumed that the resulting warming would endow the Earth with a more beneficial global climate within which humanity might thrive with less cold winters and the temperate zone extending further north. (It is perhaps worth remembering that Arrhenius lived in Sweden.) Again, this difference between Arrhenius's understanding and the modern climate science's consensus view is easily understandable when one realises what we know today. True, Arrhenius's calculations could help explain ice age glacials but Arrhenius did not know (as biosphere scientists do today) the temperature difference between an ice age glacial and the present (interglacial), nor did he know the *exact* atmospheric concentration of carbon dioxide in glacials (again, as biosphere scientists do today). Had he known these things, then armed with his calculations, he would have realised that doubling the then (late 19th-century) atmospheric carbon dioxide (rather, the warming effect of *all* human-generated greenhouse gases assuming IPCC business-as-usual scenarios) would result in the same degrees Celsius warming from the present as ice age glacials were cooler. This might have alerted him to the likely serious effect on the planet. Again, as noted earlier, he did not know how crowded the planet would become in the 21st century: so much so that it would become difficult for populations to move away from areas made less habitable by climate change and that disruptions to global agriculture might have a serious effect on food supply given today's more finely-tuned, high-input, high-productivity agricultural systems. (This likely undermining of global 'food security' is a threat of which the UN Food and Agricultural Organization (FAO) has warned in recent years [5, 6].)

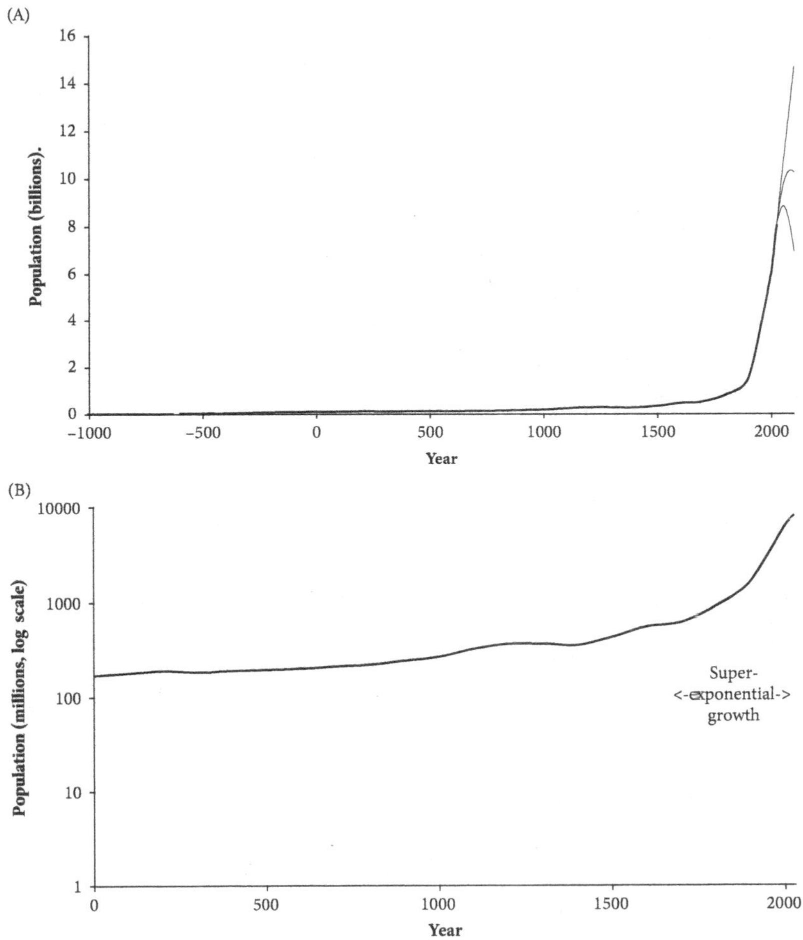


**Figure 8.2** (A) World population over 3,000 years from 1,000 BCE compiled from various demographic, historic estimates presented in a data table by Joel Cohen (Cohen, J. E., *How Many People Can the Earth Support?* Norton: New York [1995]) and future forecasts from the UN Population Division of the Department of Economic and Social Affairs (UN Population Division of the Department of Economic and Social Affairs, *World Population Prospects: The 2022 Revision.* United Nations Secretariat: New York [2022].) Actual population depicted up to 2021. Thereafter future projections are depicted. including the three lines depicting

So while Arrhenius may have been wrong regarding his calculations' implications in the timing of growth in greenhouse gas emissions, the calculations themselves were a fair attempt and, given the lack of data and restricted latitude approach (let alone that he did it all painstakingly by hand without a computer), they were quite an achievement.

Yet, often forgotten is that in the few decades following 1896, this theory was not high on many scientists' research agendas. Indeed, serious doubts arose as to the importance of the increase in atmospheric carbon dioxide in changing the Earth's global climate. However, in 1938 a steam technologist working for the British Electrical and Allied Industries Research Association, one Guy Stewart Callendar, had a paper published in the *Quarterly Journal of the Royal Meteorological Society* in which he noted that humankind had added some 150,000 tons of carbon dioxide to the Earth's atmosphere the past half century (to 1938). This, he calculated, would have warmed the atmosphere by some 0.003°C *per* year. He also looked at (a limited number of) meteorological records. These suggested that the global temperature had increased at an average rate of 0.005°C *per* year the half-century up to 1938 (*7*), which is equivalent to 0.5°C over 100 years.

Guy Callendar was not alone. Three decades later, at the end of August 1972, an atmospheric scientist, J. S. Sawyer, estimated that the warming that might be expected with a continued growth in fossil fuel emissions of carbon dioxide to the year 2000 would be 0.6°C (*8*). This was quite prescient, because less than half the total amount of carbon dioxide released into the atmosphere between the industrial revolution and 2000 was in the atmosphere by 1970. (Doubling 0.6°C gets us to 1.2°C, so less than doubling gets us to under 1.2°C by 2000.) The Intergovernmental Panel on Climate Change (IPCC) in 2023 considered that there was 1.1°C warming since the late 19th century to the decade around 2010 (*9*), so 0.6°C really is not a bad estimate for about half the carbon dioxide. What is more, for a couple of decades up to 1970 the global temperature had actually been declining, making Sawyer's prediction particularly brave as it seemingly went against the grain. So he was close (very much in the ballpark) and, with

**Figure 8.2** *(continued)* the UN's 2010 future projections: low medium and high to 2100. (B) World population from year 0 CE to the present on a logarithmic scale. Had the global population been constant then the graph would be flat (with a 0° slant). If it had grown at a constant exponential rate then this would have been represented as a rising straight line with the degree of angle representing the constant growth rate. However, what the figure shows is an upward curve between the years 1500 and 2000. During this period, the rate of growth was itself growing: this period was a period of super-exponential growth. (Data from Figure 8.2A redrawn by the author with the population on a logarithmic scale.)

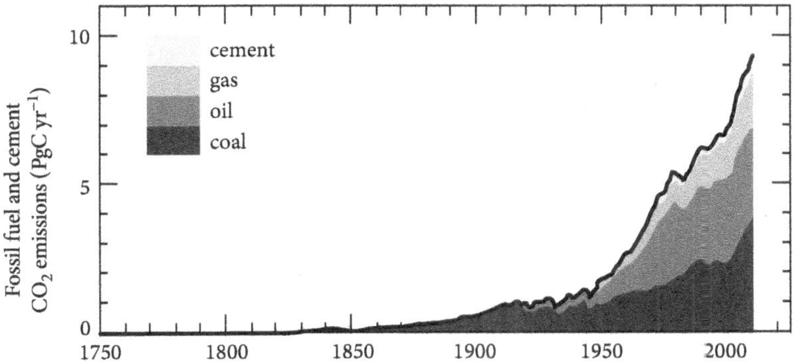


**Figure 8.3** Annual anthropogenic $CO_2$ emissions in petagrams of carbon *per* year ($PgC\ yr^{-1}$) from 1750 to 2011. Fossil fuel and cement $CO_2$ emissions by category. From Figure TS4 of IPCC Technical Summary reproduced with permission. Adapted from Figure TS.4 in Stocker, T.F., Qin, D., Plattner, G.-K., Alexander, L. V., Allen, S. K., Bindoff, N. L., Bréon, F.-M., Church, J. A., Cubasch, U., Emori, S., Forster, P., Friedlingstein, P., Gillett, N., Gregory, J. M., Hartmann, D. L., Jansen, E., Kirtman, B., Knutti, R., Krishna Kumar, K., Lemke, P., Marotzke, J., Masson-Delmotte, V., Meehl, G. A., Mokhov, I. I., Piao, S., Ramaswamy, V., Randall, D., Rhein, M., Rojas, M., Sabine, C., Shindell, D., Talley, L. D. Vaughan, D. G., & Xie, S.-P., Technical Summary. In: Stocker, T. F., Qin, D., Plattner, G.-K., Tignor, M., Allen, S. K., Boschung, J., Nauels, A., Xia, Y., Bex, V., & Midgley, P. M., eds., *Climate Change 2013: The Physical Science Basis. Contribution of Working Group I to the Fifth Assessment Report of the Intergovernmental Panel on Climate Change*. Cambridge University Press: Cambridge, United Kingdom and New York, NY, USA (2013).

the benefit of hindsight, we can now see that between the 1970s and 2000 the global temperature rose by close to 0.5°C (*9*)!

As the greenhouse theory of global warming was becoming developed and robust, so governments started to take an interest.

During US President Carter's term, there was one US greenhouse climate report of particular note: The JASON report of 1979. JASON was, and is, a semi-autonomous governmental defence think-tank, but for a while at the end of the 1970s it was asked to look at the greenhouse effect and the likelihood of climate change. Their report was titled *The Long Term Impact of Atmospheric Carbon Dioxide on Climate*. It noted that measurements were showing that atmospheric concentrations of carbon dioxide had been increasing for over a couple of decades and it estimated that with even more carbon there would be warming of a few degrees. Indeed, they predicted a 2°C–3°C rise in the global temperature with the equivalent of doubling carbon dioxide (*10*). This very much ties in with the more recent, higher greenhouse gas IPCC scenarios.

Across the Atlantic, over in the United Kingdom, the government was to begin to take an interest in global warming concerns. One to engender interest was not a politician but a civil servant—Crispin Tickell. He had spent many years working with British Ministers on relations with Europe as well as working with the Foreign Office before serving a term between 1984 and 1987 as the British permanent representative to the United Nations (UN). In short, he was familiar to leading British politicians of the 1970s and 1980s. In addition to all this, he had long had an interest in climate matters. His interest particularly developed between 1975 and 1976, when he wrote a paper on the relationship between climate and humans. This was subsequently republished much later in 1986 (*11*).

And so it was that in the mid-1980s he shared his climate concerns with the then British Prime Minister, Margaret Thatcher. She had herself graduated from university with a degree in science and, irrespective of her own climate concerns, was able to share in Tickell's apprehensions that human activity may be disturbing the global climate. How much of what followed was sparked by Tickell's influence, and how much from Thatcher's own background and beliefs, is not known, but it seems certain that conversations with Tickell did encourage her to act. At the time there was increasing UN interest in environment and development issues and Tickell was at the UN and reporting back to a government with a sympathetic Prime Minister at its helm. The UN in this period, and following a speech at the UN by Thatcher, moved to formulate a scientific perspective on human-induced greenhouse concerns. To this end, in 1986, it charged two of its agencies (the UN Environment Programme, UNEP, and the World Meteorological Organization, WMO) to establish the IPCC, which it did in 1988.

Clearly, if human-induced (anthropogenic) climate change was taking place then there would be serious economic consequences given fossil fuels' central role in powering the global economy and civilisation. There was going to be some political kickback and vested interests. The IPCC, it was hoped, would put paid to that by providing an independent, impartial scientific perspective.

Its six main reports, or *Assessments* (IPCC, 1990, 1995, 2001, 2007, 2013, and 2021), have concluded that the emissions of greenhouse gases and aerosols due to human activities continue to alter the atmosphere in ways that are expected to affect the climate (*12, 13, 14, 15, 16, 17*).

Today, with the IPCC *Assessments*, we can see that the atmosphere is indeed changing due to human activity, just as Arrhenius thought it might. Indeed, the IPCC's first *Assessment*'s business-as-usual scenario for 21st-century warming in 1990 (see Figure 8.4A) has held, in that subsequent *Assessments*' analogous scenarios (see Figure 8.4B) were all within the 1990 business-as-usual scenario's upper and lower estimates. In short, in modern times we have had a reasonable idea as to how global warming is likely to play out for the best part of a third of a century. That various IPCC *Assessments* have seen comparatively minor adjustments, as modelling and potential future scenarios have been refined, is neither here nor there: it should not matter to the political classes if scenario forecasts change by a fraction of a degree when it comes to their deciding policy. That the three decades since the first IPCC *Assessment* (1990) have seen actual global temperatures track the business-as-usual analogous scenarios testifies to their robustness and engender confidence in the IPCC scenarios.

The IPCC *Assessment Reports* for 2007, 2013, and 2021 all use slightly different emissions scenarios in what in the first IPCC *Assessment* (1900) would be the business-as-usual scenario. In 2013, the IPCC used what were called Representative Concentration Pathways (RCPs) in their future scenarios and in 2021 these became Shared Socio-economic Pathways (SSPs). Although they are nominally analogous—such as between RCP8.5 (2013) and SSP5-8.5 (2021) in Figure 8.4—they are not the same. Even so, it is not surprising that there is some difference—less than half a degree Celsius—between the IPCC 2007, 2013, and 2021 high fossil carbon scenarios: the differences are not all down to changes in climate modelling. For example, the gas-to-gas compositions differ; so the SSP5-8.5 scenario has higher carbon dioxide but lower methane concentrations compared to RCP8.5 (*16*, *17*). Yet despite this, the projected warming from RCP8.5 (2013) and SSP5-8.5 (2021) is almost identical (see Figure 8.4B).

Additionally, the IPCC scenarios largely exclude what are called 'long-term feedbacks'. As the term suggests, these are feedbacks that will come into play over the long term. They include things like the thawing of permafrosts that releases the greenhouse gas methane so furthering warming in a positive feedback manner (see Chapter 2). The response of Greenland and Antarctic ice sheets to warming and the release of carbon from soils and permafrosts are other examples of long-term feedbacks. Such feedbacks are mostly positive in that they enhance warming and sea-level rise. Because the IPCC openly says it has not included many of these feedbacks (due to difficulty in quantitatively estimating their strength) and, given that some of these feedbacks will likely come into play before this century's end and that many of them confer a positive (rather than a negative) feedback, so the IPCC's scenarios for the latter half of this century are likely to be conservative. This is very much contrary to climate deniers' claims that the IPCC's scenarios exaggerate matters. (Since 2021, the IPCC *Assessments* have begun to include long-term feedbacks, but it still warns that care needs to be taken especially with regards to ice-sheet-associated long-term feedbacks [*17*].)

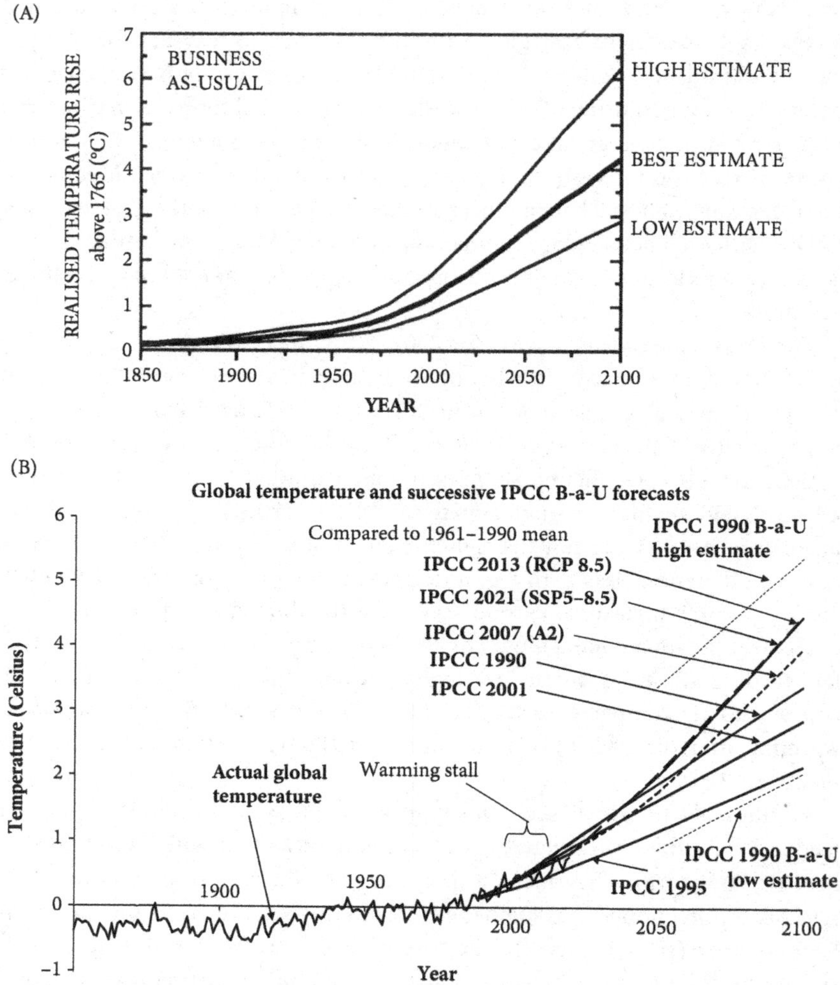


**Figure 8.4** (A) IPCC 1990 Business-as-Usual scenario. Reprinted with permission from Figure 12 in Houghton, J. T., Jenkins, G. J., & Ephraums, J. J., eds., *IPCC Scientific Assessment. Contribution of Working Group I to the First Assessment Report of the Intergovernmental Panel on Climate Change.* Cambridge University Press: Cambridge, United Kingdom, New York, NY, USA and Melbourne Sydney (1990) (ref. *12*).(B) Actual global temperature 1850–2013 (based on Climate Research Unit at East Anglia University data accessed online in 2020) together with the IPCC 1995, 2001, 2007, 2013, and 2021 scenario that are most analogous

Historical data from the Climatic Research Unit (University of East Anglia, Great Britain) and the UK Met Office, HadCRUT5 (accessed 2023) https://crudata.uea.ac.uk/cru/data/temperature/.
Note: The base line temperatures are different for panels A and B. B has a different baseline as it standardises *all* the IPCC *Assessment* scenarios.

Yet, that all the IPCC *Assessments* present analogous business-as-usual warming scenarios that are within the 1990 *Assessment*'s upper and lower business-as-usual (B-a-U) estimates, is a testimony as to how clear the science has been about the likely warming the past one-third of a century. This is the value of comparing each of the Assessments' projected warming in Figure 8.4B.

Even so, the sociopolitical kickback has been substantial and in the 21st century's first two decades—despite all the evidence—there has been much public debate as to whether the current rise in atmospheric carbon dioxide has caused the Earth to warm. An alternative climate-denier view is that the warming has been too erratic and is due to random climate variation or some other factor such as volcanic emissions of carbon and methane. Indeed, that the drunkard's, meandering walk rise in mean global temperature stalled between 1998 and 2014 fuelled deniers' arguments (see 'warming stall' in Figure 8.4B). Though, as noted earlier, the $^{13}C$ data discounts the volcanic explanation and the increase in atmospheric carbon dioxide tallies with those of emissions less the absorption by carbon sinks such as the oceans. By the early 21st century, nearly all scientists involved in climate change and related research were of a near unanimous view that humans have instigated a period of global warming (*18*) and that we are in a CIE!

In terms of a co-evolution of life and planet narrative, both the CIE and warming are a result of a short-circuiting of the deep carbon cycle and—just as life was involved in the other flips of the Earth's state—this is being undertaken by life, only in this case just one species: *Homo sapiens sapiens*. In the normal course of events (without human interference), $^{12}C$-rich fossil fuels would likely have remained geologically buried and in most instances only released when plate-tectonically carried to a subduction zone to mix with the $^{12}C$ and $^{13}C$ pool as part of the upper mantle before being volcanically returned to the atmosphere with the $^{12}C$ to $^{13}C$ ratio reset.

**Figure 8.4** *(continued)* to the IPCC 1990 Business-as-Usual (B-a-U) scenario. All have been adjusted to a zero global temperature that is equivalent to the 1961–1990 mean: the various IPCC *Assessments* have different baselines and the figure has accounted for this. The upward-sloping, two dotted lines between 2050 and 2100 are equivalent to the 1990 Business-as-Usual high and low estimates for that time period. The anomalously low IPCC 1995 scenario curve is due to the IPCC in that *Assessment* including sulphate cooling. Subsequent *Assessment* scenarios included other positively climate forcing factors. The projected warming for the 2013 and 2021 analogous Business-as-Usual scenarios (for the technically minded, called RCP 8.5 and SSP5 8.5) are almost identical and hard to distinguish. Scenario data from references *12, 13, 14, 15, 16, 17*.

The degree of this short-circuiting of the deep carbon cycle has been extremely pronounced in recent years and unparalleled in the history of human civilisation that has so far lasted many thousands of years. Throughout history, since ancient times, human civilisations have increasingly relied on energy resources as well as being affected by climate change.

The first state-wide coordination of settlements was arguably that established in Egypt over five thousand years ago. Subsequently there was the rise of the Akkadian empire in Mesopotamia. It stretched from the headwaters of the Tigris-Euphrates Rivers to the Persian Gulf during the late third millennium BC. However, there was to be an episode of significant climate change that possibly lasted for a century and which took place some 4,200 years ago. This episode ended the Akkadian empire. This abrupt climate change was likely cause by an alteration in the North Atlantic circulation (itself caused by ice rafting stalling the global surface and abyssal water conveyer) and it saw northern hemisphere cooling. Here, with less ocean evaporation and altered weather patterns, areas experienced decades of drought.

The event in the geological record is so sufficiently pronounced that in 2018 the International Commission on Stratigraphy used it to define the beginning of the age called the Meghalayan starting 4,200 years ago (4.2 kya) just prior to the beginning of the Bronze Age and continuing to the present (*19*). These droughts affected early civilisations and have specifically been attributed to the collapse of the Akkadian empire (*20*).

Since then, over four thousand years, human civilisation has slowly developed, all the while using more and more energy. Fortunately, for much of this time through to the industrial revolution of the 18th and 19th centuries, the fuel used was mainly biomass such as wood, though this supplemented manual labour and animal power. But, with industrialisation, fossil fuel consumption began to dominate. First mainly coal, then increasingly oil and natural gas.

By 2019, fossil fuels globally accounted for 492.34 EJ (exajoules), or 84%, of a total primary energy consumption that year of 583.9 EJ (*21*). There then followed a dip in global energy consumption due to the CoVID-19 pandemic's lockdowns, but this rebounded and in 2022 global annual primary energy consumption had reached 604.04 EJ (*22*). This last, if this energy was consumed in the form of oil, it would be comparable to 14,229 mtoe (million tonnes of oil equivalent).

But the perhaps more startling statistic is the recent jump in growth of energy consumption. If (skipping the earlier Egyptian and Akkadian civilisations) we take the commencement of the Meghalayan, 4.2 kya, as the starting point of human civilisation with its early nation-states, it is estimated that humanity has consumed some 14.2 ZJ (zettajoules) through its existence up to the year 1950 (*23*). The year 1950 roughly marks the time when the atomic weapons became thermonuclear, whose testing generated distinctive radioactive fallout, providing a geological marker for the dawn of a proposed new epoch (though 'stage' or 'age' within the Holocene epoch might be more appropriate) of humanity marking a global impact, the 'Anthropocene' (*24*).* Since 1950 to the beginning of 2020 it is estimated that humanity has consumed ~22 zettajoules (ZJ). This means that in the 70 years since 1950 *Homo sapiens sapiens* has consumed more energy resources than in the previous 4,200! It is a marked transition in energy use and it is the up-to-now dominance of fossil fuel (remember, 84% of a total primary energy consumption in 2019 and nearly 82% in 2022) that has been the principal factor instigating the current CIE and global warming. Yet even though we are aware of the consequential human-induced climate change, our consumption of fossil fuels continues to increase (see Figure 8.5A). The only deviations from this steady increase relate to the economic consequences of the 1973 oil crisis, the 1980 depression, and, more markedly, the 2007/8 financial crisis and the 2020 CoVID-19 global pandemic.

Recent global *per capita* energy use has also seen a steady increase from 46.6 gigajoules (GJ) in 1965 to 75.7 GJ in 1999. While 21st-century *per capita* energy use in many developed nations (such as Germany, France, the United Kingdom, and the United States) has decreased, this has been offset by a *per capita* increase by some other countries, notably China and India (see Figure 8.5B). However, here not all of the growth has been domestic. Some of the developed-nation decrease has been due to economic carbon leakage: they import goods that have been created using fossil fuels. Here, the greenhouse emissions are associated with the manufacturing nation and not the importer. Consequently, due to nations' economic carbon leakage, the picture as to which nations are driving fossil fuel emissions is not quite as clear as it might appear. Combining the steady *per capita* increase in energy consumption with global population growth (see Figure 8.2B), from 2.53 billion in 1950 to over 8 billion by the end of 2022, has served to drive the increase in energy-related greenhouse gas emissions.

However, carbon emissions are not the only way humans are affecting the global carbon cycle.

* The term 'Anthropocene' was under consideration by the International Commission on Stratigraphy for it to be formally designated as an epoch. However, there is an argument against this. The formally accepted epochs mainly last for millions of years: only the Holocene is shorter at just 11,700 years. The Holocene essentially covers our current interglacial that has seen our species spread across the globe, impacting biomes; for example, with the extinction of megafauna. Much discussion around the proposed 'Anthropocene' epoch focuses on recent human-induced climate change, radioactive markers, plastics, *etc.* Problems that, geologically speaking, are relatively transient. It could be argued that relegating the Holocene to a bit-player is unnecessary (*25*). A compromise might be to consider the 'Anthropocene' a 'stage' or 'age' within the Holocene. The International Commission on Stratigraphy rejected adopting the term in 2024.

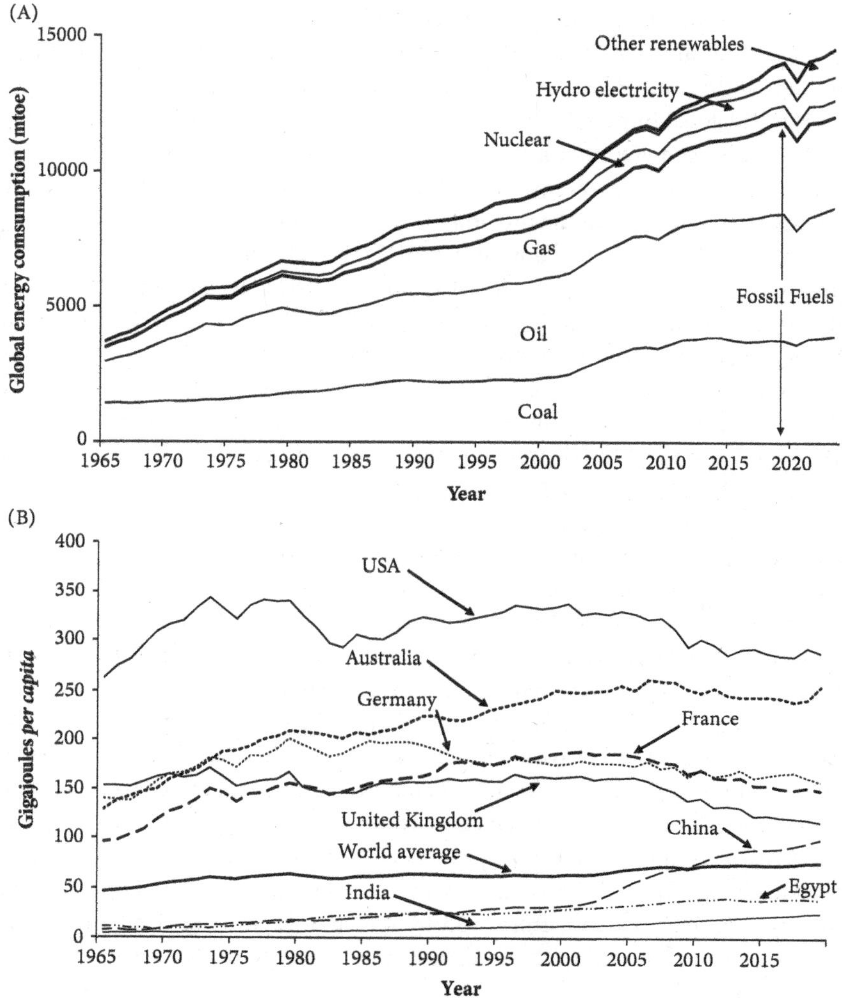


**Figure 8.5** (A) Late 20th- and early 21st-century primary energy. This is dominated by fossil fuel use. (B) Late 20th- and early 21st-century annual *per capita* energy use by selected nations. See Appendix 2 for data sources. Note, this ignores economic carbon leakage.

## Other human impacts on the carbon cycle

You will recall that the other two major flips of Earth state saw biological evolutionary innovation integral to those flips (eukaryotic oxygenic photosynthesis and terrestrial colonisation by multicellular species) and these saw some species involved boom. Respectively, oxygenic photosynthesis became the dominant form of photosynthesis and the populations of those species grew, and with multicellularity terrestrial colonisation there was the Ediacaran flourishing (Avalon boom) of species followed by the Cambrian boom (see Chapter 5). So, if we are to experience, or are in the midst of, another similarly caused flip of state then there should be the rise of a species associated with a new unique evolutionary innovation. Here, once more, humans (*Homo sapiens sapiens*) seem to fit the bill with the evolution of sentience capable of complex tool-wielding.

As evident from Figure 8.2A, the human population has grown unprecedentedly the past two and a half centuries. Indeed, if the rate of growth had been constant then population plotted on a logarithmic scale (see Figure 8.2B) would give a straight line with the degree of angle of slope reflecting the rate of growth. That there has been a period when such a plot has itself had an upward curve means that at that time the rate of population growth had itself been increasing: in other words our global population has seen a period of super-exponential growth!

That the beginning of this period of human population super-exponential growth coincided with the 17th-century scientific revolution and enlightenment, or even the late 18th-century industrial revolution onset, that itself followed the 16th-century renaissance, may be significant. With renaissance knowledge flourishing, furthered as never before with the development of the printing press, it was almost inevitable that this would lead to the industrial revolution. This and the subsequent continuation of increased scientific understanding and its application to developing technology, has only improved human well-being, longevity, and fecundity, hence population. So, as Figure 8.2B depicts, from the industrial revolution through to the mid-20th century there is an explanation for the global population's super-exponential growth rate.

Today our population is of such a size (several billion and counting) that it is affecting biomes and even the biosphere.

By the early 21st century, globally some 15% of ice-free land became dominated by human use (such as urban areas and roads) whereas 55% is part-dominated by human use (such as agriculturally dominated landscapes—agriculture takes up 11% of land—and canalised rivers). This human influence on land is so great that it is estimated that just under a third of terrestrial (land based) net primary productivity (the amount of energy, carbon, or biomass fixed by photosynthesis—see Appendix 2) is influenced by humans (*26*).

Human appropriation of this biological, net primary production, apart from leaving less for other species to use, alters the composition of the atmosphere; levels of biodiversity; energy flows within food webs; and the provision of important ecosystem services. (These last include things such as fixing carbon or trees stabilising soil on slopes and holding water to prevent flooding. They are 'services' ecosystems provide which if they didn't we would have to pay for them. So, for example, a grassy flood plain provides water storage in times of flood. If, instead, housing was built, the ecosystem service would be removed and so money would be required for alternative flood management.)

Other analyses of photosynthesis appropriation are more conservative: it depends on how the calculation is done. One study in 2004 did not take into account land-use change: such as woodland or forest being replaced by cities: clearly cities do not photosynthesise carbon nearly as much as forests do. The researchers conducting the 2004 analysis noted that if they had included the missing elements, then its estimate of net primary production appropriated by humanity would be in line with the other studies revealing human appropriation of over a third of terrestrial carbon photosynthetically fixed by biological systems across the planet!

Key features of the 2004 analysis included it having a mid-estimate of just over a fifth (20.32%) of terrestrial (land based) net primary production being consumed by humanity. The study calculated that over 4 petagrams of carbon (PgC) of primary production is taken to provide food. This includes grains consumed by us humans directly as well as for animal feed: here, very roughly, there is an equal split in net primary production sequestered. It also concluded that over 4.3 PgC is taken in the form of wood for fuel, while wood for construction accounted for 2.5 PgC. In total humans took 11.54 PgC out of 56.8 PgC fixed by photosynthesis on land across the globe (*27*).

That one species has been responsible for this substantive co-option of biological primary production, hence ecosystems, and has harnessed so much energy, arguably speaks to that species representing some sort of evolutionary milestone. We will return to this shortly. For the time being let us look at another possible indicator of an impending co-evolution of life and planet transition critical transition.

## New mineral species

In Chapter 6, when discussing the characteristics of the successive flips of the Earth's state, it was noted that each flip of state marks the beginning of a long-term period of generating new mineral species (see Figure 6.4). This is because supercontinent fragmentation associated with the two previous flips of the Earth's state (see Chapters 3 and 5) is part of the Wilson cycle of supercontinent fragmentation and formation (see Chapter 1) and because with time continents became larger (see Chapter 3). As crust gets carried to different depths (larger continents tend to be thicker) it gets exposed to different temperatures and pressures, and is also reworked, so that new mineral species are formed.

Further, as each of the previous flips of the Earth's state has seen an increase in atmospheric oxygen, so each flip has enabled new mineral species of a higher oxidation state. This too is part of the co-evolution of life and planet narrative.

Turning to the (geologically) recent rise of human sentience, we have seen the creation of human-made materials from clay pottery, bronze, and other alloys through to today's semiconductors, carbon fibres, and so forth. For example, by the early second decade of the 21st century some 9.5 million tonnes of plastic *per* year were entering the ocean (*28*).

So, in common with the past flips of the Earth's state that saw the beginning of a long-term period of the creation of new mineral species, today we are seeing new, human-made materials becoming part of the geological record.

So significant is the amount of human-made material being generated it is evidenced by the landmark year of 2020. That year was the first that saw the mass of human-made items (metals, bricks, concrete, asphalt, and so forth), at 1.1 teratonnes (1.1 exagrams), exceed the dry mass of all living biomass on the planet! Since the latter half of the 20th century, human-made (anthropogenic) mass had been doubling every 20 years from about 0.1 teratonnes around the year 1960. So, in purely mass terms, post-2020 sees one species of animal—our own—generate as much mass as all other species combined (*29*).

This manipulation of matter is all the more remarkable for a single species of animal given the dry biomass of all animals together with that of prokaryotes, protists, and fungi (non-plants), comprises about 10% of living biomass globally, with plants accounting for the remainder, around 90%. It is remarkable as our own species—*Homo sapiens sapiens*—itself comprises a paltry 0.01% of global biomass (*29*).

In living biomass terms, up to very recently the post–Snowball II Earth was in a sense a time dominated by plants. Today humans are having a very real impact on the biosphere, its composition of species, many of its landscapes globally, as well as its atmosphere, hence oceans (that are becoming more acidic), and not least the global climate.

That the present Anthropocene is characterised by extraordinary energy consumption and new mineral species was noted in 2020 by an international collaboration of 18 geoscientists. Their review appeared in the journal *Communications Earth & Environment* with the lead authors Jaia Syvitski, Colin Waters, John Day, John Milliman, and Colin Summerhayes, all of whose work addresses the question of how best to define the formerly proposed Anthropocene epoch (though as noted earlier it might serve better to view it as a 'stage' or 'age' within the Holocene). Their review not only noted the jump in global energy consumption over the seven decades since 1950 exceeds that during the entirety of the Meghalayan age (that began 4.2 kya) as cited earlier in this chapter, but that since the proposed start of the Anthropocene (1950) the number of new mineral species and synthetic compounds being created has been unprecedented. Indeed, it exceeds that depicted in Figure 6.4. They report that these recent additions now exceed 180,000, with most species created since 1950! They conclude that the global human population, its productivity and energy consumption, and most human-generated changes that impacts the global environment, are highly correlated. This extraordinary outburst of consumption and productivity demonstrates how the Earth System has departed from its (up till then) Holocene state since around 1950 CE and that this is forcing abrupt physical, chemical, and biological changes to the Earth's geological record. In other words that the Earth is entering a new state (*23*).

If the primordial anaerobic Earth was associated with prokaryotes, the partially oxygenated Earth with eukaryotes and the fully oxygenated Earth with multicellular species, then the forthcoming change of state will see one species dominating (irrespective of in an informed managed manner or uncontrollably) and affecting the biosphere.

We truly are entering what was proposed to be the Anthropocene.

Or are we?

An alternate view to there being a new epoch (or perhaps 'stage' or 'age' within the Holocene) called the Anthropocene is that it is not an epoch but an event. Just as there was Snowball I and the great oxidation and Snowball II with multicellular life on land, so could the Anthropocene be, not be an epoch but, an event? This is a musing presented by Andrew Bauer of Stanford University (*30*). So, if following Snowball I we had an Earth whose subsequent time was oxygenated due to oxygenic photosynthesis, and after Snowball II we had a planet with substantive multicellular life on land, is there anything that might characterise matters after an Anthropocene event (assuming it is best defined that way and not as an epoch)?

Well, there is another characteristic associated with the previous flips of the Earth's state and that is information processing which, around the time of the Anthropocene, has gone digital.

## Information processing and the rise of humans

The first flip of the Earth's state from a purely chemical equilibrium state was due to life perturbing that equilibrium. This is possibly evidence by, for example, the earliest banded iron formations (BIFs)—see Chapter 1. These BIFs demonstrate that something (life, it is argued) altered the oxidation state of (at least part) of the world's oceans. Life itself consists of biochemicals whose orderly function is determined by an inheritable (down the generations) suite of instructions (information) inherent in nucleic acids (DNA and RNA).

The second flip (but the first really major one) of state was determined by the evolution of oxygenic photosynthesis (OP) with cyanobacteria and also perhaps possibly the rise of eukaryotes. Some of these eukaryotes were also oxygenic photosynthesisers and, if they did not play a part (as they might have) in this first flip, their OP certainly ensured the planet remained in its new, low-level atmospheric oxygen state. That eukaryotes are the result of endosymbiosis—indeed multiple symbiogenesis events (see Chapter 4)—meant that for each event two prokaryotic species merged with one ending up as organelles: such as chloroplasts or mitochondria. These organelles have their own DNA which survives and can be seen in living eukaryotes today.

One way of looking at the rise of eukaryotes is that this was also a revolution in information processing: a number of endosymbiotic events saw the merger of information from a number of species.

The third flip of the Earth's state (or the second really major flip), proposed by the new co-evolution of life and planet narrative, is associated with the rise of terrestrial multicellular species (see Chapter 5). While the initial terrestrial colonising multicellular species were not animals (metazoans), metazoans did follow. One feature common with metazoans is that of a nerve system and neural processing organs (brains of varying degrees of complexity). And so the second major flip of Earth's state that gave an atmosphere, with close to present-day levels of oxygen, enabled there to be a revolution in biological information processing: animals think!

Information processing in terms of 'thought' has arisen many times in animals. For example, a carnivore hunting, whether alone or in a pack, has to process quite a bit of information. So we have lions, tigers, wolves, *etc.*

Information processing organs have evolved a number of times. The 'brains' of insects anatomically include the *corpora pedunculata* or 'mushroom bodies'. And significant brains are a common feature of most vertebrates (these diversified from the phylum to which the class of insects belongs, somewhere around half a billion years ago—just a score or so million years following the Ediacaran). Such animal organs have independently evolved a number of times as the last common ancestor of these lineages had only a non-centralised nerve net (with no neuron concentration). Given repeated independent evolution, more centralised information processing per se therefore seems to be an easy evolutionary step (*31*).

You may think—possibly due to owning, or knowing somebody that owns, a cat or a dog, or through seeing wildlife programmes, as well as being one yourself—that reasonably complex thinking is confined to mammal species. This is not so, it is found in completely different evolutionary branches. Here the notable example are members of the Cephalopoda—species of octopus, squid, and so forth—that have been shown to be adept problem-solvers. They also have a distinctly different nervous system to vertebrates (Vertebrata). They have a brain, but roughly a third of their neurons are distributed through their tentacles (though these are still far more concentrated than a non-centralised nerve net). They also have the ability for some memory and learning as well as some problem-solving (*32*). This demonstrates that rudimentary intelligence can arise a number of ways: species exhibiting intelligence could be due to the convergence of distinct evolutionary paths (*33*).

Biological 'intelligence' has therefore evolved independently a number of times and may, along with brains, be considered an easy evolutionary step even if it is only our own species that has the degree of intelligence to do crosswords and manifest a complex, technological society.

With the rise of our species, *Homo sapiens sapiens*, and its largest of all species' mass of brain-to-body ratio, we have also seen an unprecedented evolutionary leap in metazoan information processing. The degree of our species' ability to process information has enabled it to dominate the planet and—as mentioned in this chapter's previous section—influence it on a global scale, altering its atmosphere, creating new mineral species, *etc.* Furthermore, as we will return to shortly, it has created the means to separately record and, more recently, process information through technology: for example, you are likely reading this either as print on paper (in analogue form) or on a screen (with the book in digital form).

The evolution of our species, similar to the first three flips of the Earth's state, has involved biological interactions with the Earth system. (Not just through our fossil fuel use, short-circuiting of the deep carbon cycle, but in other ways too that we will come to shortly.) Given this, and that our species is central to changing the current state of the Earth's state—irrespective of whether or not this is a key critical transition in the evolution of life and planet—it is worth summarising human species (spp.—species plural) evolution and that of human and human-mediated information processing as well as noting that human evolution too has been driven by environmental changes mediated by the Earth system.

That there have been a number of tool-using human as well as pre-human species means that advanced intelligence was not a unique evolutionary step even if it ended up today in one extant human species: our own. So the evolution of intelligence is not a particularly difficult evolutionary step in itself. While it might not perhaps be an easy evolutionary step, it was not a particularly hard one to make.

Indeed, in the broad sweep of geological and evolutionary time, it took roughly as long (roughly the same order of magnitude of years) for humans to arise as did each of the previous flips of planetary state to take place since the previous one. They took two-thirds of a billion years to arise following Snowball II and the increase in atmospheric concentration of oxygen to near present-day levels. This oxygen increase itself was one and a half billion years following Snowball I and the first significant presence of atmospheric oxygen, albeit at just a couple of per cent. While Snowball I itself came the best part of a billion years (0.9 by) of the banded iron formations (BIFs) *likely* to be the result of photosynthesis, and Snowball I came 1.6 billion years after the earliest BIFS that *may*, possibly have been a result of an early form of photosynthesis. These milestones in planetary state (and we will return to our own species' impact on the planet in Chapter 9) each were driven by biological evolutionary innovations (the rising of life itself, photosynthesis, oxygenic photosynthesis—and possibly involving OP eukaryotes—and multicellularity respectively) interacting with the planet and which have, in a rather loose sense, been more or less evenly spaced across our planet's lifetime: they were not all clustered either early on or later in the planet's lifespan or even huddled together somewhere in the middle.

Humans are primates belonging to the biological family of Hominidae (hominids) that also comprise: chimpanzees, bonobos, gorillas, and orangutans. One candidate for the last common ancestor, or possibly the cousin of the last common ancestor, of humans and gorillas, is the fossil ape *Danuvius guggenmosi*. It lived some 11.6 million years ago (mya) and its anatomy suggests evidence of a form of bipedalism, extended limb clambering: bipedalism is common to the great apes and humans (*34*, *35*).

In our arm's-length scale of our journey through time from the formation of the Earth, this is less than half the distance from your fingertip to your finger's first joint: on the grand geological scale of our planet's history, we are very near the present.

Another candidate is *Chororapithecus abyssinicus*. ('Abyssinia' being the former name of Ethiopia, where *Chororapithecus* was discovered.) It lived over 10 mya at a time when that part of Africa was seeing geological rifting, hence a changing landscape (*36*).

Though *not* associated with a planetary flip of state, as with the flips of Earth's state, it was environmental changes of state interacting with species that drove hominid evolution! Only mutations in species that increased their survival in the new environments were genetically passed on to subsequent generations. The landscape of the African region at that time was a mosaic of patches of forest, open areas with a braided river system. *Chororapithecus*, it is thought, occupied forest environments. This complex, changing landscape of the late Miocene saw considerable hominid diversification.

The periodic changes of landscape were not only caused by local geological forces, such as the aforementioned rifting, but also planetary scale changes. You will recall in Chapter 2 that the flipping between cold glacials and warm interglacials in the Quaternary ice age was pace-makered by three cycles of Earth orbital parameters that govern how much solar energy the land-dominated northern hemisphere receives. These global climate changes were in turn amplified by changes in albedo from the waxing and waning of ice sheets. (More ice, more solar energy reflection, more cooling, and *vice versa*.) And they were further amplified by the carbon dioxide drawn down into the oceans as they cooled, and returned to the atmosphere as they warmed. Yet these cyclical orbital parameters were still operating when the Earth was warmer than in the present (cooler) Quaternary, so largely ice free, in the Miocene (23–5 mya—see Appendix 3): back then our planet did not have a significant ice sheet in Antarctica. This in turn meant that there were still cycles of less extreme global climate change that, nonetheless, could still manifest in some regional areas (such as in Ethiopia and Tanzania) as major environmental change. This continual environmental change provided, if you will, evolutionary hoops through which our ancestor species jumped.

This environmental change continued after the Miocene ended.

In 2005 Martin Trauth and colleagues reconstructed the east African climate 0.5–3 mya from ten lake-basin records in the area. These are all in the East African Rift System (EARS). The EARS is a near north-south line of rift valleys that starts with the Afar Basin in northern Ethiopia, runs through the Omo-Turkana-Sugata basin in north Kenya, and ends with the Olduvai basin in south Tanzania. The rift valley formation and deepening began in the Miocene in the north and migrated south. In turn, valley formation allowed for the formation of deep lakes provided that rainfall exceeded the rate of evaporation. So these lakes' existence and size were governed by a combination of geological and climate processes.

The researchers identified three possible hominid periods and marked wetter periods in the EARS: 2.7–2.5 mya, 1.9–1.7 mya, and 1.1–0.9 mya. The geological record shows that there were large deep lakes in the north during the earliest of these periods, transitioning through more arid times to deep lakes in the last. Between these times, the lakes dried out. There were therefore periods of ample vegetation and lakes alternated with dry periods: flips between times of biological plenty and stress. The idea is that these stress times spurred speciation while the periods of plenty enabled populations to grow. This would have had an important impact on the speciation and dispersal of mammals, including hominids (*37*).

Subsequently, the Quaternary (from 2.58 mya) and its series of glacial-interglacial cycles was the period of time when much of human evolution took place. The Earth, having gradually cooled, now saw significant ice in Antarctica as well as on Greenland. This ice waxed and waned, paced by the three orbital parameters and, as previously noted, the ice waxing and waning amplified the cycles of warming and cooling: there were now cycles of glacials and interglacials. These provided hoops through which species had to jump, hence provided further evolutionary pressure.

That the global climate, hence regional environment, went through cycles of change is evident in the EARS from lipid biomarkers and isotopic signatures from organic matter preserved in lake sediments at Olduvai Gorge 2 mya. Corroborating evidence also indicates that that environment saw recurrent ecosystem variations, where open $C_4$ grasslands abruptly transitioned to closed $C_3$ forests on a timescale of several hundreds to thousands of years. (You will recall that $C_4$ and $C_3$ plants were encountered in Chapters 6 and 7.) The geological record shows that during this time Lake Olduvai repeatedly expanded and then contracted to less than a seventh of its greatest area. The timing of the recurrent environmental variations correlate most strongly with Earth's orbital precession; this is one of the three orbital factors determining the timing of glacials and interglacials. Two mya was a key juncture in human (*Homo* spp.) evolution with the emergence, and subsequent migration of an archaic species of human, *Homo erectus* (*38*).

Ethiopia and Tanzania were not the only parts of Africa seeing cycles of environmental change the past 2 million years. The geological record from the Limpopo catchment in subtropical south-eastern Africa also shows repeated, cyclical change the past 2 million years. This change appears to be controlled by interplay between two Earth orbital factors (precession and eccentricity) as well as changes in ice volume at high latitudes. The area is known for fossils of the early hominin *Paranthropus robustus*, now considered an australopithecine, and it is thought that environmental change played a role in its eventual extinction (*39*). It is also a species associated with stone tool use (*40*).

Briefly, let's take a step back a couple of million years to see the run-up to the rise of the hominins (various human—*Homo*—species).

The australopithecines were a genus of hominids that, though having a much smaller brain than subsequent *Homo* spp., had an upright posture. They are thought to have evolved from a population of *Ardipithecus ramidus* that itself is thought to have descended from the ancestor species of the last common species of the line that led to chimpanzees: *Ardipithecus* evolved shortly (geologically speaking) after the hominid/chimpanzee divergence (*41*). Fossils of *Ardipithecus ramidus* have been dated to 4.4 mya and within a couple of hundred thousand years they led to the australopithecines. One of the australopithecine species (spp.) is thought to have included the ancestral population that eventually speciated to become the various *Homo* spp (*42*). *Ardipithecus* seems have been able to walk upright at least for a limited period as well as move in trees (*43*), and it lived in a wooded habitat (*44*).

The fossils of the early australopithecine *Australopithecus anamensis* have been dated to 4.2–3.9 mya. However, in east Africa, somewhere around 3.5 mya, the australopithecines diversified to such a degree that a number of different species co-existed at the same time, including: *A. afarensis*, *A. bahrelghazali*, and *A. deyiremeda* (*45*, *46*).

The diet of a late australopithecine, *Australopithecus sediba*, has been elucidated from an analysis of the dental plaque found in a 2-million-year-old fossil from South Africa. It included an almost exclusive $C_3$ plant diet that probably included harder foods (nuts, wood, and bark) as well as tree leaves, fruits, grasses, and sedges. Like *Ardipithecus ramidus* (approximately 4.4 mya ago) and modern savanna chimpanzees, *Australopithecus anamensis* consumed $C_3$ foods in preference to widely available $C_4$ foods (some grasses). This contrasts with later australopithecines and early *Homo* spp. that consumed variable amounts of $C_4$ foods, but their diets included more $C_4$ foods than the diets of modern chimpanzees. This suggests that they probably used their environments in different ways than do living apes today (*47*, *48*).

One population of one australopithecine went on to found our own genus, *Homo*.

The earliest fossil attributed to *Homo* is from Ethiopia and dates from nearly 2.8 mya (*49*). Considerable subsequent diversification led to various *Homo* species: among others, *H. habilis*, *H. rudolfensis*, and *H. naledi*, all in Africa. In Africa, together with later elsewhere, there was *H. erectus* (which some also consider as *H. ergaster* in Africa) and *H. heidelbergensis* (also known as *H. antecessor* outside of Africa). Outside of Africa there was a recently (2021) discovered 'Nesher Ramla *Homo*' species from remains found in Nesher Ramla, Israel, that date from 140–120 kya. This was another tool-making and -using species and it seems to be the ancestral species to *H. neanderthalensis* (the Neanderthals), and Eurasian archaic human species *H. luzonensis* and *H. floresiensis* among others (*50*, *51*). From a population of one of these (possibly *H. heidelbergensis*), and possibly interactions between other *Homo* spp., descended today's generations of *Homo sapiens sapiens*. (See Appendix 2 for the difference between *Homo sapiens sapiens* and *Homo sapiens*.)

Over the decades, the date for the oldest fossil attributed to the earliest of anatomically modern humans (*Homo sapiens*) has been pushed further back in time. This push-back is due to the Signor–Lipps effect (see Chapters 4 and 6). The oldest African fossils currently known were found in Jebel Irhoud, Morocco, and dated in 2017 along with stone implements, by two separate international collaborations of researchers. Both the fossils and implements date from 315 thousand years ago (kya) (*52*, *53*). Yet separate genetic evidence of both living populations as well as from fossils suggests that *Homo sapiens* diverged from the lines of the Eurasian Neanderthals and Denisovans, more than 500 kya. This is a lot earlier than 315 kya. So it is likely that even older fossils of the earliest *H. sapiens* have yet to be found.

There is of course a complication in that not all genes associated with modern humans give rise to anatomically modern humans: the early *H. sapiens* genetic lineage may have had features pre-dating the full emergence of what we consider looks human; the dividing line between *H. sapiens* and archaic human species could be more blurred than thought (*54*). This means that it is difficult to determine a specific birthplace and time for the first appearance of *H. sapiens* (*55*).

This blurring between *Homo* spp. is further complicated by interbreeding. Consequently, the genome of early anatomically modern humans differs from that of humans today. It is likely that not only did different semi-isolated populations of early *Homo sapiens* in Africa interbreed but that there was breeding with other *Homo* species. What is certain—as revealed by genomic analysis—is that outside of Africa *Homo sapiens* interbred with Neanderthals (*Homo neanderthalensis*). Indeed, today some east-Asian and Europeans have 1.38% and 1.15% respectively of their genome having Neanderthal ancestry. Even in present-day African Kenyans, 0.08% of their genome has a Neanderthal origin (*56*).

Evolution resulting in modern humans appears complicated. In essence, though, what was happening was that through both diversification and cross-breeding, there were plenty of opportunities for increased cognition. Indeed, there are many archaic species of hominin and, at the time of writing, another was recently discovered in China: *Homo longi*. *H. longi*, it is putatively thought, is a cousin species to both modern humans and Neanderthals (*57*, *58*, *59*). Increased cognition conferred such an evolutionary advantage that those that had it eventually out-competed other species.

Before moving on to discuss the progression from first tool-use to modern technology, it is worth looking at recent research that brings together a number of the changing environment episodes with that of the evolution of various *Homo* species.

The work of Axel Timmermann and his small collaboration of researchers largely based in South Korea working with a few European-based scientists was published in 2022 and it looked at the way climate affected archaic human species' habitats. One issue with evidence for climate-driven environmental change spurring evolution is that the terrestrial, local climate evidence, while it may span many thousands of years, does not cover the entirety of the past 2 million years. Nor do we know—because of the Signor–Lipps effect—when early *Homo* species evolved. What we do know from the good number of snapshots of hard evidence we have—some of which is discussed above—is what the climate and ecology was like for various early species of *Homo*. We therefore have a reasonably good understanding of the ecological habitat preferences of a number of *Homo* species. We also know, from dating fossils, by when these species existed even if we do not know when they first evolved. Having said that, genetic evidence enables us to at least suggest when each might have evolved.

The past three decades has seen considerable advancement in computer climate models. These can be used to recreate palaeoclimates in increasingly more regional detail, hence the regional ecology of the time.

What the Axel Timmermann team did was to use a computer climate and ecology (actually biological 'net primary production'—see Appendix 2) to model and track the regional environmental suitability over 2 million years for various species of early human: *Homo ergaster* and *Homo habilis* (which the Timmermann team considered as a single tool-using species), *Homo erectus*, *Homo heidelbergensis*, *Homo neanderthalensis*, and *Homo sapiens*. Each species had its own nuanced habitat suitability. Their model showed both at which times and where these various species could have flourished and, interestingly, when and where their respective habitat suitabilities overlapped; these being the times and places when one species was more likely to give rise to another.

Their work suggests hominin habitat overlap from approximately 400–300 thousand years ago (kya) and that climate change in southern Africa and Eurasia contributed to the evolution of *Homo heidelbergensis* into populations of *Homo sapiens* and *H. neanderthalensis*, with *H. sapiens* evolving 310–200 kya. This last is consistent with what fossil and archaeological evidence there is. The Timmermann team's work also suggests the climate-change-driven evolution of *H. heidelbergensis* from *H. ergaster*, 850–600 kya in eastern Africa. In central Europe, they suggest *H. heidelbergensis* evolved into *H. neanderthalensis* between 400–300 kya with, once more, climate change being the driver. This last is consistent with genetic evidence. Finally, they suggest that *H. heidelbergensis*' evolution into *H. sapiens* took place in southern Africa around 300–200 kya, as central Africa saw little overlap in either of these species' habit suitability whereas southern Africa did. In short, the Timmermann team have presented a plausible climate-driven human evolution scenario that chimes with what fossil, archaeological, and genetic evidence we have (*60, 61, 62*). Refining their climate-to-ecological environment model—as well as developing a more detailed model—would undoubtedly further clarify matters, especially focussing on those locales where the key evolutionary transitions took place.

In this chapter's co-evolution of life and planet narrative, the above largely focuses on timings and some of the Earth system aspects of the evolution of intelligence and our species. In turn, the following paragraphs examine some of the biological, ecological, and social/technological dimensions. This last is important, as technology development and use necessitates information and information-processing.

It would be easy to make the assumption that only *Homo sapiens* and some precursor species had the intelligence to make and use tools. This is not so, not only do other, the archaic, human spp. make and use tools, so do some animals. For example, chimpanzees (*Pan troglodytes*) have used stones to crack nuts but they do not make stones sharper through stone-on-stone chipping. It is, though, possible to teach some animals such tool-making. This has been done with captive bonobos (*Pan paniscus*), but the resulting stones are far from those associated with archaic *Homo* and their ancestor spp. (*63*).

There is one species, other than *Homo* spp. and *Paranthropus* sp., that self-learned to make stone tools and that's capuchin monkeys (*Sapajus libidinosus*) of the Serra da Capivara National Park in Brazil (*63*). They make and use these tools to break into nuts. However only human spp. and some of their ancestor species are known to have made stone tools specifically for cutting and scraping. Having said that, it is at this moment not known which human ancestral species first made stone tools. To date, the oldest evidence for tool use dates to 3.3 mya found in West Turkana, in the Kenyan part of the EARS, and this is before the first known *Homo* spp. (*64*). Of course, as contemplated earlier, it could be a Signor–Lipps effect, that we have yet to discover earlier remains of a *Homo* sp.

That other species show tool-using and tool-making capabilities demonstrates that the road to our own species' present-day, technology-wielding state is not unique, hence not just a lucky chance, even if other species have not travelled as far down this road as have we.

The reason for the need of our ancestral species having increased intelligence (hence information processing and retention) is the aforementioned cycling transitions within a longer-term shift from jungle and a heavily wooded environment to a more open landscape. Across the past half-dozen million years, we can infer the type of landscape in which they lived. It is possible to look at the $^{12}C$ and $^{13}C$ in palaeosols at hominin fossil sites on the basis that this reflects the ratio of these isotopes in organic matter (and not carbonate, which would reflect atmospheric carbon hence carbon drawdown and release—see Chapter 2). Organic matter's differing carbon isotopes indicate the types of plants present and whether they are $C_3$ trees or $C_4$ grasses. Such an analysis of 1,300 samples of ancient EARS soils reveals that hominin spp. tend to prefer landscapes with less than 40% wooded cover. Furthermore, tracking through 6 million years of time, as human spp. evolved, there was a preference for even more open landscapes. Humans in the Awash Valley of the EARS 130 kya (comparatively late in the evolution of *Homo* spp.) lived in landscapes with an average of just 10% tree cover. Conversely, those living 3–4 mya lived in landscapes with around 40% or more tree cover (*65*).

There are many theories as to 'why' human intelligence evolved; too many to go into in detail here. However, distilling these down, there are four categories of driver: (i) ecological/environmental (me *versus* nature), (ii) cooperative (us *versus* nature), (iii) competing individuals (me *versus* you), and (iv) competing groups of the same species (us *versus* them). In 2018, biologists Mauricio González-Forero and Andy Gardner of St. Andrews University modelled combinations of these factors against actual brain volumes of six *Homo* spp. factoring in metabolic and their other physiological dimensions. They found the combination of these four factors across all six *Homo* spp. saw ecological/environmental (me *versus* nature) and cooperative (us *versus* nature) were the two dominant factors, with no human species having these two factors accounting for less than 70% of evolutionary forcing. Indeed, for four species these two factors solely accounted for brain evolution. With specific regard to *H. sapiens* itself, the biologists concluded that the evolution of adult-sized brains and bodies is best explained when individuals face a combination of 60% ecological, 30% cooperative, and 10% between-group competitive challenges (*66*, *67*). Even though this is just one model, that ecology and cooperation so clearly dominate, does suggest that this is a reasonably robust conclusion. Furthermore, it is a conclusion that is easily explained.

As noted above, human evolution occurred across millions of years of repeated environmental change. It also saw archaic ancestor species, who lived in heavily wooded environments, evolve into anatomically modern humans who lived largely in more open grassland dominated environments. Open environments make it harder to hunt prey especially for a species that cannot run as fast as its prey species: in open environments prey can see predators further away and have the space to run. Overcoming these environmental problems, for predator species such as humans, necessitates skilful cooperation. Open environments also make it harder for humans to protect themselves against predators, especially when many predator species are both faster and stronger than us. Again, overcoming this problem necessitates cooperation. This, in turn, necessitates information exchange and intelligence.

There are many theories as to 'how' human intelligence evolved; again too many to go into in detail here. However, in essence, evolving larger brains necessarily involves other, concomitant anatomical changes.

Brains consume considerable energy; the bigger the brain the greater the energy consumption. The human brain is about three times larger than that of our closest living relative, the chimpanzee, and so requires much more energy. Adult human brains constitute around 2%–3% of an adult's body mass yet account for 20%–25% of energy intake (*68*). Yet, relative whole-body energy consumption rates of individuals at rest are about equal in the two species, which raises the question, how do humans manage to cover the energetic requirements of their much-enlarged brains? One hypothesis is that there is an evolutionary trade-off between the size of the brain and that of the digestive tract in anthropoid primates. The idea is that if other processes have reduced a species' energetic needs of digestion, it should be able to evolve a relatively larger brain (*69*).

One way to do this is to switch from an almost vegetarian diet to one more omnivorous: meat (protein and fat) is an energy rich food. Another energy-rich food source is below-ground plant tubers. Today, in the mixed woodland and open savanna landscape of Tanzania, there are some 40,000 kilograms of tubers *per* square kilometre. That there were tubers in the landscapes of early hominins is evidenced by the presence of fossil mole rats (*Tachyoryctes* spp.) that have been found alongside fossils of hominins: mole rats subsist almost entirely on tubers (*70*).

There is, though, a problem. By 3.3 mya *H. erectus* had relatively smaller teeth, reduced chewing muscles, weaker maximum bite force capabilities, and a relatively smaller gut compared to earlier *Homo* spp. as well as their ancestral species. Subsequent *H. sapiens* furthered this trend. So how did *H. erectus* and its smaller gut cope?

Today we process our food typically through mashing (*e.g.* mince and vegetable puree) and cooking, as well as undertaking both together (*e.g.* soup). Yet fire was not commonly used until about 500 kya. The answer appears to be that the selection for smaller masticatory eating features in *H. erectus* and *H. sapiens* was made possible by the combination of using stone tools to slice and pound both meat and tubers (*71*). As previously note, stone tools were in use by 3.3 mya, which is before *Homo* spp. arose and so allowing gut-shortening traits to arise prior to the evolution of our species (*72, 73*). Indeed, bones have been found with stone-created marks along with fossils of *Australopithecus afarensis* dating from at least 3.39 mya (*74*). So if stone tool use was already employed by the time of the first *Homo* sp. then, with increased dietary meat and tuber consumption, the evolutionary selective pressure to maintain both larger teeth, jaw muscles, and a long gut was removed. This freed up energy intake that could then go towards further brain development.

The other method to process food was cooking. There is unambiguous evidence—in the form of burned bone and ashed plant remains—of cooking by hominins dating from a million years ago in the Wonderwerk Cave, Northern Cape province, South Africa (*75*). This dating makes it likely that it was *H. erectus* that first cooked food (*76*).

As with there being some non-hominins that can appreciate and use stone tools, so there are non-human species that appreciate cooked food and are sufficiently aware of the cooking process that they will defer immediately available raw food now for cooked food later. This has been demonstrated in a remarkable study by two psychologists at Harvard University, Cambridge, Massachusetts, Felix Warneken and Alexandra Rosati. They looked at one of humans' closest relatives, chimpanzees (*Pan troglodytes*). Across nine experiments, they showed that they: (i) prefer cooked foods, (ii) comprehend the transformation of raw food that occurs with cooking, (iii) will defer food now for foods later, (iv) were willing to actively give up possession of raw foods in order to transform them, and (v) could transport raw food as well as save their raw food in anticipation of future opportunities to cook. They conclude that several of the fundamental psychological abilities necessary to engage in cooking may have been shared with the last common ancestor of apes and humans, predating the control of fire (*77*). Given this it would seem that the evolutionary leap for the ability of humans to cook food was as if pushing at an open door: it just took sufficient cognitive ability—in part enabled by the stone processing of meat and tubers—for the use of fire for cooking. This was a threshold moment in which *Homo* spp. transitioned to an evolutionary positive feedback state that furthered the development of intelligence.

With further cognitive ability, as evidence by increased cranial volume in successive *Homo* spp., so Stone Age technology developed further. Stone points identified as spear tips have been found in South Africa dating from 500 kya (*78*). And small stone artefacts identified as arrow heads found in South Africa have been dated to around 71 kya. This is within a time period that saw *Homo sapiens* migrate out of Africa. Had they been armed with bows and arrows, they would have been virtually unstoppable (*79*).

We are now well into the time of *Homo sapiens* and it is now worth pausing to reflect on the Earth system drivers of environmental change. Just as over the past few millions of years the Earth overall cooled towards what would be the Quaternary ice age of cyclical glacials and interglacials (see Figure 2.3A) that repeatedly challenged archaic hominins and *Homo* spp., so within the glacial and interglacial cycles there were still (smaller) jumps in global temperature which translated in regional shifting patterns of varying precipitation and aridity. The peak of the previous interglacial to the present spans some 130,000 years. Within this the deuterium record (a proxy for temperature, see Chapter 2) of both the Antarctic (southern hemisphere) and Greenland (northern hemisphere) shows a broadly similar pattern of change and so we can be confident that these reflect changes in the global climate.

Figure 8.6 shows the Antarctic deuterium record for the past glacial and interglacial cycle. The short, sharp jumps between states, though smaller than the interglacial-to-glacial and back flips of state, would still nonetheless have a regional environmental impact and so challenge *Homo sapiens*, spurring both its continued biological and cultural evolution.

Furthermore, the coolest parts of glacials would have seen more water locked up in continental ice sheets. This would lower sea levels, creating land bridges, so facilitating migration. Indeed, the coolest parts of the last glacial (22–30 kya and 60–72 kya) were when leaving Africa was easiest.

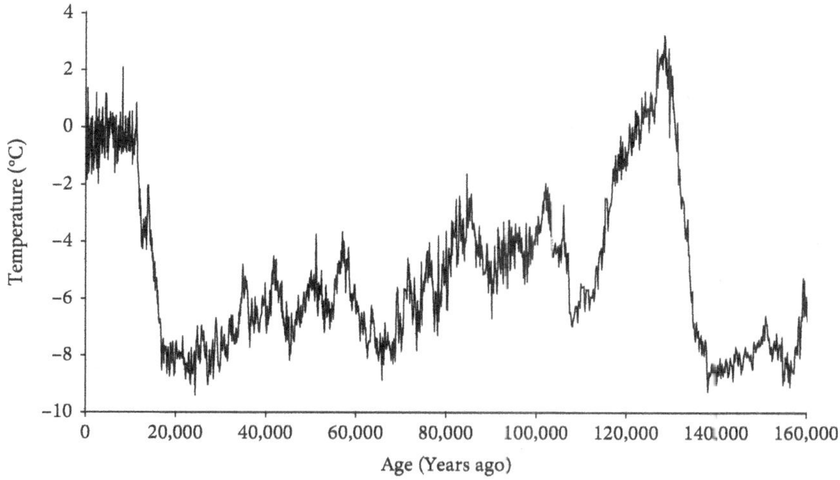


**Figure 8.6** The latest glacial–interglacial cycle with temperature determined by the Antarctic deuterium record. Data from National Oceanic and Atmospheric Administration as used in Cowie, J. *Climate Change: Biological and Human Aspects*, 2nd expanded ed. Cambridge University Press: Cambridge, Great Britain (2013). See also Figure 2.3A. The changing global climate saw changing regional environmental conditions in Africa spurring hominid and hominin evolution.

Here, in 2021 an international team led by Huw Groucutt, Abdullah Alsharekh, and Michael Petraglia, published their correlation of stone age archaeological sites in Arabia. They noted that these dated from times that coincided with brief 'green' windows of wetter times, where Arabia saw lakes and rivers, approximately 400, 300, 200, 130–75, and 55 thousand years ago. Their idea was that glacial times of low sea level enabled entry into Arabia from Africa and these 'green' windows permitted these populations to briefly flourish and for Arabia to be far more easily traversed so facilitating migration of these thriving populations into Eurasia (*80, 81*). Once more, it appears that regional effects arising from changes in the Earth system affect human ecology.

Increased intelligence did not just manifest improved Stone Age technology, there were cultural developments too that involved information exchange. One of these, of which some evidence survives through to today, is cave art. Abstract cave art has been found at the 73 kya Middle Stone Age levels at Blombos Cave, South Africa. It is the earliest cave art so far found (*82*). Subsequent, more developed cave art consisting of a figure appearing to be half human and half animal from the Fumane Cave, northwest of Verona, Italy, dates to between 32 kya and 36.5 kya. This makes it older than the art in Grotte Chauvet in southern France estimated to date from 32 kya (*83*). Further away, cave art depicting a hunting scene has been discovered in the limestone cave of Leang Bulu' Sipong 4, Sulawesi, Indonesia, that portrays several figures hunting wild pigs and small bovids (members of the biological family of cows, sheep, bison, and oxen). These date to at least 43.9 kya (*84*). All these examples date from the depth of the last glacial.

Up to and through this time *H. sapiens* led a hunter-gatherer existence. However, human ecology was to transform with settled farming: species would not be harvested from the wild but nurtured. This defines the Mesolithic-to-Neolithic transition; a transition that took place at different times in different places.

It is not certain what prompted this move. One theory, applying to the rise of farming outside of Africa, has it that it was the development of bread making: it would be easier to farm cereals for a commonly used food than to seek them out in the wild in a happenstance manner. What is certain is that bread making preceded this agrarian shift. The archaeological remains of charred food has been found in ancient fireplaces in Jordan and identified as being bread-like products dating to 14.2–14.4 kya. This precedes the emergence of agriculture by a few thousand years (*85*).

Alternatively, regarding dairying (which was also first practiced prior to the agrarian transition) it had been undertaken by nomadic communities. However, it too was easier to undertake in a settled agrarian setting. Certainly, dairying was practiced in North Africa 7 kya. Again, this precedes the emergence of agriculture in that region by a few thousand years (*86*).

Another hypothesis is that small changes in the Earth system resulted in significant local/regional changes in climate that made foraging in some places less feasible compared to the careful nurturing of crops. This spurred a move to regions supporting more vegetation, at which point communities became more settled (*87*, *88*).

Here, in Africa, a little background is required. You will recall that by around 11,000 years ago the Earth system had moved to a warmer interglacial (the Holocene) from a previous cold glacial mode. However, while the Holocene epoch has been one where the average global temperature has been comparatively stable, seasonality (the degree as to how varied each season was across the year) changed. This was because the northern hemisphere's summer solar energy received (insolation), having peaked early in the interglacial, began to decline. This in turn affected the African monsoon, hence the amount of water delivered to north Africa (*89*). And so a few thousand years into the Holocene the rainfall patterns in north Africa changed.

Leaving seasonality aside for a moment, the overall comparative stability of the global climate across the Holocene has only recently been plausibly accounted for. Research shows that the summer northern hemisphere energy received from the Sun was just a little higher early in the Holocene from 11 kya and then very slightly declined to 7 kya before slightly rising again. The same was also true of greenhouse gases (carbon dioxide, methane, and nitrous oxide): both these factors served to provide warmth early in the Holocene. Conversely, the remnants of the glacial ice-sheets served to cool the Earth: their still-large area immediately after the glacial reflected more sunlight. Meanwhile, seasonal effects early in the Holocene served to counteract this cooling.

Later in the Holocene matters reversed. There was declining seasonal energy received, which provided a cooling effect and, conversely, a slow, slight increase in greenhouse gas warming from 6 kya together with the now smaller ice-sheet reflection that both had a warming effect: the later Holocene saw reduced seasonal warming but less cooling from reduced icesheets so serving to cancel each other, while the rise in greenhouse gases from the mid-Holocene up to before the industrial revolution were more or less constant (with only a very minimal rise). So at the global level, the Holocene has had a comparatively stable climate (*90, 91*). This leaves the change in seasonality that had regional climate effects.

Early in the Holocene, the area that is now the Sahara Desert was much wetter, with rivers and lakes, and supporting grassland. This was the time of the green Sahara in the African Humid Period. It was the latest of series of more humid N. African periods of varying degree that took place every 20,000 years and going back 9 million years (*92*). With seasonality change, the African Humid Period ended, somewhere around 5–6 kya, in a shift—a transition—towards the arid Sahara we see today. It is possible that this began earlier, 7–8 kya, but either way once begun, as the vegetation was lost, albedo effects kicked in: sand is more reflective than vegetation. The effect of this was to further alter atmospheric circulation, promoting the reduction in the African monsoon, so amplifying the transition (*93*). Beyond North Africa, a similar shift in climate took place in East Africa, so the transition had impacts elsewhere (*94*).

With declining rainfall, humans would have abandoned the arid areas and confined themselves to those restricted areas that still had water. This, the argument goes, encouraged the establishment of permanent settlements and farming (*95*).

The first farms we know of were established outside of Africa. These date from 11.7 kya. This was right at the start of the Holocene following the last glacial maximum (LGM). You will recall that the LGM and its lower sea levels, some 22–30 kya, saw land bridges that made it easier to leave Africa and this time was the latest of the African exoduses made by various *Homo* spp. So *Homo sapiens* was already firmly established beyond Africa by 11.7 kya.

At this time there is archaeological evidence of early farming in what is today Iran (at Chogha Golan and Sheikh-e Abad). These farmers grew wild barley (*Hordeum spontaneum*) and other wild progenitor species of modern crops, such as lentil and pea. Subsequently at that site domesticated-type emmer wheat appears 9.8 kya. Between 11.7 and 11 kya farming became established in an arc from Jericho in the southern Levant through to northern Iran (*96*).

This pattern of first farming wild, and then developing domesticated, species was the typical model repeated elsewhere.

Nine to five thousand years ago saw the domestication of many crops in many locations across the globe. For example, rice (*Oryza* spp.) was domesticated independently twice: once in Africa (*Oryza glaberrima*) and once in Asia (*Oryza sativa*), both from different ancestral species, *O. barthii* and *O. rufipogon*, respectively. Genomic analysis indicates that the population of wild African rice was in decline beginning from 10 kya to reach a minimum about 3.8 kya. This decline is thought to have spurred domestication in the Inner Niger Delta of West Africa. The analysis supports the hypothesis that the drying of the Sahara, with the end of the African Humid Period, directly depleted wild resources and eventually led to domestication (*97*).

The domestication of wheat and barley subsequently took place in Turkey and the Middle East around 10–9 kya. And beyond Africa and Asia, Peru saw the domestication of cotton (*Gossypium barbadense*), squash (*Cucurbita* spp.), chilli (*Capsium* spp.), beans (*Phaseolus vulgaris* and *Phaseolus lunatus*), lucuma (*Pouteria lucuma*, formerly known as *Lucuma obovata*), guava (*Psidium guajava*), pacay (*Inga feuillei*), camote (*Ipomoea batatas*), avocado (*Persea Americana*), maize (*Zea mays*), and achira (*Canna edulis*). There are many other instances of species domestication early in the Holocene (*98*). That species domestication took place independently on different continents demonstrates that farming and the domestication of many species was a very successful innovation and an 'easy' step.

So successful was the shift from a hunter-gather to agrarian society that the increased food yields enabled support of a cadre of the population not involved in food production. For example, in Egypt from 4.78–3.625 kya (the age of the Pharaohs), it is estimated that Egypt had a population of 3 million and had a food surplus that freed about 5% of the population for other activities including belonging to the ruling elite and the construction of the pyramids. It has been calculated that the construction of the pyramid of Cheops alone would have necessitated the consumption of Egypt's entire agricultural surplus for the couple or so of decades required. Later, following the peak period of pyramid construction, some of this surplus was used to support a substantive military enabling ancient Egypt for a while to control its neighbours (*99*).

In 2019 the Archaeoglobe Project synthesised contributions of over 250 archaeologists as to land use globally over the past 10,000 years. It concluded that by 3,000 years ago the planet was largely transformed by hunter-gatherers, farmers, and pastoralists (*100*).

This seemingly flies in the face of the more accepted view of a recent transformation. Even 12,000 years ago, nearly three-quarters of Earth's land was inhabited and therefore shaped by human societies, including more than 95% of temperate and 90% of tropical woodlands. Having said that, while large-scale environmental change may not have been as recent as commonly thought, the intensification of land-use (intensive agricultural production through to high population concentrations of urban development) is a very recent phenomenon to which we will return in the next chapter (*101*).

Over three thousand years ago, moving from a hunter-gatherer to a settled agrarian existence also necessitated information and information exchange. Farming took more cooperation and planning: cooperation in terms of who was agreed to grow what, where, and planning for exactly when in the year to sow and reap. If you yourself are an active gardener, then you will know that (without a greenhouse) it is vital to plant at a specific time to within a week: not too early to prevent frost kill and not too late so as not to waste what biologists call the 'thermal growing season'.

Consequently, from the outset when contemplating an agrarian lifestyle, communities had to have their 'wise' men and women to preserve and disseminate information: it is vital to be able to instruct the community, for example, to plant this crop two or three full Moons after the winter solstice. This is why even with more primitive societies than Egypt's at that time, there was a need to accurately time the agricultural year and for the community to spend considerable effort to ensure this information processing and conveyance. For example, famously when in England, 4.3 kya, Stonehenge was constructed, it was done with a design so as to be able to identify certain times of the year: specifically, Stonehenge had stones that align with the winter and summer solstices. Similarly, in Ireland, the famous Newgrange passage tomb was constructed around 5.1 kya at a time when farming was becoming established. This mound incorporates over 200,000 tonnes of earth and stone. Notably, the passage itself was constructed to be aligned in a way so that the Sun shines along it only on the winter solstice (*102*). Solstice-aligned passage tombs, not dissimilar to Newgrange, were also constructed in Wales, Orkney, and Brittany (*103*). All these were places and at times where farming had become established.

Such 'wise' men and women would have been part of, or at least associated with, the ruling elite and there was often an understandable desire to ensure that this status was passed down to members of the elite's offspring. Hence, it is common for Royalty to be dynastic. Indeed, with specific regards to the Newgrange passage tomb, a genomic analysis of the remains of 44 individuals across a number of tombs from this time is particularly revealing. The genome of one individual from Newgrange was the adult son of a first-degree incestuous union of another. Others were relations including relatives of this individual within two other major complexes of passage tombs 150 km to the west of Newgrange. As this time was a period of sea-going western seaboard European colonisation, in-breeding due to population isolation is precluded (*104*).

These are just a few examples of the value of information processing to early agrarian societies. Today, the timing of sowing or planting and then reaping different crops at different times of the year is just as critical so as to obtain the greatest yield. Having said that, even without humans being able to appreciate and use annual cycle timing intellectually (through being able to process and hold information), there are many examples of timing being important in biology and used without being mediated through information communication. You will recall many species, especially outside of the tropics, follow an annual cycle. Indeed, covered earlier (see Chapter 1) it was noted, among other examples, that the polychaeta worm, *Platyneriseis dumerilii*, relies on the Lunar-cycle-mediated monthly tides to time its reproduction.

Clearly, sentience or intelligence, with its concomitant quality of processing and retaining information, has conferred an unprecedented advantage for our species eventually enabling it (as noted earlier this Chapter) to flourish in terms of numbers, impact on other species, harness energy, create new materials, and even (unintentionally) alter global atmospheric and ocean chemistry. However, though we can chart the evolution of human intelligence through, for example, brain-to-body mass ratios across species, it is difficult using cave art or the construction of annual cycle time-keeping artefacts (stone circles and burial mounds, even if such constructs also serve other societal purposes) to quantify the amount of information being handled such as in terms of bytes. Fortunately, a subsequent development makes this easier.

Being able to support a proportion of the community not involved in food production, and recognising the value of information exchange, virtually inevitably led to writing.

According to Mesopotamian legend, it was a high priest from the city of Uruk (in what is today Iraq) who invented writing around 5,400 years ago give or take a century or so: Uruk being the largest settlement of its time in Babylonia that used writing mainly for administrative/commercial purposes. Indeed archaeologists in the latter half of the 20th century did find the then-oldest examples of writing—cuneiform—in the region in the ancient middle east and beyond (*105*). However, towards the end of the 20th century earlier examples of writing were found. (This is another instance of the Signor–Lipps effect: see earlier this chapter as well as Chapters 4 and 6.) Storage pots have been found in Iran with markings on them that date from 9 to 4 kya. In Egypt inscriptions on pots have been found dating to around 5.5 kya. In Pakistan the Harappan civilisation used a script from at least 4.8 kya (*106*).

While it is not known if the concept of writing began in one place and was then conveyed by traders, or whether it arose a number of times independently (the latter being more compelling, given the diversity of ancient marking systems), it does appear that at least in one place (possibly many) there was a distinct chain of events. Hunter-gathers became settled agrarians, the resulting food surplus enabled a more sophisticated community, this in turn led to proto- as well as proper civilisations, and these shortly developed a form of writing.

You will be well aware of what comes next: the development of block printing, followed by print using moveable type, and finally, text stored, transmitted, and displayed electronically.

One useful property of writing is that we can quantify it in a number of ways, such as the number of letters or words. Here we might use as a unit of quantification the byte. Further, we might consider writing and print as analogue storage of data and that on computers and data servers as digital storage.

As previously noted, there has been increase in 'information' associated with each of the flips of Earth state. We previously (in this chapter and also Chapter 6) touched on new mineral species rising following each of the major flips of the Earth's state and so that the information needed to define the content of the Earth's crust had to increase. But biology also relies on information, and when the Earth shifted from a purely chemical (non-biological) equilibrium to one perturbed by biology it was necessary to have 'information' inherent in species' genetic code.

Now, before continuing, let us unpack this genetic code information concept and then see how it relates to the way we humans process information using technology today. This will be done in the next chapter subsection.

Meanwhile, this chapter subsection has spent a considerable time reviewing human evolution. This is important as, while humans prior to the industrial revolution did not have a significant impact on the Earth system (compared to today), critical transitions within the components of the Earth system (especially with regards to Africa's climate) were important to our species' evolution. This contrasts with major, fundamental, innovations in biological evolution (such as oxygenic photosynthesis and multicellularity) that did have a major impact on the Earth system causing it to pass through a critical transition: some critical transitions in the Earth system were driven by biological innovation and other Earth system critical transitions impacted on biological evolution. Both take place and we need to be aware of them and to distinguish between the two.

Now let us return to biological information and how it compares with information generated by technology.

## Biological information and information technology

At the end of the day, genetic information is information albeit information in chemical form. Indeed, it is possible to use DNA to store information that normally we would do as writing. Not only that but DNA can be used to encode a digital PDF and even JPEG pictures. In 2013, researchers from the European Bioinformatics Institute in England, did just that and successfully stored all 154 of Shakespeare's sonnets (as ASCII text), a classic scientific paper (in PDF format), a medium-resolution colour photograph of the European Bioinformatics Institute (JPEG 2000 format), and a 26-second excerpt from Martin Luther King's 1963 'I have a dream' speech (in MP3 format), coded in the form of DNA. The researchers also presented the case for using DNA as an archival storage medium. They noted that, as with any storage system, a large-scale DNA archive would need stable management. But whereas current digital schemes for archiving require active and continuing maintenance and regular transferring between storage media, the DNA-based storage medium requires no active maintenance other than a cold, dry and dark environment (such as the Global Crop Diversity Trust's Svalbard Global Seed Vault, which has no permanent on-site staff yet remains viable for thousands of years even by conservative estimates) (*107*).

So, if considering nucleic acid as an information store, then not only was the evolution of prokaryotes the establishment of life, it was also the start of cellular bio-information.

Moving on to the second flip of Earth's state, assuming that eukaryotes evolved prior to Snowball I (remember this has not been proved but is a distinct possibility) and that eukaryotic oxygenic photosynthesisers were involved in the first oxygenation event associated with Snowball Earth I, then again information increase (with eukaryotic DNA) was involved: prokaryotic DNA was endosymbiotically incorporated into the eukaryotic cell (see Chapter 4). Even if eukaryotes evolved after the Snowball I transition, cyanobacteria certainly had evolved before then and the bolting together of two photosystems (see Chapter 3) would have necessitated additional DNA: either way, more nucleic acid information was needed to enable oxygenic photosynthesis.

Next, Snowball Earth II and the second rise in atmospheric oxygen saw the involvement of multicellular species and with the colonisation of land

and rise of larger species, there was more nucleic acid, hence more bio-information, in the biosphere. Further, multicellularity subsequently saw brains evolve with their ability to process and store information. This ultimately led to the evolution of *H. sapiens sapiens* and finally, we have present-day computing: our species, its own evolution and subsequent technology, is part of a new increase in information processing unparalleled in the Earth's history.

We can now bring this information dimension to each of the flips of Earth state all together. Fortunately, here much of the heavy lifting has been done by the collaboration of Michael Gillings and Darrell Kemp from Macquarie University, Australia, and Martin Hilbert of the University of California. In 2016, they reviewed the amount of information in the biosphere in biological as well as analogue and digital forms. As noted, relating analogue and digital information to Bytes is straightforward and so is considering biological information in terms of the number of bytes. With regard to nucleic acid (RNA and DNA) four base pairs will code for one amino acid, and this can be considered a Byte or alternatively a symbol (letter, character, or single digit number).

Michael Gillings and Darrell Kemp note that the accumulation of digital information is happening at an unprecedented speed. After RNA genomes were replaced with DNA (assuming you buy into the RNA world hypothesis), it then took a billion years for eukaryotes to appear, and roughly another 2 billion for multicellular organisms with a nervous system. It then only took an additional half billion years to develop neural systems capable of forming languages. From there, it took roughly just 100,000 years to develop written language, and only a further 4,500 years before the invention of printing presses capable of rapid replication of this written information. The recent digitalisation of technologically mediated information has taken less than 30 years: less than 1% of information was in digital format in the mid-1980s, growing to more than 99% by the mid-2010s.

Their review of the academic literature provided estimates for the amount of information held in different forms: both biological and non-biological. They took as the value of the total amount of DNA contained in all the cells on Earth to be about $5.3 \times 10^{37}$ base pairs, equivalent to $1.325 \times 10^{37}$ bytes of information. With regards to just human DNA, they took a 2015 global population estimate of 7.2 billion, each with a genome of 6.2 billion base pairs, so in total being equivalent to $1 \times 10^{19}$ bytes. They also consider the amount of information in analogue form as print as well as electronic in digital form. Their estimates inspire Figure 8.7, which is an estimate of various forms of biological and non-biological information across time (*108*).

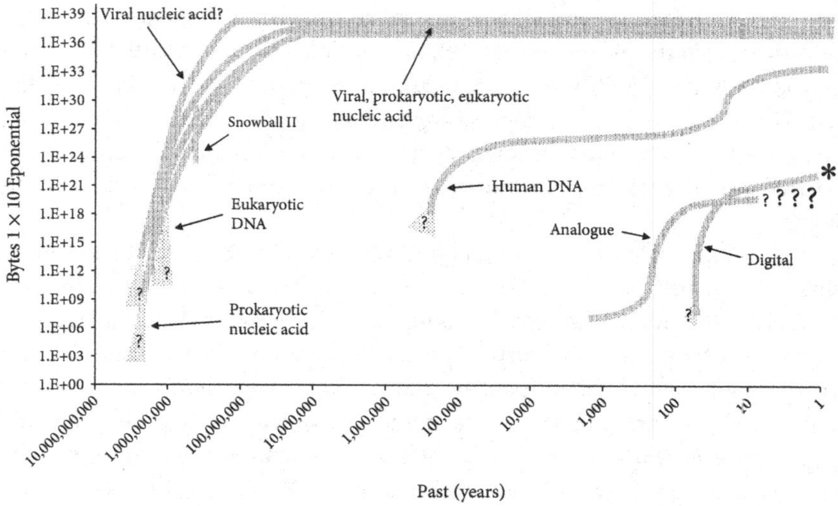


**Figure 8.7** The approximate amount of different types of information across deep time. Note, both the axes are logarithmic. That is, they increase by exponentials: 1, 10, 100, 1,000 (not 1, 2, 3, 4 . . .). 1.E + 03 = $10^3$. The figure is inspired by ref. *108*, Gillings, M. R., Hilbert, M., & Kemp, D. J., Information in the biosphere: Biological and digital worlds. *Trends in Ecology and Evolution*, 31 (3), 100–108 (2016). The early estimates and timings are largely the author's based on the discussion in Chapters 1–4.

*Note*: There is uncertainty, hence the thickness of these lines which—due to both the logarithmic axis—covers a range of values on the left and top of the graph. The viral genetic information is highly speculative, hence the question mark by its label. Here there is also uncertainty as to the rise of information in prokaryotic nucleic acid compared to viruses as to whether they were commensurate or whether there was a significant lag of the former behind the latter, hence also some of the lines' basal question marks. Similarly, there is debate as to the timing of the rise of eukaryotes: here the figure's basal question mark straddles Snowball Earth 1 (not labelled, as where it would be is smothered by other lines due to the logarithmic scale). So I fully accept that some may (will) disagree with a few of this graph's finer details. However, we do know that, following land colonisation, plants (which are of course eukaryotes) dominated global biomass. We also know that for the past couple of decades the global digital (electronic) information capacity has grown at a more or less steady exponential rate: hence the straight line. Here Rob Grant (see this book's Introduction and Acknowledgements) kindly provided the rough estimate for 2025, which is denoted by an asterisk. Conversely, there is little in the academic literature as to amount of human analogue (writing and print) information in recent years (which may be growing or even decreasing in the face of digital's rise) and this uncertainty is represented by the string of question marks

Given the rise of digital information in the 20th century and its continued growth in the 21st, we appear to be on a digital information trajectory that may, in time, begin to rival some forms of biological information. Drawing parallels of this, with the evolution of information modes associated with the major flips of Earth state, is tempting.

## The prospect of a new flip

It is now worth revisiting the characteristics associated with a biological evolution mediated flip of the Earth's state. You will recall it is possible to distinguish between the flips of state mediated by prior biological evolution (albeit timed due to geological factors) from other (non-evolution-related) transient events in the Earth system, such as the end-Cretaceous extinction that did in the dinosaurs. (See Chapter 6.) We can now modify Table 6.1, taking out these other transient events and adding the current set of circumstances. This is portrayed in Table 8.1.

Other than life's beginning affecting the global climate being unproven (we simply do not have the data from that far back), it almost inevitably occurred: once established, a significant population of prokaryotes would both draw-down atmospheric carbon and alter the atmospheric mix of methane and carbon dioxide from that determined by purely chemical (non-biological) equilibrium, hence in turn the atmosphere's greenhouse properties (global climate change). Today we see all the hallmarks of biological evolution once more resulting in the Earth flipping state. Other than the caveats noted earlier this chapter regarding

**Table 8.1** The principal characteristics of the major co-evolution of life and planet transitions.

| Event | Carbon cycle perturbation (CIE) | Climate change | Evolution development | New environment colonised | Information processing enhanced |
|---|---|---|---|---|---|
| Life begins | ✓ | ? (Likely) | ✓ (Life arises) | ✓ (Hot springs/vents) | ✓ (First nucleic acid) |
| Snowball Earth I | ✓ | ✓ | ✓ (Eukaryotic OP) | ✓ (Sea surface) | ✓ (Incorporation of prokaryotic DNA) |
| Snowball Earth II | ✓ | ✓ | ✓ (Multicellularity) | ✓ (Land) | ✓ (development of animal brains) |
| Human technology | ✓ | ✓ | ✓ (Sentient species) | ✓ (Earth orbit +) | ✓ (computing/AI) |

eukaryotes colonising new environments (prokaryotes did inhabit the sea surface, but oxygenic-photosynthesising [OP] species were the ones that could harvest more of the light spectrum and the first boom of [if not] OP eukaryotes [then certainly cyanobacteria] tipped the balance, pushing the Earth system across a critical threshold), all the features of the previous flips of Earth state—the perturbation of the carbon cycle, a global climate change event, the evolution of a novel biological innovation, a new environment colonised, and enhanced information processing—are exhibited today!

The evolution of a species (us), capable of creating and wielding sophisticated technology, is driving the current transition. We humans have short-circuited the deep carbon cycle in using fossil fuels causing a carbon isotope excursion (see Figure 8.1). Here, there is concomitant global warming, and we are now beginning to inhabit a new environment. With a sustained presence beyond the biosphere in Earth orbit and the prospect of a permanent presence on the Lunar surface, our new environment is space. Finally, with the development in turn of complex language, writing, printing, digital storage of information, and processing thereof (which now includes artificial intelligence), human technology has seen new ways for information to be processed.

We are now at the end of our arm's-length journey through the Earth's existence and it is possible to chart the key events in the co-evolution of life and planet narrative. This is summarised in a simplified, but to-scale, timeline (see Figure 8.8).

Starting with the formation of the Earth, Figure 8.8 charts the key developments in the intertwining strands of our planet's geological and biological evolution. Excluded for clarity's sake are the times of commencement of long-term periods of new mineral species arising (see Chapter 6). Suffice it to say that they all are associated with flips of Earth state and that today humanity is creating new materials that are entering the geological record as novel mineral species.

The italic text across the top of Figure 8.8 relates to the state of the Earth system.

The Earth system was originally in chemical equilibrium. At some unknown point (exemplified by the question marks), but relatively shortly after (or possibly during the Late Heavy Bombardment—LHB?) life arose and then biologically perturbed the Earth system with the life-driven component to the carbon cycle. There was then a critical transition during Snowball I from an anaerobic biosphere to one partially oxygenated. Another critical transition at Snowball Earth II saw the Earth system move into the fully oxygenated state seen today.

The previous flips of Earth state have all seen the Earth system go through a critical transition. Other than the original rise of life, these transitions have each seen extinctions. The two oxygenation events saw anaerobic species retreat.

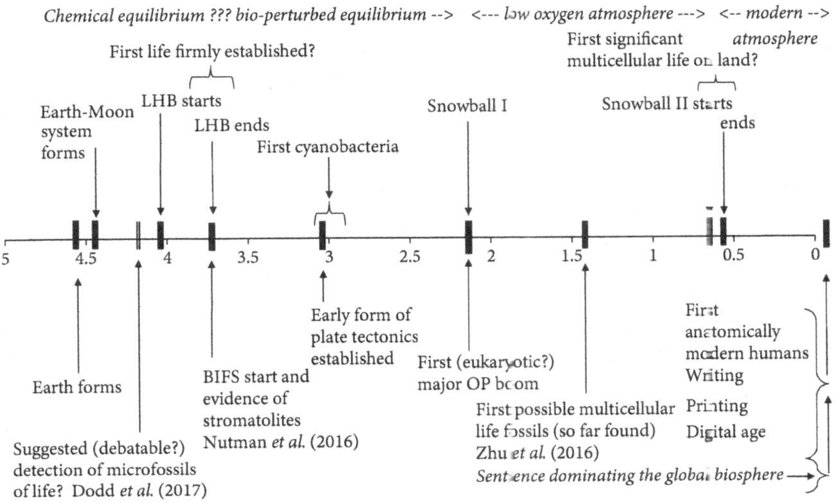


**Figure 8.8** A to-scale simplified timeline of the key co-evolution of life and planet elements across the Earth's existence. Italic text relates to the state of the Earth system. Note: Estimates for first cyanobacteria, first eukaryotic boom, and first multicellular life are the author's: readers should consult the academic literature for the thoughts of others. (For in-diagram references see Chapter 1 references *28* & *30*, and Zhu, S., Zhu, M., Knoll, A. H., *et al.*, Decimetre-scale multicellular eukaryotes from the 1.56-billion-year-old Gaoyuzhuang Formation in North China. *Nature Communications*, 7, 11500 (2016).)

The rise of a sentient, technology-wielding species similarly has seen, and is seeing, a wave of species extinction. This began with the prehistoric extinction of megafauna (large animals) outside of Africa as these species had not co-evolved with *Homo* spp. It is continuing today with the urbanisation and the loss of wild habitat to agriculture as our population increases. It is also continuing due to global warming and, relatedly, the acidification of the oceans.

These previous flips saw a dramatic change in the global environment, driven by the rise of new species.

In ecology, when a species shapes its environment, this is known as 'niche construction' and species that do this are 'ecosystem engineers'. One of the classic examples of niche construction by ecosystem engineers are beavers, whose dams alter their river environment and biodiversity (*109*). With the rise of oxygenic photosynthesising species, and then multicellular species, there was ecosystem engineering on a global level. Our own species, *Homo sapiens*, has also been engaging in ecosystem engineering and niche construction for thousands of years. It is easy to point to the negative impacts of human niche construction such as biodiversity loss, megafaunal extinctions, and so forth. However, the benefits to our species have been considerable and include the creation of landscapes capable of supporting increasingly dense human populations and the creation of biomes that enhance human food supplies and reduce exposure to predators and natural dangers (*110*). As we shall see in the next chapter, our own species is also now affecting the planet on a near global level. (The far right of Figure 8.8 represents where we are imminently heading if not, arguably, are.)

Of course, on a pedantic level, that current times do feature the hallmarks of past major transitions of the Earth system is not in itself actual proof that we are currently undergoing a transition of such significance: it could be just coincidence. This, I contend, is a moot point. With climate change, we are seeing a change in the Earth system. We are replacing highly diverse natural ecosystems with managed ecosystems many of which are dominated by agricultural monocultures and are of low biodiversity (*111*). Therefore, irrespective of whether events now are comparable to the evolution of oxygenic photosynthesis and the rise of eukaryote, or akin in significance to multicellular life evolving, these environment and human development issues need addressing most urgently. Understanding how the Earth system responds to perturbations, and knowing how biological and geological evolution have interacted, instigating critical transitions, can only help in our being able to recognise and manage the changes that are already beginning to take place.

## References

1. Caldeira, K., & Kasting, J. F., The life-span of the biosphere revisited, *Nature*, **360**, 721–723 (1992).
2. Fourier, J., Mémoire générale sur les températures du globe terrestre et des espaces planétaires. *Mémoires de l'Académie Royal des Sciences*, **7**, 569–604 (1827).
3. Tyndall, J., On the absorption and radiation of heat by gases and vapours, and on the physical connexion of radiation, absorption and conduction. *The London, Edinburgh, and Dublin Philosophical Magazine and Journal of Science*, Series 4, **22**(146), 169–194, and **22**(147), 273–285 (1861).
4. Arrhenius, S., On the influence of carbonic acid in the air and upon the temperature of the ground of the absorption and radiation of heat by gases and vapours, and on the physical connection of radiation, absorption and conduction. *Philosophical Magazine*, **41**, 237–276 (1896).
5. Food and Agriculture Organization, *The State of Food Insecurity in the World* [various annual reports]. Rome: UN Food and Agriculture Organization (2001, 2003, 2004, 2008, 2009, 2010, 2011).
6. Food and Agriculture Organization, *The State of Food and Agriculture: Climate Change, Agriculture and Food Security*. Rome: UN Food and Agriculture Organization (2016).
7. Callendar, G. S., The artificial production of carbon dioxide and its influence on temperature. *Quarterly Journal of the Royal Meteorological Society*, **64**, 223–237 (1938).

8. Sawyer, J. S., Man-made carbon dioxide and the greenhouse effect. *Nature*, **239**, 23–26 (1972).
9. Intergovernmental Panel on Climate Change, *Climate Change 2023: Synthesis Report.* Intergovernmental Panel on Climate Change: Geneva, Switzerland (2023).
10. JASON Defence Advisory Group, *The Long Term Impact of Atmospheric Carbon Dioxide on Climate JSR-78-07.* Mitre Corporation: Virginia, USA (1979).
11. Tickell, C., *Climatic Change and World Affairs*, 2nd edition. University Press of America: Lanham, Maryland, US (1986).
12. Intergovernmental Panel on Climate Change, *Climate Change: The IPCC Scientific Assessment.* Cambridge University Press: Cambridge, Great Britain (1990).
13. Intergovernmental Panel on Climate Change, *Climate Change 1995: The Science of Climate Change—Summary for Policymakers and Technical Summary of the Working Group I Report.* Cambridge University Press: Cambridge, Great Britain (1995).
14. Intergovernmental Panel on Climate Change, *Climate Change 2001: The Scientific Basis—Summary for Policymakers and Technical Summary of the Working Group I Report.* Cambridge University Press: Cambridge, Great Britain (2001).
15. Intergovernmental Panel on Climate Change, *Climate Change 2007: The Physical Science Basis—Working Group I Contribution to the Fourth Assessment Report of the Intergovernmental Panel on Climate Change.* Cambridge University Press: Cambridge, Great Britain (2007).
16. Intergovernmental Panel on Climate Change, *Climate Change 2013: The Physical Science Basis. Contribution of Working Group I to the Fifth Assessment Report of the Intergovernmental Panel on Climate Change.* Cambridge University Press: Cambridge, Great Britain (2013).
17. Intergovernmental Panel on Climate Change, *Climate Change 2021: The Physical Science Basis. Contribution of Working Group I to the Sixth Assessment Report of the Intergovernmental Panel on Climate Change.* Cambridge University Press: Cambridge, Great Britain (2021).
18. Oreskes, N., The scientific consensus on climate change. *Science*, **306**, 1686 (2004).
19. International Commission on Stratigraphy, *International Chronostratigraphic Chart v2018/07* (2018). Available from http://www.stratigraphy.org/ICSchart/ChronostratChart2018-07.pdf.
20. Cullen, H. M., deMenocal, P. B., Hemming, S., *et al.*, Climate change and the collapse of the Akkadian empire: Evidence from the deep sea. *Geology*, **28** (4), 379–382 (2000).
21. British Petroleum, *BP Statistical Review of World Energy 2020.* British Petroleum, London, Great Britain (2020).
22. Energy Institute, *Energy Institute Statistical Review of World Energy 2023.* Energy Institute, London, Great Britain (2023).
23. Syvitski, J., Waters, C. N., Day, J., *et al.*, Extraordinary human energy consumption and resultant geological impacts beginning around 1950 CE initiated the proposed Anthropocene Epoch. *Communications Earth & Environment*, **1** (32), 1-13,ff10.1038/s43247-020-00029-yff (2020).
24. Waters, C. N., Zalasiewicz, J., Colin Summerhayes, C., *et al.*, The Anthropocene is functionally and stratigraphically distinct from the Holocene. *Science*, **351**, aad2622 (2016).
25. Owen, P., Unwanted epoch. *Geoscientist*, **31** (4), 11 (2021).
26. Vitousek, P. M., Ehrlich, P., Ehrlich, A., & Matson, P. M., Human appropriation of the products of photosynthesis. *Bioscience*, **36**, 368–373 (1986).
27. Imhoff, M. L., Bounaoua, L., Ricketts, T., Loucks, C., Harriss, R., & Lawrence, W. T., Global patterns in human consumption of net primary production. *Nature*, **429**, 870–873 (2004).

28. Lau, W. W. Y., Shiran, Y., Bailey, R. M., *et al.*, Evaluating scenarios toward zero plastic pollution. *Science*, **369**, 1455–1461 (2020).
29. Elhacham, E., Ben-Uri, L., Grozovski, J., *et al.*, Global human-made mass exceeds all living biomass, *Nature*, **588**, 442–444 (2020).
30. Bauer, A. M., Correspondence—Anthropocene: Event or epoch? *Nature*, **597**, 332 (2021).
31. Roth G., Convergent evolution of complex brains and high intelligence. *Philosophical Transactions (B) of the Royal Society*, **370**, 20150049 (2015).
32. Hochner, B., Shomrat, T., & Fiorito, G., The octopus: A model for a comparative analysis of the evolution of learning and memory mechanisms. *Biological Bulletin*, **210**, 308–317 (2006).
33. Godfrey-Smith, P., Cephalopods and the evolution of the mind. *Pacific Conservation Biology*, **19**, 4–9 (2013).
34. Böhme, M., Spassov, N., Fuss, J., *et al.*, A new Miocene ape and locomotion in the ancestor of great apes and humans. *Nature*, **575**, 489–493 (2019).
35. Kivell, T. L., Fossil ape hints at how bipedal walking evolved. *Nature*, **575**, 445–446 (2019).
36. Suwa, G., Kono, R. T., Katoh, S., *et al.*, A new species of great ape from the late Miocene epoch in Ethiopia. *Nature*, **448**, 921–924 (2007).
37. Trauth, M. H., Maslin, M. A., Deino, A., & Strecker, M. R., Late Cenozoic moisture history of East Africa. *Science*, **309**, 2051–2053 (2005).
38. Magill, C. R., Ashley, G. M., & Freeman, K. H., Ecosystem variability and early human habitats in eastern Africa. *Proceedings of the National Academies of Sciences USA*, **110** (4), 1167–1174 (2013).
39. Caley, T., Extier, T., Collins, J. A., *et al.*, A two-million-year-long hydroclimatic context for hominin evolution in southeastern Africa. *Nature*, **560**, 76–79 (2018).
40. Plummer, T. W., Oliver, J. S., Finestone, E. M., *et al.*, Expanded geographic distribution and dietary strategies of the earliest Oldowan hominins and Paranthropus. *Science*, **379**, 561–566 (2023).
41. Lovejoy, C. O., Reexamining human origins in light of *Ardipithecus ramidus*. *Science*, **326**, 74–74e8 (2009).
42. White, T. D., Asfaw, B., Beyene, Y., *et al.*, *Ardipithecus ramidus* and the paleobiology of early hominids. *Science*, **326**, 75–86 (2009).
43. Lovejoy, C. O., Suwa, G., Spurlock, L., *et al.*, The pelvis and femur of *Ardipithecus ramidus*: The emergence of upright walking. *Science*, **326**, 71e1–71e6 (2009).
44. Louchart, A., Wesselman, H., Blumenschine, R. J., *et al.*, Taphonomic, avian, and small-vertebrate indicators of *Ardipithecus ramidus* habitat. *Science*, **326**, 66–66e4 (2020).
45. Haile-Selassie, Y., Gibert, L., Melillo, S. M., *et al.*, New species from Ethiopia further expands Middle Pliocene hominin diversity. *Nature*, **521**, 483–488 (2015).
46. Spoor, F., The middle Pliocene gets crowded. *Nature*, **521**, 432–433 (2015).
47. Henry, A. G., Ungar, P. S., Passey, B. H., *et al.*, The diet of *Australopithecus sediba*. *Nature*, **487**, 90–93 (2012).
48. Schoeninger, M. J., The ancestral dinner table. *Nature*, **487**, 42–43 (2012).
49. Villmoare, B., Kimbel, W. H., Seyoum, C., *et al.*, Early *Homo* at 2.8 Ma from Ledi-Geraru, Afar, Ethiopia. *Science*, **347**, 1352–1355 (2015).
50. Zaidner, Y., Centi, L., Prévost, M., *et al.*, Middle Pleistocene *Homo* behavior and culture at 140,000 to 120,000 years ago and interactions with *Homo sapiens*. *Science* **372**, 1,429–1,433 (2021).
51. Hershkovitz, I., May, H., Sarig, R., *et al.*, A Middle Pleistocene *Homo* from Nesher Ramla, Israel. *Science*, **372**, 1424–1428 (2021).
52. Hublin, J.-J., Ben-Ncer, A., Bailey, S. E., *et al.*, New fossils from Jebel Irhoud, Morocco and the pan-African origin of *Homo sapiens*. *Nature*, **546**, 289–292 (2017).

53. Richter, D., Grün, R., Joannes-Boyau, R., Steele, T. E., The age of the hominin fossils from Jebel Irhoud, Morocco, and the origins of the Middle Stone Age. *Nature*, **546**, 293–296 (2017).
54. Stringer, C., & Galway-Witham, J., On the origin of our species. *Nature*, **546**, 212–213 (2017).
55. Bergström, A., Stringer, C., Hajdinjak, M., *et al.*, Origins of modern human ancestry. *Nature*, **590**, 229–237 (2021).
56. Sankararaman, S., Mallick, S., Dannemann, M., *et al.*, The genomic landscape of Neanderthal ancestry in present-day humans. *Nature*, **507**, 354–357 (2014).
57. Ni, X, Ji, Q., Wu, W., *et al.*, Massive cranium from Harbin establishes a new Middle Pleistocene human lineage in China. *The Innovation*, **2**, 100130 (2021).
58. Shao, Q., Ge, J., Ji, Q., *et al.*, Geochemical locating and direct dating of the Harbin archaic human cranium. *The Innovation*, **2**, 100131 (2021).
59. Ji, Q., Wensheng, W., Yannan, J., *et al.*, Late Middle Pleistocene Harbin cranium represents a new Homo species. *The Innovation*, **2**, 100132 (2021).
60. Timmermann, A., Yun, K.-S., Raia, P., *et al.*, Climate effects on archaic human habitats and species successions. *Nature*, **604**, 495–501 (2022).
61. Petraglia, M. D., A lengthy look at climate and hominin evolution. *Nature*, **604**, 430–431 (2022).
62. Kreier, F., Giant simulation hints at how climate shaped human migration. *Nature*, **604**, 610 (2022).
63. Proffitt, T., Luncz, L. V., Falotico, T., *et al.*, Wild monkeys flake stone tools. *Nature*, **539**, 85–87 (2016).
64. Harman, S., Lewis, J. E., Feibel, C. S., *et al.*, 3.3-million-year-old stone tools from Lomekwi 3, West Turkana, Kenya. *Nature*, **521**, 310–313 (2015).
65. Cerling, T. E., Wynn, J. G., Andanje, S. A., *et al.*, Woody cover and hominin environments in the past 6 million years. *Nature*, **476**, 51–56 (2011).
66. González-Forero, M., & Andy Gardner, A., Inference of ecological and social drivers of human brain-size evolution. *Nature*, **557**, 554–557 (2020).
67. McElreath, R., Sizing up human brain evolution. *Nature*, **557**, 496–497 (2020).
68. Potts, R., Big brains explained. *Nature*, **480**, 43–44 (2011).
69. Aiello, L. C., & Wheeler, P., The expensive-tissue hypothesis—the brain and the digestive system in human and primate evolution. *Current Anthropology*, **36**, 199–221 (1995).
70. Pennisi, E., Did cooked tubers spur the evolution of big brains? *Science*, **283**, 2004–2005 (1999).
71. Zink, K. D., & Lieberman, D. E., Impact of meat and Lower Palaeolithic food processing techniques on chewing in humans. *Nature*, **531**, 500–503 (2016).
72. Harmand, S., Lewis, J. E., Feibal, C. S., *et al.*, 3.3-million-year-old stone tools from Lomekwi 3, West Turkana, Kenya. *Nature*, **521**, 310–315 (2015).
73. Hovers, E., Tools go back in time. *Nature*, **521**, 294–295 (2015).
74. McPherron, S. P., Alemseged, Z., Marean, C. W., *et al.*, Evidence for stone-tool assisted consumption of animal tissues before 3.39 million years ago at Dikika, Ethiopia. *Nature*, **466**, 857–860 (2010).
75. Berna, F., Goldberg, P, Horwitz, L. K., *et al.*, Microstratigraphic evidence of *in situ* fire in the Acheulean strata of Wonderwerk Cave, Northern Cape province, South Africa. *Proceedings of the National Academies of Sciences USA*, **109**, E1215–E1220 (2012).
76. Roberts, R. G., Homo 'incendius'. *Nature*, **485**, 586–587 (2012).
77. Warneken, F., & Rosati, A. G., Cognitive capacities for cooking in chimpanzees. *Proc. R. Soc. B*, **282**, 20150229 (2015).
78. Wilkins, J., Schoville, B. S., Brown, K. S., & Chazan, M., Evidence for early hafted hunting technology. *Science*, **338**, 942–946 (2012).
79. Brown, K. S., Marean, C. W., Jacobs, Z., *et al.*, An early and enduring advanced technology originating 71,000 years ago in South Africa. *Nature* 491, 590–593 (2012).

80. Groucutt, H. S., White, T. S., Scerri, E. M. L., *et al.*, Multiple hominin dispersals into Southwest Asia over the past 400,000 years. *Nature*, **597**, 376–480 (2021).
81. Dennell, R., Traces of a series of human dispersals through Arabia. *Nature*, **597**, 338–339 (2021).
82. Henshilwood, C. S., d'Errico, F., van Niekerk, K. L., *et al.*, An abstract drawing from the 73,000-year-old levels at Blombos Cave, South Africa. *Nature*, **562**, 115–118 (2018).
83. Balter, M., Paintings in Italian cave may be oldest yet. *Science*, **290**, 419–420 (2000).
84. Aubert, M., Lebe, R., Oktaviana, A. A., *et al.*, Earliest hunting scene in prehistoric art. *Nature*, **576**, 442–445 (2019).
85. Arranz-Otaeguia, A., Carretero, L. G., Ramsey, M. N., *et al.*, Archaeobotanical evidence reveals the origins of bread 14,400 years ago in northeastern Jordan. *Proceedings of the National Academies of Sciences USA*, **115** (31), 7925–7930 (2018).
86. Dunne, J., Evershed, R. P., Salque, M., *et al.*, First dairying in green Saharan Africa in the fifth millennium BC. *Nature*, **486**, 390–394 (2012).
87. Kuper, R., & Kröpelin, S., Climate-controlled Holocene occupation in the Sahara: Motor of Africa's evolution. *Science*, **313**, 803–807 (2006).
88. Gatto, M.C., & Zerboni, A., Holocene supra-regional environmental changes as trigger for major socio-cultural processes in northeastern Africa and the Sahara. *Afr. Archaeol. Rev.*, **32**, 301–333 (2015).
89. Kutzbach, J. E., & Otto-Bliesner, B. L., The sensitivity of the African-Asian monsoonal climate to orbital parameter changes for 9000 years B.P. in a low-resolution general circulation model. *Journal of Atmospheric Science*, **39**, 1177–1188 (1982)
90. Bova, S., Rosenthal, Y., Liu, Z., *et al.*, Seasonal origin of the thermal maxima at the Holocene and the last interglacial. *Nature*, **589**, 548–553 (2021).
91. Hertzberg, J. A., seasonal solution to a palaeoclimate puzzle. *Nature*, **589**, 521–522 (2021).
92. Menocal, P. B., & Tierney, J. E., Green Sahara: African humid periods paced by Earth's orbital changes. *Nature Education Knowledge*, **3** (10), 12 (2012).
93. Kutzbach, J. E., Bartlein, P. J., Foley, J. A., *et al.*, Potential role of vegetation feedback in the climate sensitivity of high-latitude regions: A case study at 6000 years BP. *Global Biogeochemical Cycles*, 10, 727–736 (1996).
94. Tierney, J. E., & deMenocal, P. B., Abrupt shifts in Horn of Africa hydroclimate since the Last Glacial Maximum. *Science*, **342**, 843–846 (2013).
95. Drake, N. A., Blench, R. M., Armitage, S. J., *et al.*, Ancient watercourses and biogeography of the Sahara explain the peopling of the desert. *Proceedings of the National Academies of Sciences USA*, **108**, 358–462 (2011).
96. Riehl, S., Zeidi, M., & Conard, N. J., Emergence of agriculture in the foothills of the Zagros Mountains of Iran. *Science*, **341**, 65–67 (2013).
97. Cubry, P., Tranchant-Dubreuil, C., Thuillet, A.-C., *et al.*, The rise and fall of African rice cultivation revealed by analysis of 246 new genomes. *Current Biology*, **28**, 1–9 (2018).
98. Cowie, J., *Climate Change: Biological and Human Aspects*. Cambridge University Press, Cambridge, Great Britain (2013).
99. Pimental, D., & Pimental, M., *Food, Energy and Society*. Edward Arnold: London, Great Britain (1979).
100. Stephens, L., Fuller, D., Boivin, N., *et al.*, Archaeological assessment reveals Earth's early transformation through land use. *Science*, **365**, 897–902 (2019).
101. Ellis, E. C., Gauthier, N., Goldewijk, K. K., *et al.*, People have shaped most of terrestrial nature for at least 12,000 years. *Proceedings of the National Academies of Sciences USA*, **118** (17), e2023483118 (2021).
102. Sheridan, A., Incest uncovered at elite prehistoric Irish burial site. *Nature*, **582**, 347–349 (2020).
103. Hensey, R., in Papadopoulos, C., & Moyes, H., eds, *The Oxford Handbook of Light in Archaeology*. Oxford University Press, Oxford, Great Britain (2017).

104. Cassidy, L. M., Maoldúin, R. Ó., Kador, T., *et al.*, A dynastic elite in monumental Neolithic society. *Nature*, **582**, 384–388 (2020).
105. Woods, C., *Visible Language: Inventions of Writing.* Oriental Institute Museum Publications, No. 32. University of Chicago: Chicago, USA (2010).
106. Lawler, A., Writing gets a rewrite. *Science*, **292**, 2418–2420 (2001).
107. Goldman, N., Bertone, P., Chen, S., *et al.*, Towards practical, high-capacity, low-maintenance information storage in synthesized DNA. *Nature*, **494**, 77–80 (2013).
108. Gillings, M. R., Hilbert, M., & Kemp, D. J., Information in the biosphere: Biological and digital worlds. *Trends in Ecology and Evolution*, **31** (3), 100–108 (2016).
109. Wright, J. P., Jones, C. G., & Flecker, A. S., An ecosystem engineer, the beaver, increases species richness at the landscape scale. *Ecosystems Ecology*, **132** (1), 96–101 (2002).
110. Kareiva, P., Watts, S., McDonald, R., & Boucher, T., Domesticated nature: Shaping landscapes and ecosystems for human welfare. *Science*, **316**, 1866–1869 (2007).
111. International Union for the Conservation of Nature, World Wildlife Fund & United Nations Environment Programme, *The World Conservation Strategy.* IUCN: Gland, Switzerland (1980).

# 9
# Implications

Many implications arise from the co-evolution of life and planet narrative. These relate to both the current prospective transition as well as the possibilities for life elsewhere. The specific narrative I have presented includes that there is increased information processing with successive transitions. This is a nuance that some may have difficulty accepting. Yet, if it has any merit then it should shed light on where we, as a species, are going. Regardless of this caveat, an Earth system science (or biosphere science) approach to current climate change concerns is informative. Projected warming may not be as smooth as in the IPCC's hypothetical, future scenarios and, indeed, the IPCC itself warns that: 'Of great significance here is an assessment of the likelihood of the critical thresholds at which natural and human systems exhibit large-scale, abrupt, or irreversible changes in their response to a changing climate' (*1*). They are telling us to take heed.

## A current critical transition

Have no doubt, irrespective of the information-processing dimension to successive flips of planetary state, with human-induced warming (see Chapter 8) we are currently on a trajectory that could see us cross a number of thresholds—a number of small critical transitions of subsystems within the Earth system—that could trigger others, causing a major transition and flip the whole Earth system into a warm state. This warm state, or 'hyperthermal', would likely last two to three hundred thousand years, much like the IETM (Initial Eocene Thermal Maximum)/PETM (Palaeocene-Eocene Thermal Maximum) hyperthermal at the beginning of the Eocene 56 million years ago (see Chapter 6). It would only be shorter if we humans actively drew-down and sequestered carbon greenhouse gases.

There are a number of similarities with the current trajectory to the end of the 21st century (assuming we continue to be unable to change matters) and the IETM/PETM. (Though, the current transition is biologically/human driven whereas the IETM/PETM was a non-biologically driven flip—see Chapter 6.) Proxies in the geological record inform as to peak ocean acidity 56 mya: during that carbon isotope excursion (CIE), pH appears to have decreased by

*Earthâs Flips of State*. Jonathan Cowie, Oxford University Press. © Oxford University Press (2026).
DOI: 10.1093/9780197825402.003.0009

0.3 units. Having an indication of ocean acidity in turn enables us to estimate how much extra carbon dioxide would need to be dissolved in the oceans to create this extra acidity hence, again in turn, how much would need to be added to the atmosphere. This gives an estimate for the total carbon involved in the IETM/PETM—in both the volcanic carbon trigger (rich in $^{13}C$) and the subsequent disturbed carbon rich in $^{12}C$ rich ($^{13}C$ depleted) carbon reservoirs—to be between at least 3,000 petagrams (Pg, or 3,000 gigatonnes, Gt) to in excess of 10,000 Pg (or 10,000 Gt) of carbon (*2*).

Remember, this 3,000–10,000 Pg (or Gt) of carbon relates to *both* the volcanic trigger *and* the disturbed $^{12}C$. This means that the event's trigger *alone* must be smaller.

The UN's Intergovernmental Panel on Climate Change (IPCC) estimates that, assuming business-as-usual, the total 21st-century fossil carbon release (so ignoring carbon release from deforestation, cement production, *etc.*) is likely to be between 1,000 and 2,100 GtC (*3*). And, discounting the 11-year overlap (2000–2011), this is without adding in the IPCC-estimated 335–395 GtC released by fossil fuels alone between 1750 and 2011 (*4*).

Looking at the above figures and, for simplicity's sake, taking the total carbon released by human action through fossil fuels alone (ignoring land-use release such as through deforestation) under a business-as usual scenario (the higher end of the estimates) then we seem to be heading closer to 2,100 GtC by the end of the 21st century as we need to include pre-21st century releases. This 2,100 GtC itself is two-thirds of the aforementioned low end 3,000 PgC (GtC) of the *total* carbon involved in the IETM/PETM. And remember (it cannot be emphasised enough), this fossil fuel carbon is ignoring current land-use carbon releases. And again, remember that this IETM/PETM's 3,000–10,000 PgC (GtC) covers both the $^{13}C$ carbon trigger and the $^{12}C$ carbon response: the trigger alone would be less than this. In this context, it is very easy to contemplate that 2,100 GtC by the end of the 21st century could be enough for a hyperthermal trigger alone.

It would be good if we could get a better idea of the various carbon contributions to the IETM/PETM carbon isotope excursion to see how much was needed to trigger the event.

Actually, we have a way to get a direct handle as to how much carbon was involved in the trigger. While the change in ocean acidity gives us the total carbon (trigger and disturbed $^{12}C$), the carbon isotope excursion (CIE) itself—the change in the $^{13}C$ to $^{12}C$ ratio—gives us the disturbed $^{12}C$ involved: so to determine the carbon involved with the trigger is simply a matter of subtraction. From the CIE it is estimated that the $^{12}C$ involved was between 1,100 and 2,100 GtC (*5*). And we estimate that both the $^{12}C$ and $^{13}C$ involved constituted the aforementioned 3,000–10,000 GtC.

Immediately, you can see the first concern: this 1,100 and 2,100 GtC of $^{12}C$ involved in the IETM/PETM 56 million years ago is *exactly* the same as the IPCC considers we may release from fossil fuel alone in the 21st century!

Subtracting the $^{12}C$ involved from the total (both $^{12}C$ and $^{13}C$) in the IETM/PETM, we can see that the carbon involved with the trigger alone could be anywhere between 900 GtC and 8,900 GtC: a wide range because it is difficult to estimate matters with great accuracy from geological proxies 56 million years old.

Our potentially releasing 2,100 GtC of fossil carbon by the end of the 21st century clearly exceeds this lower 900 GtC limit for the hyperthermal trigger.

We, today, are playing Russian roulette with the future.

The trajectory we are on is a gamble that might instigate an IETM/PETM-type critical transition and—with the release of other carbon from various biosphere reservoirs—a concomitant hyperthermal event. The IETM/PETM event is perhaps a close analogue to potential future climate change (6)!

Other than that it was *not* due to a leap in biological evolution, the one major (and worrying) difference from what happened the in the early Eocene and now—which *is* due to a leap in biological evolution—is that with the IETM/PETM the trigger release of carbon was at least ten times slower than today. In other words, we are heading towards IETM/PETM-style of warming at least ten times faster than happened with that past event. This makes the current CIE quite distinct, even if in other respects where we are heading has certain similarities with the early Eocene hyperthermal (7).

In other words, the risk is that self-reinforcing feedbacks could push the Earth system towards a planetary threshold that, if crossed, could prevent stabilisation of the climate at an intermediate temperature and so cause continued warming on a 'hothouse Earth' pathway—flipping our planet into a hyperthermal state—even as annual human carbon greenhouse emissions are reduced. Crossing this threshold would lead to a hyperthermal Earth with a much higher global average temperature: degrees higher than any interglacial in the past 1.2 million years, and also significantly higher sea levels: much, much higher over the coming centuries than at any time in our current Holocene and past Quaternary interglacials. Decided human action is required to steer the Earth system away from such a potential threshold and stabilise it in a present-day habitable, Holocene interglacial-like state (8).

The situation we are in is rendered, in simplified form, with Figure 9.1. During the Quaternary, the past 2 million years or so, the Earth system has been in an ice age oscillating roughly every 100,000 years between long, cold glacials and shorter, warmer interglacials (see Chapters 2 and 8). Figure 9.1A depicts two environmental parameters: global temperature and sea level. As the Earth cools so an increasing volume of the planet's water is locked up in ice sheets on land, which lowers the sea level as the Earth system heads towards a glacial maximum. Milankovitch orbital factors then return the Earth system—melting the ice sheets and raising sea levels—towards an interglacial state. The pre-20th century interglacial state, in the simplified Figure 9.1, would see the Earth positioned over the axes' intersection. The Earth's current position is slightly off-centre to reflect the warming that has taken place since the dawn of the industrial revolution, which saw the Earth in the 20th century begin to approach its warmest of any previous Quaternary interglacial. If we warm further towards 2°C above pre-industrial levels then we will begin to exceed the warmest of any previous Quaternary interglacial.

Worryingly, the level of atmospheric carbon dioxide has already exceeded the highest levels of any previous Quaternary interglacial, it is just that inertia has yet to see the Earth system react and temperature catch up. (The Earth system is large and it is 'slow'—in the human-scaled sense of the word—to react.) At this level of carbon dioxide the risk is very real that a series of Earth subsystem positive feedbacks, some releasing greenhouse gases from various natural biosphere carbon reservoirs (examples of feedbacks were given in Figure 2.2), could push the Earth system, flipping it into a hyperthermal (hothouse) state. This state is likely to be warmer than that seen in the mid-Miocene epoch over 10 million years ago! (See Appendix 3.)

## Critical transitions and Lorenz systems

A brief aside for those with a mathematical bent (others can skip this text box), as they may note a similarity of Figure 9.1 with the graphical depiction of a Lorenz system—named after the mathematician Edward Lorenz—shifting between two semi-stable states about which the system moves. One state sees the Earth system in a Quaternary glacial–interglacial mode. The other state might be described as a Paleogene period (66–23 mya) mode: to be more exact, a late Paleogene mode. The Earth moved away from this mode during the Neogene period (23–5 mya) driven by a number of factors. These include a period of mountain building, that included the rise of the Himalayas and Alps, resulting in silicate weathering hence carbon dioxide drawdown (see Chapter 5). Another was the expansion of new $C_4$ plants (see Chapters 6 and 8) that are more capable of removing carbon dioxide from the air than $C_3$ plants. The Neogene period also saw the closure of the seaway between North and South America that reduced the ocean transport of heat to higher latitudes. Today, by kicking the Earth system, with an injection of carbon dioxide to the atmosphere, we may temporarily knock it into a hyperthermal state similar to the Paleogene. (Search the net for 'Lorenz system' and 'Lorenz attractor' for more information.)

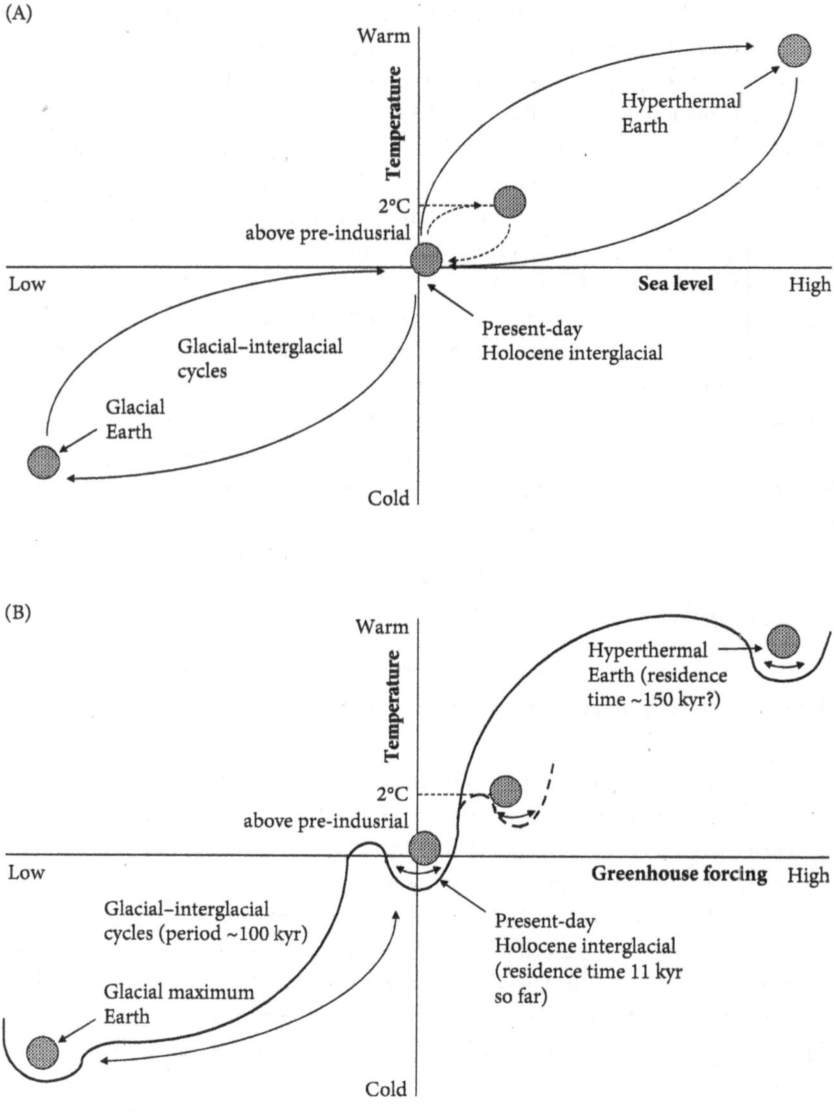


**Figure 9.1** (A) A simplified cartoon outline of two dimensions of the Earth system: average global temperature and sea level. The Quaternary glacial–interglacial cycle is depicted in the diagram's lower left. Two possible options for the future are depicted in the upper right. See text. The figure is adapted from Steffen, W., Rockström, J., Richardson, K., *et al.*, Trajectories of the Earth system in the Anthropocene. *Proceedings of the National Academies of Sciences USA*, **15** (33), 8252–8259 (2018).

Here some clarification is required. The Earth's journey from the later part of the Paleogene period (66–33 mya) to today involved mountain building (especially the collision between India and Asia to form the Himalayas), plate tectonic closing of ocean gateways (especially between North and South America), and the expansion of $C_4$ grasses. So a return to the Paleogene mode will not see the Earth system stuck there for millions of years. Instead, the Earth is likely to flip to there for a much shorter time (100–300 kyr) analogous to that that initiated the Eocene epoch: the early Eocene hyperthermal—the IETM/PETM—within the longer Palaeogene (see Appendix 3 and Chapter 6).

The Figure 9.1B visual cartoon takes a system stability perspective similar to that of Figure 2.5. As noted earlier, if the current Earth system is given a greenhouse forcing kick then it could—depending on the severity of the kick—push it into a hyperthermal state. Again, if this hyperthermal is loosely analogous to the IETM /PETM carbon isotope excursion (CIE) then it might stay in that hyperthermal state for somewhere around 150,000 years (100–300 kyr), give or take a few tens of thousands of years, as the various biosphere reservoirs of carbon adjust *via* the atmosphere.

Conversely, the transition time to this new hyperthermal state would be shorter than the transition time into the earlier IETM/PETM: centuries as opposed to a few thousand years back then. As noted above, we are today adding greenhouse forcing at a rate that is roughly ten times faster than the transition to the early Eocene hyperthermal state and so it is therefore likely to take just two or three centuries to get there, not a few thousand years! Nonetheless, the return from the hyperthermal state will ultimately rely on the natural process, such as carbonate sedimentation and silicate weathering, and so is likely to last as long as it did with the IETM/PETM (and other large) CIE(s): which means thousands of years unless we (humans) actively facilitate carbon drawdown.

Much depends on how hard the Earth system is kicked by us: the smaller the greenhouse forcing, the smaller the hyperthermal, triggering feedbacks notwithstanding. Here, keeping the Earth below 2°C above pre-industrial levels is important, as we know that earlier in the current Quaternary ice age the Earth

---

**Figure 9.1** *(continued)* (B) The same represented in a form akin to Figure 2.5 with the Earth represented as a ball in the landscape. This is more of a cartoon as, for example, in the sweep down from an interglacial to a glacial maximum there are a number of minor dips and rises. This will also be true of the rise towards becoming a hyperthermal. The exaggerated cups at the extremes and present (the current interglacial) serve to represent semi-stable states, at a hypothetical fixed upper level of greenhouse forcing, with carbon cycling between various biosphere reservoirs *via* the atmosphere in the form of carbon dioxide and methane.

system returned to glacial from interglacial modes despite parts of some previous interglacials being that warm. Though we are currently (in the first quarter of the 21st century) below this temperature (though the degree of current warming is most worrying), as said, carbon dioxide levels today are much higher than in past Holocene interglacials: they actually correspond to a more distant time (a few million years ago) in the geological past when the Earth was warmer. As noted earlier, all that is holding this warmth back is the inertia of the Earth system: it will take centuries for the volume of the oceans to warm, but warm they will and already we may be passing the Paris Accord limit (we will come to this shortly). Consequently, not only is it important to reduce emissions of greenhouse gases close to zero—carbon drawdown will also be needed. Fail to do that in good time and we are almost certain to end up in a hyperthermal somewhat analogous to the IETM/PETM.

Figure 9.1A and B are simplified representations; matters in real life are far more complicated. As mentioned at the beginning of this section, along the road of this overall, potential, critical transition of the Earth system, there are a number of smaller, potential critical transitions *within* the Earth system, yet these smaller potential transitions are connected. When one of these lesser thresholds is crossed, there is a burst of greenhouse gas release that has a positive feedback effect on the Earth system as a whole, pushing it towards the next threshold. This is much like a row of dominos: one domino falling over will knock the next and so on. As each of these smaller critical transitions within the Earth system take place, so it becomes harder for us to halt the trajectory of the whole system cascading towards a larger critical transition that flips the Earth into a hyperthermal state.

Just a few of the Earth system's subsystems that are capable of undergoing a critical transition, so-called tipping points, are depicted in Figure 9.2, together with some of their connections. Some of these have been mentioned before: permafrosts thawing (Chapters 6 and 8), methane clathrate release (Chapter 6), Amazon fires (Chapter 2), and Greenland and Antarctic ice-sheet melt (Chapters 2 and 8). (A certain artistic licence goes with the Figure. For example, it should be noted that marine methane clathrate decomposition is not as localised as depicted in Figure 9.2 but across the globe at depths and pressures conducive to clathrate stability.)

Arctic warming relates to the albedo feedbacks discussed in Chapter 2: it has long been known that the Earth's poles (especially the North Pole) are warming faster than elsewhere and computer models do not fully reflect this (though they have been continually getting better). In the north, this faster warming is sometimes called Arctic amplification or polar amplification. The major reason for Arctic amplification is the albedo difference between ice and water (see Chapter 2). Ice is highly reflective (it has a high albedo), and sea water is

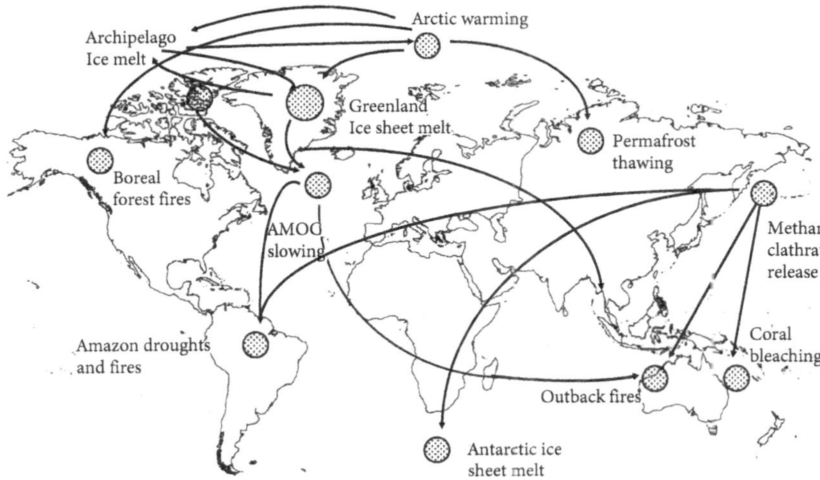


**Figure 9.2** Some 'tipping points' and connections between them. Note, not all the connections and feedbacks are shown. Inspired by, adapted, and modified from Lenton, T. M., Rockström, J., Gaffney, O., *et al.*, Climate tipping points—Too risky to bet against. *Nature*, **575**, 592–595 (2019).

not. So when Arctic ice melts in the summer, less Solar energy is reflected, which in turn means more is absorbed so enhancing warming.

Archipelago ice melt, depicted in Figure 9.2, relates to the Canadian Arctic Archipelago, which is located off the north-western shore of Greenland with its two largest islands of Baffin and Ellesmere. The archipelago contains one-third of the global volume of land ice outside the ice sheets of Greenland and Antarctica. With global warming, ice on and around the archipelago's islands melts. Where this melt exposes the sea surface, albedo change means that more heat is absorbed than reflected. This enhances warming, so facilitating further melt in a positive feedback. Research has shown that between 2004–6 and separately 2007–9, the rate of ice loss here sharply increased from 31 ± 8 gigatonnes *per* year (Gt $yr^{-1}$) to 92 ± 12 Gt $yr^{-1}$. In the 21st century's first decade, the increase in the rate of ice loss makes the Archipelago the single largest contributor to sea-level rise outside of Greenland and Antarctica (9). It looks like the Canadian Arctic Archipelago is undergoing a transition that further reinforces Arctic warming which also furthers Greenland ice sheet melt which again positively feedbacks to the Canadian Arctic Archipelago.

Finally, a warmer Arctic makes for a warmer, and at times drier, summer in the surrounding boreal forests. This increases forest fire risk hence the return of forest carbon to the atmosphere increasing carbon dioxide and furthering global warming. In this way, there is a 'connection' between Arctic warming and boreal forest fires. (Again, see Figure 9.2.) This is why the Earth system is akin to a line of (subsystem) dominos: when one subsystem crosses a threshold (flips state) so it is more likely that the connected subsystem will also flip.

The term 'AMOC' in Figure 9.2 relates to the 'Atlantic meridional overturning circulation', and here a little explanation is required. The AMOC is part of a global system, a global conveyor that carries water along both surface and abyssal currents about the Earth. This is the Broecker thermohaline circulation elucidated by Wallace Broecker in 1991 (*10*). See Figure 9.3.

The importance of this conveyor is that it carries both heat and carbon dioxide from the surface waters to the abyssal depths. Very roughly, half the carbon dioxide released due to human action is (currently) absorbed by the oceans, so if that absorption were to decline, global warming would be exacerbated. The conveyor also helps transfer heat away from the tropics to higher latitudes.

(Here a quick caveat concerning the amount of carbon dioxide 'currently' absorbed by the oceans. As the decades and centuries pass, so the oceans become more saturated with carbon dioxide and this will impede their ability to absorb

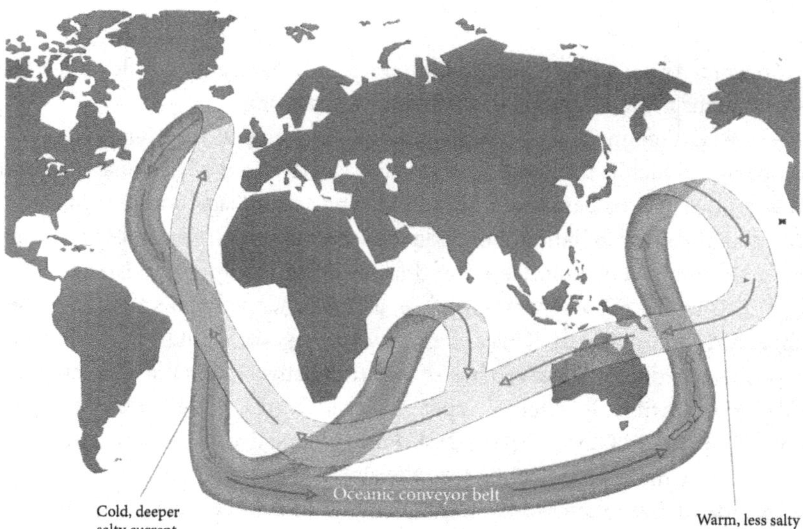


**Figure 9.3** The Broecker global conveyor of surface and abyssal currents moving water about the planet, greatly simplified. Reproduced from Cowie, J., *Climate and Human Change: Disaster or Opportunity?* Parthenon Publishing: London, Great Britain (1998), having been adapted from Broecker, W. S., The great ocean conveyor. *Oceanography*, **4** (2), 79–89 (1991).

more. In short, as atmospheric carbon dioxide increases, this absorption will eventually cease.)

The essential elements to this conveyor are its thermohaline drivers at a number of points. One of these thermohaline drivers is in the North Atlantic and is the aforementioned 'AMOC'.

It works like this. Warm water from the Gulf Stream is carried north by the North Atlantic Drift current. As it does so, not only does it take heat from the tropical Caribbean to warm NW Europe (notably the British Isles, western France, and Spain) but as it goes water evaporates and so the current becomes saltier hence more dense. It also becomes denser because it becomes cooler. There therefore comes a point at which it becomes much heavier than the northern ocean waters into which it is entering. So it sinks, taking with it any residual heat, and any excess carbon dioxide absorbed from the atmosphere, into the ocean's abyssal depths. The current then continues beneath the sea all the while losing its excess salt. Eventually, an ocean away, it returns to the surface. Both these surface and abyssal currents form one giant current or conveyor (a simple depiction of which is given in Figure 9.3).

Here's the thing. If weather patterns change, such as seeing higher rainfall in the North Atlantic, or, more significantly, freshwater melt from Greenland and the Canadian Arctic Archipelago enters the North Atlantic, this freshwater may dilute the saltier water to such an extent that it reduces its ability to sink. Should that happen, the conveyor will slow, with less heat and carbon dioxide being carried away from the surface into the ocean's abyssal depths.

The implications of the AMOC slowing are significant in quite different ways, both regionally and globally. First, if the AMOC slows then the North Atlantic Drift will encounter a blockage of water in the North Atlantic and so alter its course, moving further south forming a more enhanced current off the coast of north-western Africa. With northwest Europe receiving less heat it will become cooler and its winters colder. The winter heat northwest Europe currently receives from the conveyor is illustrated by the average January temperature found in London (Great Britain) and Kyiv (Ukraine). Both are at roughly the same latitude (actually Kyiv is a little more southern), yet Kyiv's January temperature is typically 10°C cooler than London's.

Conversely, globally a slowing of the AMOC enhances global warming. First, less heat is carried away from the ocean surface into the abyssal depths, as well as from lower latitudes, and so the tropics warm. Secondly, less carbon dioxide is transported into the ocean depths and so with more carbon dioxide in the atmosphere than there would otherwise be, greenhouse warming is enhanced. Finally, this warming of equatorial and near equatorial latitudes can contribute to seasonal drought, hence forest wildfire, in Amazonia and other places (possibly Australia): hence, the connections depicted in Figure 9.2.

So, the question becomes one of whether the AMOC is, or is not, currently slowing? Is it nearing a threshold?

There is an increasing body of work relating to this. Illustrative are two fairly recent papers. One of the two research groups was led by Levke Caesar and Stefan Rahmstorf (2018) and the other by David Thornalley and Delia Oppo (2018) (*11, 12*). The former used computer models mapping them onto real observations since 1870 to then infer change. The latter looked at 1,600 years of abyssal sediment. If they found larger grains of sediment they inferred a stronger current flowed at the time than at other times when sediment of smaller grains was laid. They inputted this real-life AMOC data inference into a high-resolution climate model and this replicated the climate change trends found in meteorological records going back to the end of the 19th century. This also replicated temperature changes in the South Atlantic: less heat was being transferred north and the North Atlantic currents shifted during times of a weaker AMOC bringing warm water back heading south down the West African coast. The study showed that with warming the AMOC must have slowed by around 15% since the mid-20th century.

The latter study also looked at ocean sediments (but at a different Atlantic location to the former, though one also associated with the AMOC) and inputted this into a climate model. However, as noted, it looked at sediments laid down over a longer period dating as far back as 1,600 years. The picture that emerged was that of the past 1,600 years the AMOC was largely stable: it only slowed over the past 150 years or so (since 1850) and also that it has continued to weaken at a faster rate more recently. The researchers attribute this long-term trend of slow weakening to the melting of ice as the Earth came out of the Little Ice Age (a cool period roughly between 1350 and 1850). The more recent faster weakening of the AMOC they attribute to human-induced global warming.

Both the studies complement each other, even if there is some difference in the exact time on onset of circulation declines. (This last might be a site-specific thing as the studies looked at different locations?) Both studies attribute the weakening to freshwater (from ice melt) entering the North Atlantic and so diluting the cold salty water so hindering its sinking. And finally, both studies attribute this last to global warming. That the AMOC has weakened in recent times is of concern as it has less of a way to go before it possibly completely shuts down. Finally, a further weakened AMOC may itself lead to considerable changes in climate and precipitation patterns throughout the northern hemisphere (*13*).

The question then arises as to if and when the AMOC may shut down completely? In 2021, the Intergovernmental Panel on Climate Change (IPCC) said that a collapse in the 21st century would be very unlikely (with 'medium confidence') (*14*). But in 2023 Danish mathematicians, Peter and Susanne Ditlevsen from the University of Copenhagen, provided an autocorrelation analysis (see Chapter 2) of historic sea surface temperatures where the North Atlantic downwelling takes place. Their results estimate a collapse of the AMOC to occur around the mid-21st century under the current scenario of future emissions. Their results generated some debate in climate science circles not least because historic data from more than several decades ago is less reliable than that more recently gathered. In addition, they themselves say that they 'cannot rule out that other mechanisms are at play'. But they also say that 'given the importance of the AMOC for the climate system, we ought not to ignore such clear indicators of an imminent collapse' (*15*).

With regards to permafrosts (see Chapter 6), there are signs that substantive such areas are likely subject to thaw releasing carbon dioxide and, the even more potent greenhouse gas, methane. Currently, northern peatlands hold ~80% of the global peatland carbon. The carbon in both northern peatland as well as frozen peatland (permafrost) is more or less in balance: as peatlands develop, they draw down carbon dioxide, whereas as permafrosts thaw they release it. At the moment, the system globally provides a small net sink for carbon of 0.10 ± 0.02 $PgCyr^{-1}$ (or gigatonnes carbon *per* year). However, a recent study, that looked at some 7,000 northern peatland cores, identified that the potential for peatlands to remain long-term carbon sinks may reverse due to increased permafrost thaw. The study estimates the northern permafrost store of carbon to be 185 ± 70 Pg (or gigatonnes) over a present-day area of 1.7 million $km^2$. If the global climate stabilises at 2°C warming, they estimate that only a half of the pre-industrial permafrost peatland extent will remain; but at +6°C warming, peatland permafrost essentially disappears (*16*). This would add roughly 185 PgC (or GtC) to the atmospheric reservoir. This compares with an early 21st century atmospheric reservoir of roughly 830 Pg. So an additional 185 PgC represents an additional 22%. In a purely carbon context, this would provide a significant jolt to the Earth system. However, in a greenhouse context it is worse: most of the current atmospheric carbon is in the form of carbon dioxide but much of the carbon released from permafrost thaw would be in the form of the far more (albeit shorter-term) greenhouse-potent methane gas.

The implications of all this are that should we head towards a hyperthermal state then permafrost thaw would be one of a number of factors keeping the planet in that warm state until natural carbon drawdown began to restore matters. If (note the 'if') such a hyperthermal were analogous to the IETM/PETM (see Chapter 6), then, as discussed earlier, we might expect the Earth system to remain in that elevated state for around 100,000 to 200,000 years as carbon cascaded between various biosphere reservoirs often *via* the atmosphere in the form of greenhouse gases, unless humans actively augmented carbon drawdown. This is why there is a cup in the warm/high greenhouse-forcing quadrant of Figure 9.1B cartoon.

Permafrost is just one reservoir of soil carbon. Globally it is estimated that the total soil reservoir of carbon is between 3,500 and 4,800 PgC (*4*). This soil carbon ranges from carbon in cold permafrosts through to carbon in tropical forest soils. This last can also be released with warming.

It was thought that permafrosts might see the greatest loss of soil carbon to the atmosphere, with little being lost from tropical soils: tropical soils currently accumulate carbon—they are currently a carbon sink. However, the past decade or so has seen attention address soils other than permafrosts. Here, it had been thought that carbon in tropical soils—global tropically some 502 Pg C—was relatively stable, but a number of studies are challenging this view. Of these recently, in 2020, Andrew Nottingham, Patrick Meir, Esther Velasquez, and Benjamin Turner published their study of likely carbon release from such tropical forest soils (*17*). They had an experimental area of lowland tropical forest soil on Barro Colorado Island, Panama, and wanted to see how the soils would respond to warming. Their experiment looked at the carbon dynamics of four pairs of five-metre diameter circular plots over two years. The soil of one of each pair was electrically warmed to 4°C above the natural temperature, while the other was not heated to serve as a control. 4°C is the warming the tropics, it is thought, will experience under an IPCC business-as-usual scenario by the end of the century. They then ran their experiment.

Their results chimed with satellite observations of carbon loss.

Now, you may wonder how satellites can see a 4°C future? Well, of course they can't: at least, not in terms of the entire planet. The thing is that currently the annual variation in the regional tropical temperature in some places can be as much as 4°C: in different places there are warm years and there are cool years even if overall the planet is not 4°C warmer. What satellites can do is detect the carbon coming off these tropical regions when they are warmer. The researchers found from their plots that warming increased carbon loss by 55% and that not all of this loss was replaced by carbon accumulating in the plants and with carbon returning to the soil in plant roots. Tropical soils, they conclude, with warming would stop being a carbon sink and become a carbon source. Extrapolating their results to tropical soils planet-wide, and assuming 4°C warming by the end of the century, they estimate a global loss of more than 65 Pg C. This is equivalent to 7% or 8% of atmospheric carbon present in the mid first decade of the 21st century.

So, with some of the Earth's subsystems potentially crossing critical thresholds, how likely is it that the planet as a whole might undergo a major transition, shift, and flip into a new state? In 2012 Anthony Barnosky, of the University of California, and colleagues looked into this in a review paper for the journal *Nature*. They concluded that a 'global-scale state shift is highly plausible within decades to centuries, if it has not already been initiated' (*18*).

One recent telling is that parts of the Amazon rain forest have ceased being a carbon sink (ceased absorbing more carbon each year than it emits) and have become a carbon source. Luciana Gatti, Luana Basso, and colleagues, mainly based at the São José dos Campos in Brazil, have directly measured the atmosphere in four regions across Amazonia for nine years (2010–2018). They also used data from several sites on remote islands and coastal headlands around the South Atlantic Ocean to establish background concentrations of gases. They found that south-eastern Amazonia, in particular, now acts as a net carbon source (total carbon flux minus fire emissions) to the atmosphere especially in the dry season. The overall trend for deforestation, warmer and drier dry seasons, drought stress, fire, and carbon release in eastern Amazonia critically threatens the Amazon carbon sink (*19*).

All this brings us to the question of whether this trend, of a part of Amazonia ceasing to become a carbon sink and instead becoming a carbon source, relates to the broader Amazon basin as a whole?

You will recall (Chapter 2) that an oscillating system that takes longer to return to its average state and, indeed, increasingly deviates from its former average state, can be quantified by the mathematical and statistical measure of 'autocorrelation'. In 2022 Chris Boulton, Timothy Lenton, and Niklas Boers from the University of Exeter looked at historic data of satellite microwave data of the Amazon basin for over a quarter of a century from 1991. This microwave data specifically looked at vegetation optical depth (how thick is the vegetation cover) and also how 'green' is the cover (a measure of photosynthetic activity). They then ran an autocorrelation analysis of the data. What they found was that autocorrelation declined from 1995 to 2003 (you would expect it to go up and down with natural variability and not just go in one direction) but, not only did it increase from 2003, by 2006 it was beginning to exceed 1995 levels and this increase continued so that it almost doubled the 1995 levels by 2016! Indeed, that this change was due to human activity they demonstrated as the increase in autocorrelation was greatest in those parts of the Amazon basin in areas of greatest human land use and roads, and least in areas further away (*20*). It therefore looks as if the Amazon basin as a whole is heading for a critical transition and may shift to becoming a net carbon source, as opposed to a sink, just as south-eastern Amazonia has become.

Here there is an added twist. Overall, globally forests act as a carbon sink. Worldwide, the land carbon sink has absorbed around 25% of all fossil-fuel emissions since 1960 to near the end of the 2010 decade. This is in part because increased atmospheric carbon dioxide encourages further photosynthetic drawdown of carbon. In higher latitudes, the increased botanical growing season with warming further encourages drawdown. An additional factor for mid-latitude forests is increased nutrients from human activities have spurred growth. However for tropical forests it is the extra carbon dioxide alone that has spurred drawdown (as too hot a temperature serves to impede plant growth). Here's the thing, if we cease adding carbon dioxide to the atmosphere (which we undoubtedly should) then this means that tropical photosynthetic rates (hence carbon drawdown) are unlikely to increase (*21*). Factor this in with the Gatti–Basso team's results (driven by land-use, climate change, and fire) and the University of Exeter team's results, and it is most likely that much, if not all, of Amazonia will become a net carbon source later this century.

The following year another small collaboration, led by Giovanni Forzieri and Alessandro Cescatti of the European Commission's, Joint Research Centre at Ispra in Italy, used satellite data of tree mortality over 20 years to 2020 across the globe. They ran an autocorrelation analysis (again see Chapter 2) and found that globally between the two decades 2000–2010 and 2011–2020, the global forest autocorrelation hardly changed. However, by breaking down the global tree cover by key biomes—tropical, arid, temperate, and boreal—they saw that that there was a noticeable difference. Tropical, arid, and temperate forests all saw increased autocorrelation indicating that they were heading towards a critical transition: that is to say they would cease being the type of forest they had been, or even cease to be forest at all. Conversely, boreal forests saw a decreased autocorrelation indicating an average increasing trend in resilience. From a global perspective, the more resilient boreal forests were offsetting the decline in resilience of tropical, arid, and temperate forests. Finally, the researchers note that approximately 23% of intact undisturbed forests, corresponding to 3.32 PgC (GtC) of annual gross primary productivity, have already reached a critical threshold and are experiencing a further degradation in resilience. This has implications for forests' carbon storage ecosystem function, hence the way forests need to be managed (*22*).

Worryingly, in 2022 a European collaboration of ten researchers published, in the leading journal *Science*, a review of components of the Earth system that have a potential to cross a threshold and undergo a critical transition. Its lead authors were David McKay and Arie Staal of Stockholm University, and others in the collaboration were from the University of Exeter and the Potsdam Institute for Climate Impact Research. They looked at these components individually, each in turn, and concluded that several might be triggered with warming within the 1.5°C to 2.0°C warming above pre-industrial times, the warming limit as stipulated in the 2015 UN Paris Climate Accord: in other words, achieving the Paris Climate Accord's weaker 2°C goal might not prevent components of the Earth system from drastically changing (*23*). Here, perhaps the component of greatest concern—it is one of the most likely with the least warming—is that of boreal permafrosts (as we discussed a page or so ago). Not only do they consider it quite likely that they will undergo a transition at 1.5°C warming above pre-industrial (which is only 0.5°C above the global temperature around 2020) but that in transitioning, the permafrost would release methane, the potent greenhouse gas that would exacerbate warming and make it even more likely that other thresholds would be crossed.

And this is the rub. This assessment examined each component *individually* in turn but, as already has been noted (Figure 9.2), in some cases, such as with permafrosts, crossing one threshold makes it more likely that there will be a critical transition in another component of the Earth system. The authors of the Stockholm-Exeter-Potsdam assessment were acutely aware of this. Their conclusion was that current policies leading to ~2°C to 3°C warming are unsafe because they would likely trigger multiple climate tipping points. Their assessment of climate tipping points provides strong support for the Paris Agreement to limit global warming to the stricter limit of 1.5°C.

To sum up, there are a number of subsystems within the Earth system that could undergo their own individual critical transition. Further, as each one does, so it increases the likelihood that another subsystem will undergo a transition. The more of these systems that cross a critical threshold, the higher the chance, with the continued addition of greenhouse gases, of the *whole* Earth system flipping into a hyperthermal state somewhat analogous to the Initial Eocene Thermal Maximum/Palaeocene-Eocene Thermal Maximum hyperthermal and a similar carbon isotope excursion.

Given that we are most likely on a trajectory that will take our world across a climate threshold and critical transition that will flip the planetary state, what, if anything, should we do? Or alternatively, given that life on Earth has survived past CIE hyperthermals, need we do anything?

## The need to manage the future

The Earth system has experienced a number of hyperthermal events, including the IETM/PETM 56 million years ago (mya). For example, as noted earlier (see Chapter 6), another hyperthermal carbon isotope excursion (CIE) took place 183 mya in the Toarcian age of the Jurassic period. This too had a peak lasting the order of roughly a couple of hundred thousand years. Yet, though both CIEs were pronounced and distinct climate events, other than certain shelled (calcareous) marine species, neither was a major extinction event (though there were extinctions of these calcareous—calcium shelled—marine species due to ocean acidity), let alone an extinction of the order of the end-Permian event or that of the end-Cretaceous (again see Chapter 6). This is not to say that both these events lacked ecological change: both saw biome shift and ecosystem re-ordering with global temperature change (*24*). Therefore the hyperthermal to which we are currently heading will not be by itself a major climate-induced extinction event even if there will be (very) major ecological and environmental impacts. However, in addition to the climate dimension, humanity is currently imposing a heavy global footprint on natural systems due to a combination of increasing *per capita* resource demand multiplied by a global population exceeding 8 billion and growing. This factor, added to climate change, is likely to result in extinctions.

Broadly speaking, doing nothing or doing something are the two options we have. It is up to us to ensure that the decision process (the policymaking) for the option we go for is an educated one.

This last point is not one I wish to go into in any depth here, but in essence, for the democratic nations at least, this means ensuring our politicians are informed. The problem is that most politicians do not have a skill set that includes science, technology, engineering, maths/medicine (STEM) subjects. This has implications when events take place that relate to STEM topics.

Here, the 2020/1 SARS-CoV2/CoVID-19 pandemic was revealing: some nations previously rated highly for pandemic preparedness by the Global Health Security Index, actually performed poorly when the pandemic came. One was the UK, in terms of deaths as a proportion of the population: compared to European Union nations, by the end of the first wave, other than Belgium, the UK had the highest CoVID mortality rate (*25*). At the time of writing, here in the UK, there will be an inquiry into the response to the pandemic, and I do not wish to pre-empt that. However, it is worth noting that of over a score of politicians in that nation's governmental Cabinet at that time (*26*), most had qualifications relating to politics, philosophy and economics, English literature, law, and history. Just two held a STEM degree, and a third member had studied at agricultural college before working on a farm. Perhaps with its predominantly humanities expertise, the then UK Cabinet members did not have the best skill-mix to deal with a pandemic let alone address issues arising in a scientific age with economies becoming increasingly technological and knowledge-based. (For fairness and political balance, it should be noted that when the UK's current opposition party was last in power, its Cabinet had a similar skill mix: there has inherently been a lack of STEM skills in the leaderships on both sides of the House of Commons [*27*]: I am not making a party political point.)

But it is not just politicians: there is a lack of STEM in the UK's Civil Service too. When the UK Chief Scientific Advisor (the scientist advising the Prime Minister), Professor Sir Patrick Vallance, came into office, one of the first questions he asked was how many civil servants had a STEM degree? Alas, the Civil Service does not collate that data, but it does have a fast-track cohort of their most promising staff who are slated for rapid advancement, and here they did have a figure. Just 10% of fast-track civil servants held a STEM degree! Professor Sir Patrick has made it clear that while he did not want the reverse matters—and have 90% of civil servants to be STEM qualified—the then current balance was clearly very wrong (*28*).

Over in the United States there are similar problems too. Only in 2021 for the first time was the director of the White House Office of Science and Technology Policy appointed to the President's Cabinet. And so geneticist Eric Lander became the first US scientific adviser to have such Presidential access with his promotion by the then newly elected President Joe Biden. The wonder is that it took so long for such a move given the United States' economy's dependence on science and technology and its knowledge base. Eric Lander faced a number of challenges not least because Biden's predecessor, Donald Trump, had cut funding from key areas of science (which Trump repeated in his second term following Biden's Presidency). One concern, Eric Lander saw, was that there are large parts of the United States that do not have a science high school or a science-based industry. This makes it difficult to enable all Americans, should they wish, to participate in science even if they are really interested in it (*29*). Subsequently, with Donald Trump's second Presidential terms, there are once more cuts in US science investment.

The political class, in both Britain and the United States, and likely elsewhere, it would seem for the most part simply does not have the STEM skill set to handle the science-related issues of the day let alone to manage the likely global transition this century will see. As noted, the foreseeable global policy changes required, to either avoid or manage an entire Earth system transition, are considerable.

Speaking of considerable future global change, at this point it is worth pausing and briefly returning to some of the topics discussed in the last chapter relating to how the Earth changed in the past from the depth of the last glacial maximum (the LGM) over 18,000 years ago (18 kya) to the start of the Holocene interglacial some 11 kya, through the hunter-gatherer to agrarian transition that began over 3 kya (Chapter 8) and how these compare to current global vegetation changes.

Importantly, how do current ecological changes measure up to those between the LGM and the start of the Holocene interglacial? This will give us an additional perspective on the scale of the current transition problem.

In 2021, Ondřej Mottl and Suzette Flantua of the University of Bergen, in Norway, led an international collaboration including some US scientists from the University of Wisconsin–Madison, the home of the Neotoma Palaeoecology Database—of pollen in lake sediment cores from around the world. Knowing the species of pollen in a sediment laid down many years ago—even thousands of years ago—it is possible to elucidate what species lived nearby at that time. The collaboration looked at 1,181 pollen sequences spanning all continents, except Antarctica, to understand how plant ecosystems have changed since the end of the depth of the last glacial maximum (LGM) about 18,000 years ago, and how quickly this change occurred.

They discovered that the rate of botanical change initially peaked between 16,000 and 8,000 years ago; the time of the LGM-to-Holocene transition. Ecosystems subsequently stabilised but not until about 4,000 years ago. Then, as separately noted in the last Chapter, the rate of change substantially increased at a pace that continued to rise and which is even faster today, when most plant ecosystems are changing at least as fast as they did at the peak of LGM-to-Holocene transition. Given that compared to the LGM-to-Holocene transition, the subsequent global climate during the Holocene has been comparatively stable the past 10,000 years or so, this means that any substantive change the past 4,000 years is non-climatic and so attributable to human land-use: ecosystem engineering (particularly agriculture and forestry) and niche construction. Here, the scale of human effects on land ecosystems has exceeded even the climate-driven transformations of the last deglaciation! This in turn suggests that the greater still changes seen over the past two centuries—due to the intensification of land use—is the tip of a deeper trend we will see in the years to come (*30*).

Again, as previously noted, in essence we have a choice as to whether we want to manage the future, or whether we will continue with business-as-usual to affect the Earth system in an unprecedented way and end up with a full-blown hyperthermal? Here, equally as much depends as to what a full-blown hyperthermal would be like (there is a lot still to understand about past hyperthermals) as well as the other human ecology implications of business-as-usual. After all (again as previously noted), it might be hypothetically argued, that as the Earth system has successfully endured past CIEs without a major extinction 'doing nothing' might well be an option?

Make no mistake—'doing nothing' (if not the almost equivalent of 'not doing enough') is a very real option and may be one we drift into through inaction despite political rhetoric (i.e. the aforementioned political skill set).

The various IPCC *Assessments* conjecture under business-as-usual scenarios that the Earth will warm by 4°C–5°C above pre-industrial temperatures by the year 2100 (see also Figure 8.4). Now, even if in the year 2100 humanity wanted to halt, and at that point cease, all greenhouse gas emissions, the IPCC speculate that inertia of the Earth system would see warming continue by about a degree or more over the next nine centuries to the year 3,000 (*31*). However, this speculation specifically does not take into account long-term feedbacks (see Chapter 8) and, notwithstanding this—as the thought that under a business-as-usual scenario humanity would suddenly cease all emissions in 2100 is clearly ludicrous—it would be prudent to consider business-as-usual warming for the 23rd century to be at least several degrees warmer than today. Such a consideration might focus political minds before we get to such warming?

Unfortunately, the academic literature as to the biological and human impacts on so great a warming is very sparse: more has been written on 4°C–5°C warmer scenarios. But, and fortunately, there has been considerable palaeoecological study of the last glacial maximum (LGM) 18–25 thousand years ago (kya) when the Earth was at least, at its coldest, 5°C or more cooler than during our current (Holocene) interglacial that began 11.7 kya (*32*). So by comparing the two—the LGM and the Holocene—we can see the degree of ecological change planet-wide an average temperature difference of about 5°C (or perhaps a little more) makes, even if the future is one of potential warming (not cooling) from where we are today by that amount. The heavy lifting of teasing apart to summarise this volume of palaeoecological science has already been done by Connor Nolan and Jonathan Overpeck, of the University of Arizona. They worked with an international team two score strong in research departments across the globe.

The Nolan–Overpeck collaboration had their research findings published in 2018 in the journal *Science* (*33*). They looked at 594 palaeoecological studies of the LGM 21,000 years ago, at nearly 600 sites worldwide (with the exception of Antarctica). From these, they determined vegetation responses to temperature changes of 4°C to 7°C (the glacial-to-interglacial temperature difference varies in different parts of the globe). They found that found that ecosystems planet-wide experienced major change. Between then (the LGM) and the start of the Holocene interglacial, all the sites showed species change and many (about 70%) showed biome change—for example, tundra becoming forest. Most of the remainder (26% of the total) exhibited significant but more moderate change. These last were mainly in the tropics.

Even though that study compared today's interglacial with a cooler past, an ecological comparison between today and a comparably warmer future would most likely (almost certainly) show a similar pattern of extensive change. The tropical biomes would be least affected, though likely expand in range, while higher latitudes would see greater warming. Indeed, as mentioned earlier, we are already seeing signs of this last.

Again as previously noted, much study has been devoted to a warmer world of just four or five degrees centigrade. Indeed, in 2011 the Royal Society published the proceedings of a symposium the world of 'Four Degrees and Beyond' (*34*). This pulled together much of the scientific understanding of the 21st century's first decade.

So, what might a 4°C world look like?

The first thing to note, as alluded to above, is that with a 4°C warmer world not everywhere will be equally warmer than today: the higher latitudes will warm more than the tropics. Northern Alaska, the Canadian Arctic Archipelago, the northern coast of Finland and that of Russia, as well as parts of Antarctica will warm by nearly 8°C or more. This is not to say that the tropics will escape extreme warmth: they are already the warmest parts of the planet and on land they too will warm by 5°C–7°C. Europe, parts of the United States, southern

Argentina and Chile, and Australia will warm by an estimated 4°C–6°C. The areas that will warm the least will be the oceans, and remember the oceans make up most of the Earth's surface. It is thought that much of the oceans will warm by only 3°C–4°C. This is because a warmer world in, say a century or so's time, will still see surface waters being subducted to abyssal depths by the Broecker global thermohaline conveyor and being replaced by cooler abyssal water. Also, because water has a higher heat capacity than earth and rock (it takes more energy to raise its temperature). With regards to rainfall, while global average precipitation is projected to increase, most areas that are currently arid and semi-arid are projected to dry further or have more arid seasons, while the moist tropics and mid-latitudes are projected to become wetter (*34*).

With major biome change, we will have to change the way we farm. In some cases this will be *just* altering the types of crops grown in a particular place (not that 'just' is an appropriate term given the near global scale of adjustment needed). In other cases, with warmer temperatures and in areas with less rainfall, it may be that farming will no longer be possible. However, there will be still other areas where previously farming or pasture was not possible but will become so in a 4°C warmer world. From Figure 9.4, depicting current agricultural land and the rainfall changes expected in a warmer world, losers will include: north-western United States, southern Europe, north-western African coast, southern Africa, north-eastern South America, and northern and south-eastern Australia, among others. Winners will include: the eastern Canada-US border region, northern Europe, north-eastern Russia, mid-Africa, and western Australia, among others. Bear in mind that adapting the way the world supplies its food as it enters a 4°C world will be ongoing as there will be further changes with further warming and also that this will be all set against a backdrop of a larger global population.

And, of course, it is not just climate that will affect a warmer world. Not least among the non-climate (but nonetheless warming-induced) changes will be sea level. In a world that warms by 4°C by 2100, global sea level will increase by no less than half a metre and possibly by up to as much as two metres. This last was the estimate arising out of the Royal Society 'Four Degrees and Beyond' symposium whose proceedings were published in 2011 (*35*). Since then there was the news of the aforementioned melt from the Canadian Arctic Archipelago. Also, only recently published (2020) were the results of a comprehensive analysis of the Greenland ice sheet. That analysis concluded that, since the 1990s to 2018, the cumulative ice loss from Greenland as a whole, and by itself (not including the Canadian Arctic Archipelago), have been close to the rates predicted by the Intergovernmental Panel on Climate Change for their high-end climate warming scenario (*36*). With this subsequent knowledge, the 'Four Degrees and Beyond' symposium sea level rise estimate certainly seems conservative.

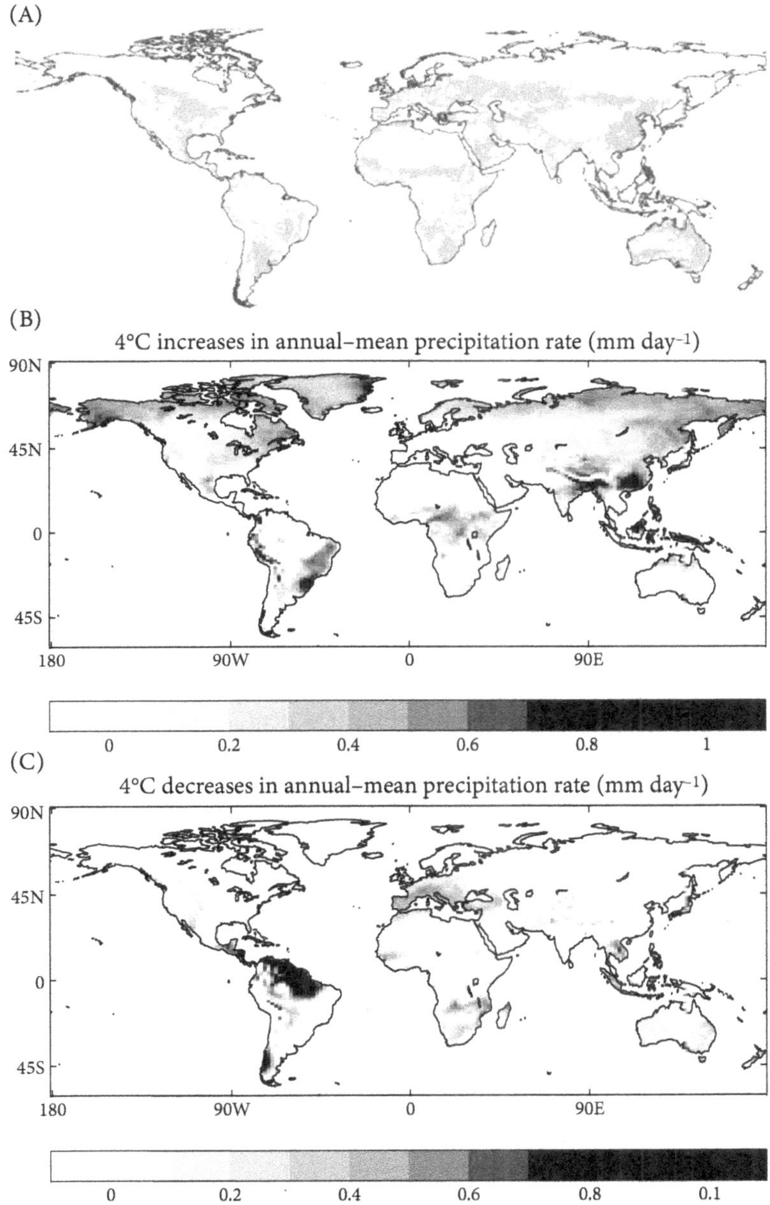


**Figure 9.4** (A) Distribution of present-day agricultural land. (B) Areas where precipitation is predicted to increase in a 4°C world. (C) Areas where precipitation is predicted to decrease in a 4°C world. Crop and pasture regions of the world. Panel A is adapted with permission from Figure 2.15 in Forster, P., Ramaswamy, V., Artaxo, P., Berntsen, T., Betts, R., Fahey, D. W., Haywood, J., Lean, J., Lowe, D. C., Myhre, G., Nganga, J., Prinn, R., Raga, G., Schulz, M., & Van Dorland, R., Changes in atmospheric constituents and in radiative forcing. In Solomon, S., Qin, D., Manning, M., Chen, Z., Marquis, M., Averyt, K. B., Tignor M., & Miller,

In the southern hemisphere, Antarctica's potential contribution to sea-level rise is worrying. More than half of Earth's freshwater resources are held by the Antarctic Ice Sheet which, with global warming, represents by far the largest potential source for global sea-level rise. Its coastal ice shelves have been melting faster. In 2015, a study in the journal *Science* showed that the continent's ice shelves as a whole transitioned from losing (or with snowfall even gaining) 25 ± 64 cubic kilometres *per* year for 1994–2003 to a rapid loss mode of 310 ± 74 cubic kilometres *per* year for 2003–2012. West Antarctic losses increased by ~70% in the decade to 2012. The concern is that once a coastal ice shelf is gone then it is easier for the ice sheet on the land to move towards the sea (*37*).

With warming, Antarctica itself seems to be crossing, or about to cross a number of thresholds. In 2020, Ricarda Winkelmann and colleagues from the Potsdam Institute for Climate Impact Research in Germany, have modelled a number of these. One for instance, is that as surface ice melts, so the ice sheet surface lowers. As lowlands are warmer than uplands, so the rate of melt increases. Another is that ice flow increases under warmer conditions. Yet another is the aforementioned loss of marine ice shelves that buttress the ice on the land: removing the buttress frees the landed ice to move towards the sea. These are all positive feedback cycles, the progression of which accelerates melting.

There are also negative feedback cycles that slow melting. One of these is that as the ice sheet melts, so the weight of ice that depresses the land lessens and this (over many years) allows the land to rise, taking the surface to a cooler level. However, such negative feedbacks are not strong enough to overcome the positive feedbacks of accelerated ice melt. The Potsdam study's results imply that, if the 2015 UN Paris Climate Accord to limit global warming to well below 2°C above pre-industrial temperatures is not met, one or more critical Antarctica thresholds might be subsequently crossed, committing us to long-term, possibly irreversible, sea-level rise of up to dozens of metres (*38*).

---

**Figure 9.4** *(continued)* H. L., eds., *Climate Change 2007: The Physical Science Basis. Contribution of Working Group I to the Fourth Assessment Report of the Intergovernmental Panel on Climate Change.* Cambridge University Press: Cambridge, United Kingdom and New York, NY, USA (2007).

Panels B and C kindly provided by the Met Office (UK) in 2012. © 2011 Crown copyright, Met Office and for outside the UK © British Crown copyright. For technical details as to the HadGEM2-ES model used for the CMIP5 centennial assumptions see Jones, C. D., Hughes, J. K., Bellouin, N., *et al.*, The HadGEM2-ES implementation of CMIP5 centennial simulations. *Geoscientific Model Development*, **4**, 543–570 (2011), and for those technically minded wishing to construct a more refined version, based on the more recent HadCRUT5, should see the update Morice, C. P., Kennedy, J. J., Rayner, N. A., *et al.*, An updated assessment of near-surface temperature change from 1850: The HadCRUT5 dataset. *Journal of Geophysical Research*, **126**, e2019JD032361 (2021).

As to what it would be like for humans to live in a 4°C or 5°C warmer world, there has been considerable study and informed conjecture; far too much to go into here. Yet the extremes predicted for a possible 4°C or 5°C world are telling. Here three separate and recent studies, using three different methodologies, all come to the same conclusions as to where in a warmer world envisaged under the IPCC business-as-usual scenarios it would be simply too hot for humans to live, or at any rate too hot to live without air-conditioning.

The first study was by a US collaboration based in Hawai'i and published in 2017. It reviewed academic papers published between 1980 and 2014, and found 783 cases of an excess of the normal death-rate, these were excesses associated with heatwaves in 36 countries. Based on the climatic conditions of these lethal heatwave events, they deduced a threshold beyond which daily mean surface air temperature and relative humidity becomes deadly. Around 30% of the world's population is currently exposed to climatic conditions exceeding this deadly threshold for at least 20 days a year. By 2100, this percentage is projected to be ~74% under a scenario of growing greenhouse emissions. An increasing threat to human life from heatwaves now seems almost inevitable, but will be far worse if greenhouse gases are not considerably reduced (*39*).

The second research paper was by Eun-Soon Im, Jeremy Pal, and Elfatih Eltahir and also published in 2017. Based on biomedical evidence, it took 35°C for 6 hours as an upper limit on human survivability in a natural (not air-conditioned) environment. It then used climate models to predict which parts of the world would experience such warmth by the end of the century under business-as-usual scenarios. They identified regions that today are where about a fifth of the world population reside and particularly in densely populated agricultural regions of South Asia (*40*).

The third paper was by an international collaboration in a fascinating study designed by Chi Xu and Marten Scheffer and published in 2020. It took regions as having a mean annual temperature of over 29°C as being the edge of the temperature niche for human viability. It showed that for thousands of years, humans have concentrated in a surprisingly narrow subset of Earth's available climates, characterised by mean annual temperatures around ~13°C. (With regards to a mean annual temperature of 29°C being the limit of human viability and the previous, Im, Pal, and Eltahir study, note that a climate of with an average of 29°C will see both colder and warmer episodes.) This third paper noted that currently the region of the world where the mean annual temperature is over 29°C occupies only 0.8% of the Earth's land surface, and is mostly concentrated in the Sahara. However, by 2070 under IPCC business-as-usual scenarios, it will cover 19% of the planet's land (*41*)!

All three studies came to the same conclusion. Two studies—the first and the third—noted that parts of the Amazon and equatorial African rain forest as well as the northernmost parts of Australia would become too warm and uninhabitable for humans. (This was left out of the second study—the Im, Pal, and Eltahir paper—as that work focussed on the most populated regions at risk.) All three papers agreed that the populous regions at greatest risk were the Indian subcontinent and South-East Asia, southern China, and the nations of the Malay peninsula as well as around the Gulf of Siam and also the East Indies. This means that 1 to 3 billion people would be affected! The resulting migration pressure in the decades to come contrasts with that of the early 21st century migration crisis (largely war and economically driven) that by comparison measures in a paltry few millions.

Finally, this idea of a climate too hot for humans is not something for the future; there are events today that see increased mortality. As I write a first draft of this chapter in the northern hemisphere summer of 2021, over in North America, in the Pacific Northwest region, there is a heatwave that has seen records broken in multiple cities by several degrees Celsius with peak daily temperatures far above 40°C including setting a new all-time Canadian temperature record in the village of Lytton, at which a temperature of 49.6°C was measured on 29th June and where wildfires spread the following day. Given that the observed temperatures were so far outside historical experiences, and in a region with only about 50% of dwellings having air conditioning, there was excess mortality over what would normally be expected.

With such major and record-breaking heatwaves, the question asked is whether or not they are part of weather's natural variability, or were part of climate change? In the case of this event, an international collaboration of climate scientists, called the World Weather Attribution, provided an assessment. They specialise in attributing events to natural variation, climate change or a combination. Based on the historic record and climate modelling they concluded that the 2021 North American event was virtually impossible without human-caused climate change: the observed temperatures were so extreme that they lay far outside the range of historically observed temperatures. This made it hard to quantify with confidence how rare the event was. In what they considered the most realistic statistical analysis, they estimated it to be about a 1 in 1,000-year event in today's climate terms.

Interestingly, as discussed here earlier, their conclusions touched on the possibility of the climate system shifting in a non-linear way to a new state. They noted that there were two possible sources of this extreme jump in peak temperatures. The first was that this heatwave was a very low probability event, even in the current climate which already includes about 1.2°C of global warming above pre-industrial temperatures—the statistical equivalent of really bad luck, albeit aggravated by climate change. The second option was that non-linear interactions in the climate system had substantially increased the probability of such extreme heat, much beyond the gradual increase in heat extremes that had previously been observed. This option had not been included in the climate models of the time. Such was the extremity of the event that looking into the future, in a world with 2°C of global warming above pre-industrial (that might be reached as early as the 2040s), this event would have been another degree hotter. An event like this (as noted above) was currently estimated to occur only once every 1,000 years, but would occur roughly every 5 to 10 years in a future 2°C above pre-industrial (or just 0.8°C warmer than 2021) (*42*).

Clearly, to address this section's titular topic, 'The need to manage the future', we do need to manage the future and address climate change concerns. We should own up to our (humanity's) role in transforming ecosystems and embrace responsible policies befitting a species that has engaged in millennia of ecosystem engineering and human niche construction (*43*). If we do not, the prognosis, for our global society is most unlikely to be good.

## Problems in avoiding or managing the transition

The key to managing the transition to a new planetary state, one dominated by our species' actions, will be to avoid those aspects of the transition we consider unwelcome. Previous (natural) transitions across climate thresholds have seen a change in the nature of the carbon cycle and, during the actual transitions, fluctuating atmospheric carbon and concomitant climate change. However, with these past slower transitions, species were more easily able to migrate: today, with today's agricultural and urban landscapes, this is less so. With the current (human-induced) transition threat, it is avoiding global warming of 2°C (preferably 1.5°C) above the mean, pre-industrial global temperature that is crucial.

One way of looking at the problem is considering our emissions of greenhouse gases and to curb these. Much has been written about this issue but the bottom line is sadly that, since the first IPCC report in 1990, emissions have so far not fallen, nor stabilised, but have continued to grow (Figure 8.3) along with increased consumption of fossil fuels (Figure 8.5) and emissions associated with land use.

Another way of looking at the problem would be to see what is the level of atmospheric carbon that must not be exceeded to keep the Earth below 2°C (preferably 1.5°C) warming above pre-industrial temperatures?

In 2009 a small, largely British-based, research collaboration led by Myles Allen and David Frame crunched the numbers. They concluded that the total amount of fossil carbon past, present, and future released by humans had to be kept below 1 trillion tonnes of carbon or 1,000 gigatonnes of carbon (1,000 GtC) which in turn is equivalent to 3.67 trillion tonnes of carbon dioxide. Yet, since industrialisation began, by 2008 a little over half of this had already been emitted (*44*).

At first, it might seem as if our still having roughly half the 'carbon allowance' (500 GtC) that would confer an additional 'cumulative warming commitment' of 2°C might not seem problematic. After all, humanity took the best part of two and a half centuries to emit half a trillion tonnes of carbon. However, this ignores the power of exponential growth: a water lily doubling in size each day to cover a pond in 50 days, covers half the pond on day 49! You will recall from earlier in this chapter that in 2001 the IPCC estimated the then current trends would see the emission of between 1,000 and 2,100 GtC in the 21st century alone (*1*). The excess between this and less than 500 GtC is considerable: the need to curb emissions is urgent. Indeed, the Myles Allen and David Frame team's 2009 base-case estimate is that the total emissions from 2008 to 2050 need to stay below 190 GtC for us to have a good chance (75%) of keeping warming below 2°C.

The Myles Allen and David Frame team's study appeared in the journal *Nature*. In the same edition there was a complementary study led by Malte Meinshausen and Nicolai Meinshausen. That the studies chimed so well was in no small part because a number of the researchers were involved in both works including Myles Allen and David Frame themselves. This second study took a probability approach based on both models and real world observations of past climate change and atmospheric greenhouse gases. They aimed to determine the likely limits to greenhouse gas emissions by 2050 that should not be exceeded to keep 21st century warming below 2°C. It concluded that limiting cumulative carbon dioxide emissions between 2000 and 2050 to 273 GtC (equivalent to 1,000 Gt $CO_2$) yields a 75% probability of keeping warming below 2°C (*45*).

Yet between the start of 2001 and the end of 2019 global emissions were already 164 GtC (or 600 Gt $CO_2$) having increased by 45%. In short, by the end of 2019, 21st-century emissions were over 60% of the 2000–2050 273 GtC emissions allowance limit (source: see Appendix 2, 'Energy statistics and data') leaving a 109 GtC allowance for 2020–2050.

Even back in 2009 it was clear that the bottom line of both these studies was that dangerous climate change (2°C warming above pre-industrial) is going to be hard to avoid. Unless emissions begin to decline very soon, severe disruption to the climate system will entail expensive adaptation measures and may eventually require cleaning up the mess by actively removing carbon dioxide from the atmosphere (*46*). Today (as I write in 2023), given the emissions the past couple of decades, the problem has become more acute.

Clearly, problem number one, in avoiding damaging warming associated with the current transition, is that our greenhouse emission trends are both too great and also going in the wrong direction (increasing annually, not decreasing).

So, what are our chances of limiting emissions?

Obviously much depends on nations' political will. However, this has to overcome the inertia of existing trends. Here, a key dimension is the fossil fuel infrastructure (for example, gas and coal power stations) that are currently being constructed or are on the drawing board having the investment in place for their construction.

In 2019, a group of researchers largely based in California, looked at existing global fossil-fuel infrastructures and, assuming recent performance, the greenhouse emissions they are likely to generate over their planned lifetimes. They estimated that, if operated as historically, they will cumulatively emit about 179 GtC (or 658 Gt $CO_2$). More than half of these emissions were predicted to come from the electricity sector; infrastructure in China, the United States, and the 28 member states of the European Union represents approximately 41%, 9%, and 7% of this 179 GtC total, respectively.

With regards to proposed and infrastructure being constructed in 2018, if built and then operational for their planned lifetimes, proposed power plants and those under construction would emit roughly an extra 51 GtC (188 Gt $CO_2$).

Together, existing, under construction, and planned/proposed fossil fuel infrastructures over their proposed lifetimes altogether would release 231 GtC (about 846 Gt $CO_2$). This is greater than the aforementioned, entire carbon emissions budget allowance of 109 GtC that remains for 2020–2050 if mean warming is to be limited to 1.5°C and perhaps two-thirds of the remaining carbon budget if mean warming is to be limited to less than 2°C. Their estimates suggest that: no new fossil carbon-emitting infrastructure can be commissioned; proposed plans must be scrapped; and that existing infrastructure may need to be retired early (or be retrofitted with carbon capture and storage technology) in order to meet the Paris Accord climate goals (*47*).

Cancelling all fossil carbon infrastructure to curb subsequent fossil carbon emissions has a corollary, and that is to leave fossil fuels in the ground. Here, as Dan Welsby, James Price, Steve Pye, and Paul Ekins of University of London calculated in 2021, the amount of fossil fuel that would need to be left in the ground, so as to meet the Paris Accord limit of 1.5°C warming above pre-industrial, is not trifling: we would need to leave most of it in the ground. We would have to leave nearly 90% of economically viable global coal reserves and nearly 60% of oil and gas reserves in the ground to have even a 50% chance of hitting the Paris Accord target (*48, 49*)! But, do we have the will to turn our back on this economic potential? (Honest answers only, please.)

In short, given the limits to the total level of human-generated carbon in the atmosphere necessary to keep warming below 2°C (let alone the more ambitious Paris Accord's target of below 1.5°C), in the context of 21st century's first two decades of emission trends, and not to mention the anticipated emissions from existing, proposed and fossil fuel infrastructure under construction, we will (with certainty) miss the warming targets unless drastic (hence costly) action is taken. It could, though, be possible to meet these targets if sufficient measures are taken!

One real likelihood is that we take action, and avert the most serious impact of business-as-usual, but not sufficient action to meet the 2015 UN Paris Accord targets, hence still incur significant climate impacts. In this circumstance we will not only have made the necessary expenditure to switch the global economy to one with less fossil carbon emissions, but also have to pay for the costs incurred through climate change impacts in part outlined in the previous chapter section. This might happen even if we transfer to a truly zero fossil carbon economy further down the line. If so, it looks like we could be in for a bumpy ride.

That we are in for a mid-21st-century bumpy ride is not new news as there have been ample studies predicting this over the past half century—both quantitative and qualitative studies. I do not wish to dwell too long on these as futurology is (rightly) questionable: after all, who can predict the future with accuracy? Nonetheless, a good number of past studies have predicted problems for our global society in the mid-21st century. The thing is that if any study, forecast, or prediction is noticed and action taken, then the prediction automatically becomes undermined. For example, predict that that cliff-top path will see walkers fall off then the simple solution of erecting a warning sign, or a fence, will greatly reduce injuries and fatalities. However, because now with a sign, or fence, there are hardly any injuries or fatalities does not mean that the cost of a sign or a fence was a wasted investment. In short, even if successful, prediction is doomed to forecast failure with mitigating investment, this does not mean such prediction should be disregarded.

One of the earliest attempts to mathematically model future global human ecology was by Jay Forrester of the Massachusetts Institute of Technology (MIT). He had devised a computer model of the global human population and its resource consumption, which incorporated not just population levels, but industrial output, food availability, pollution, and finite resource levels. Again, by today's standards it was a very simple model. Then, in 1972, an MIT team led by Dennis Meadows produced a report based on runs of the Forrester World Model for the Club of Rome. This was *The Limits to Growth* report (*50*). The report itself attracted much attention, including by the media, sold over 30 million copies, and helped provide the impetus in establishing the United Nations Environment Programme (*51*). But it was also heavily criticised. For example, according to the report, by now we should have exhausted the global supply of copper. The report failed to recognise both the growth in copper recycling and improved mining techniques. Indeed, by the early 1980s techniques had improved to such an extent that the spent tailings from former copper mines were being re-mined. It must also be noted that Jay Forrester himself was not an author of *The Limits to Growth*: the model and the more seriously flawed *The Limits to Growth* report are separate entities.

The Forrester 'World' model (actually, a more developed, 1973 version of the original World model was used by *Limits*), in essence, consisted of some separate, but interacting, subsections namely:

- the food system (agriculture and food production)
- the industrial system
- the population system
- the non-renewable resources system
- the pollution system

Each of these in turn lead to outputs of: food *per capita*, population, resources used, pollution, and industrial output *per capita*. They were graphically plotted with 1900 to 1970 plots being real world and, going into the future, the calculated forecast.

The evolved version of the World model used to underpin *The Limits to Growth* had around 200 equations, so it was back-of-the-envelope stuff: even by the end of that decade (the 1970s) other such models had around 100,000 equations (*52*). Nonetheless, from 1972, subsequent years have shown the model's standard run to be somewhat prescient.

Over a third of a century later Graham Turner, of CSIRO Sustainable Ecosystems in Australia, looked at the standard run forecast and the real world data for the subsequent three decades following the original publication of the standard run. It showed that the standard run outputs for all five variables—for the then future, and now in the past, three decades—compared favourably with the real world data: the model was by all accounts a successful predictor, at least for three decades (*53*). Indeed, track real world population and the World model standard run through to 2020 and it can be seen that for half a century since 1970 there is considerable agreement (see Figure 9.5a). (Ironically, less so with more developed subsequent versions of the model.)

The World model's standard run outputs show exponential growth in population, food *per* capita, and industrial output *per* capita through various parts of the mid-21st century and a decline in each thereafter. Figure 9.5A shows just the model's food *per* capita and population outputs together with the actual population to 2018 for comparison.

This change from exponential growth to a mid-21st-century peak followed by decline suggests a transition of state of the system and as such chimes with this chapter's discussion so far.

Figure 9.5B depicts UN data of actual population to 2021 and its high, medium, and low scenario variants thereafter to 2100. If the World model's standard run has any merit for later in the 21st century, then the human population would be tracking more the UN's low variant. However, this is a big 'if'. As the

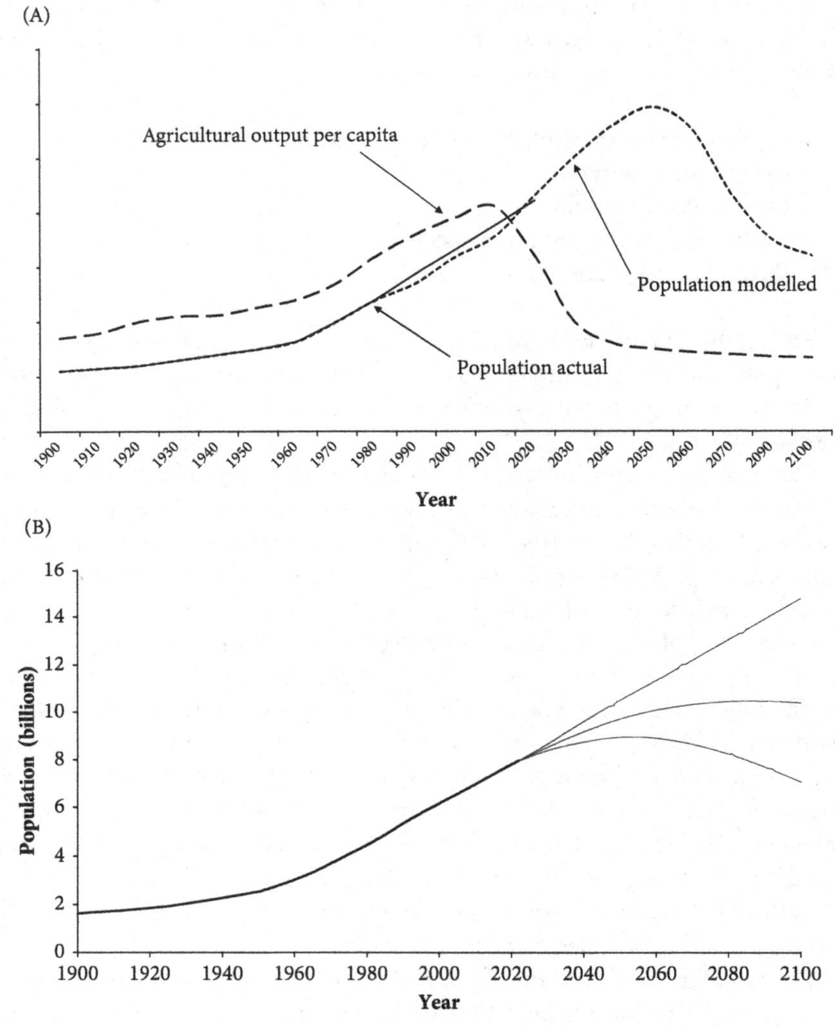


**Figure 9.5** (A) The modelled agricultural output *per* capita, population actual and population modelled by the World model as used in *The Limits to Growth* (1972 [*50*]). There are no units, as the graph depicts how these variables change with time compared to between 1900 and 1970 as here the lines are based on real-world data. The subsequent-to-1970 model forecasts are proportional to that. The actual population data from 1900 to 2018 is based on UN data and forecasts.
(B) World population actual to 2021 and then UN high, medium, and low variants to 2100. (Data from: United Nations, Department of Economic and Social Affairs, Population Division (2022). *World Population Prospects 2022*, Online Edition. Rev. 1. [2022]).

World model's standard run moves further into the future, so the constants in its equations are likely to drift from reality: that they have done so well, so far, is remarkable. Indeed, with regard to critical transitions and the discussion in Chapter 2, this is particularly true as the system enters the zone of transition (see Figure 2.4B) and a more chaotic (in the mathematical sense) time. In short, we should not take the World model's standard run as an accurate prediction for the middle-to-late 21st century but just a possible rough guide.

Having said all this, it should be emphasised that the World model's qualified success, such as it is, with regards to population and agricultural production as used in *The Limits to Growth*, is in the early 21st century beginning to see some drift from reality. This, though, is not the point I am making.

Though it may not be an accurate forecast of times further into the future, the World model does suggest some form of mid-21st-century transition. Indeed, simply tweaking its equations' constants just a little does not erase this transition—that's the point! I found this out, back in the late 1970s when working on the World model, how difficult it was to alter the model to avoid a severe transition: many of the model's equations' constants had to be changed (the equivalent to assuming the successful implementation of substantive hypothetical policies) quite markedly and even then a transition was still there, it was just not so pronounced.

The Forrester World model is not the only early computer model whose outputs exhibit a certain robustness with time; many of the early climate models do too.

Models rely on two key factors: their structure and the quantitative estimates they make for their components. Simplifying matters right down, a global climate change computer model's structure will not only relate greenhouse emissions to warming but also rely on assumptions as to how much emissions will be generated and things like the climate forcing associated with a unit of a particular greenhouse gas. Let the years roll by and, with the benefit of hindsight, we can see how accurate, or robust, a model was in predicting warming. Yet it could be that a model's design—no matter how good—could be undermined by poor estimates. In 2019, a study of 17 projections arising out of 14 computer models developed between 1970 and 2007 found that 10 projections were consistent with observations but, correcting the models' original components estimates with the latest and re-running those projections, 14 of the 17 agreed with observations (*54, 55*). These chimed with the successive IPCC's *Assessments' Scientific Basis* reports' business-as-usual (b-a-u) analogue scenario forecasts all falling within the 1990 *Assessment*'s b-a-u upper and lower limits discussed in the preceding chapter (see Figure 8.4B). So with models, among much else, we have had ample warning and time to address some of the issues we now face, which makes it all the more lamentable that so little has been done.

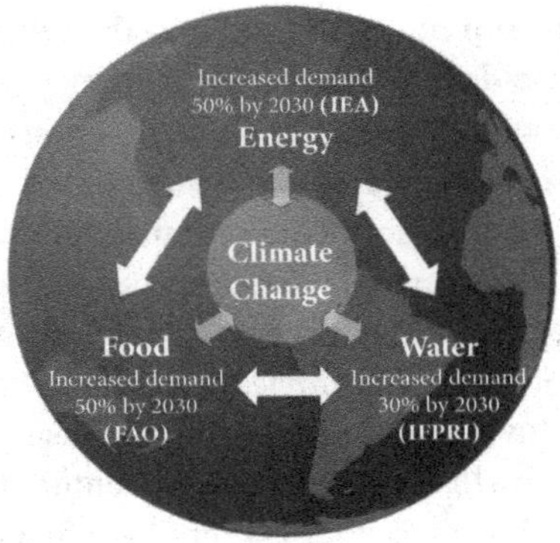


**Figure 9.6** A pictorial summary of John Beddington's mid-21st century 'Perfect Storm' (2009 [56]). Reproduced with permission.

Other than number crunching (a quantitative) approach to futurology, it is possible to identify what may be some of the critical factors that will determine the future (a qualitative approach). In 2009, the United Kingdom's then Chief Scientific Advisor, Prof. Sir John Beddington, did just this in his 'Perfect Storm' paper (56). He noted that the mid-21st century will see the need to sustainably manage the world's rapidly growing demand for energy, food and water to service a growing world population and all this against a backdrop of climate change. Further, he emphasised that each of these issues was interrelated (see Figure 9.6) and that they come together to create a 'perfect storm' of global events.

In managing the transition to a sustainable world, John Beddington presents five challenges:

- Can 9 billion people be fed equitably, healthily, and sustainably?
- Can we cope with the future demands on water?
- Can we provide enough energy to supply the growing population coming out of poverty?
- Can we mitigate and adapt to climate change?
- Can we do all this in the context of redressing the decline in biodiversity and preserving ecosystems?

These, Beddington contends, will require a mix of behavioural change and technological solutions from many scientific disciplines, ranging from biotechnology and engineering to newer fields such as nanotechnology. He concludes: 'This vital contribution from science will not happen by default'.

There is little point going into all the interactions Beddington discusses here (there is no space) but they do all add, in a very obvious way, to the picture presented in this chapter section that there are significant problems and difficulties to overcome in managing the transition to sustainability. This is a multifaceted issue, which means that we have to change many aspects of our current global lifestyle to a more sustainable one. Indeed, it also means that on the way striving for a more sustainable future we do not inadvertently create new problems. For example, just one issue of recent concern is the rise of plastic pollution: plastic use having risen in no small part to reduce food spoilage and provide a low-energy way of packaging drink.

In 2021, work was published from an international collaboration of researchers led by Carmen Morales-Caselles and Andrés Cózar of the University of Cadiz that looked at inventories from scientific surveys and civil society networks to create a database of millions of litter items found across seven major aquatic environments, ranging from rivers to shorelines and the open ocean, and from the surface to the seafloor. They had data on more than 12 million litter items confirming overwhelming prevalence of plastic items (which here exclude micro-particle pollutants). On average, 80% of the items were made of plastic, followed by metal (7%), glass (5%), and fabric (3%). The largest share of plastic was found in surface waters (95% of surface water litter items), followed by shorelines (83%). Below the surface, the proportion of plastic in total litter increased progressively from 49% on riverbeds to 64% on near-shore bottoms and 77% on deep seafloors. Of the plastic found, the top three major contributors were: bags (14% across all environments), plastic bottles (12%), food containers and cutlery (9%), and wrappers (9%) (*57*).

This plastic pollution is comparatively recent: plastic was not manufactured in any scale before 1950. Yet by 1980, global production *per annum* had risen to over three-quarters of a billion tonnes and then more than double this figure by 1990. By 2015 global plastic production *per annum* had nearly reached nearly 8 billion tonnes, over half of which was discarded as waste, with approximately just 6% recycled and 8.5% incinerated. Roughly, around 3% of global plastic waste enters the oceans each year, which in 2015 was around 8 million tonnes (*58*). If current trends continue (note the 'if') then global production will exceed 25 billion tonnes by 2050. And if production and waste management trends up to 2017 continue, roughly 12 billion tonnes of plastic waste will be in landfills or in the natural environment by 2050 (*59*).

Addressing plastic waste is a sustainability issue that we as a global society have to address. It is also connected with the current increase in mineral species (see Chapters 6 and 8) for some of this plastic, in some form, will enter the geological record.

There are undoubtedly problems in avoiding or managing our species' transition to a sustainable mode of existence and it is not going to be easy, but, as discussed in the previous chapter section, we need to manage the future if we are to avoid much human suffering through destruction of the natural systems and climate that makes human life equable.

It is certainly not going to be easy, even if we do succeed in adopting a near-zero fossil carbon existence, conserve natural systems, and the other things that need to be done. Part of achieving this will be to address the issue of the global human population. Yet the demographics suggest that the inertia (which includes factors such as people living longer) in the system will make it hard to go below the low end of the UN low variant scenario (see Figure 9.5B). Regionally, it is likely that Africa and South Asia will experience the greatest human pressures on future ecosystems (notwithstanding the climate change pressures discussed earlier). Humanity's large demographic momentum means that there are no easy policy levers to change the size of the human population substantially in the coming decades, short of extreme and rapid reductions in human fecundity (*60*).

In addition to our global society needing to become environmentally sustainable, we must remember that the forthcoming transition that our human system component of the larger Earth system will itself enter the zone of transition (see Figure 2.4B) within which there is much uncertainty. This is not a place within which the human system needs a kick. Yet, for example, we are currently (2022) seeing with Russia's war on Ukraine disruption of the global food supply, hence rising prices and the risk of starvation. Further, as we develop the human system's electronic and digital dimension, so we must ensure that it is secure and stable as well as have back-ups (non-digital options) in place. Here, there is much that can be said, but one example is the move away from a cash economy to electronic financial transfers. But what happens if the latter fails? Software failures, cyber-attacks, and even something as mundane as a power cut, effectively curtails day-to-day digital transactions: a daily shop would be nullified. In 2023, some 10,000 properties in the small English town of Tring (a reasonably wealthy area north-west of London) experienced a three-day internet outage. As it happened, this was caused by rats destroying a key internet junction. After the first day, most of the local businesses had to close as the local population's cash ran out: cash machines could not help as they too rely on the internet (*61*).

So, the afore example was small and local: a large, non-local, serious issue could not be a problem, surely? Well, Sir Oliver Letwin would beg to disagree. He was a Cabinet minister and for 6 years was a member of the UK's National Security Council as the Minister with responsibility for the nation's resilience. He presented his concerns of Britain not having alternate back-ups to digital failure in his 2020 book *Apocalypse How? Technology and the Threat of Disaster* (62). In it, he details the issue—so I am not going to duplicate that here—and laments the number of safeguarding proposals that have been rejected by government.

We need back-ups: striving for a pluralistic, mixed cash and digital economy would be prudent. A similar lesson for pluralistic and back-up resources is again currently (2022) being learned by western Europe, which is struggling—amid the Russian war against Ukraine—due to its reliance on Russian natural gas that is undermining economic sanctions western Europe imposed due to the war with Ukraine.

Another system failure arising from that conflict is the global wheat crisis due to Ukraine being unable to export wheat. Russia and Ukraine together export about one-third of the world's wheat, and more than half of the world's sunflower oil and seeds. Alison Bentley, director of the Global Wheat Program at the International Maize and Wheat Improvement Center in Texcoco, Mexico, notes the folly of having 2.5 billion people depend so heavily on three main regions of wheat production and export. She advises: spreading our bets; real-time monitoring (including by satellites of agricultural production); and support for women in agricultural areas, citing the UN's Food and Agriculture Organization saying that this would increase yields by 20%–30% (63). Meanwhile, Michel Ghanem and colleagues highlight the need to diversify food systems: again, to spread our bets. African countries should focus instead on their diverse indigenous and traditional cereals, roots, tubers, oil crops, vegetables, and fruit. Investment in historically grown, nutritious African cereals such as sorghum, fonio, and teff as well as barley and legumes would, they say, be transformative (64).

The key thing is that back-ups are needed for resilience against system failure whatever the cause.

Finally, given that a good proportion of this book was written in 2020 during the CoVID-19 pandemic, there is one other impact of our growing human population on natural systems worthy of mention: natural systems can impact back.

As noted in the previous chapter, our species has altered over half of Earth's habitable land to meet the needs of its burgeoning population. The transformation of forests, grasslands, and deserts into cities, suburbs, and agricultural land has caused many species to decline or disappear, but others—you may be surprised to hear—have prospered. Here, as to which species suffer and which thrive, there is a trend: the winners tend to be generalist species that are small and abundant and that have 'fast', short lives, such as rats and starlings; the losers tend to be ecological specialists, such as rhinoceros or ostriches, that have highly specific feeding or habitat requirements and that are comparatively larger, rarer, and longer-lived than are non-specialists.

There is also a second trend: winners tend to be species that contribute to disease in humans, either directly or indirectly through an intermediate host (*65*). These diseases humans catch from animals are called zoonotic.

SARS-CoV-2 is a zoonotic virus that gives rise to CoVID-19. It has been known for a few years now that SARS-related viruses circulate in some populations of bats. A large number of SARS-related coronaviruses have been detected in horseshoe bats since 2005 in different areas of China. In 2017 came news of 11 SARS-related viruses from a single cave in Yunnan Province, China, that was inhabited by many horseshoe bat species (*Rhinolophus* spp.), several of which were infected with SARS-related viruses (*66*). Since then, in 2020, a dead bat that had been captured back in 2010, but stored in a freezer in Cambodia, was found to have been infected with a virus that was 99% genetically identical with SARS-CoV-2 (*67*). It is thought that this virus, or one similar to it, passed through an intermediate host species becoming a form that made it easier to infect humans. Malayan pangolins (*Manis javanica*) are considered the likely possible intermediate species (*68*). (There is also a laboratory-release hypothesis for SARS-CoV-2's rising, but China has not been as open with the World Health Organisation as it arguably might have been, so we do not have evidence either way: I am not going to go down that particular rabbit hole.)

In 2020, in the midst of the SARS-CoV-2/CoVID-19 pandemic, research was published in *Nature* from a small team of British-based bioscientists with lead authors Rory Gibb and David Redding of University College London. Now, while this came out during the pandemic, it should be noted that the actual conception of the research and its undertaking was done *before* the world heard of CoVID-19. Indeed the journal received their paper's manuscript in January 2019. This is worthy of note because the research cast a direct link between the human encroachment on natural systems and zoonotic disease.

The Rory Gibb and David Redding collaboration drew on the US-funded PREDICTS (Projecting Responses of Ecological Diversity In Changing Terrestrial Systems) project database. It contains more than 3.2 million records from 666 studies that counted animals, noting their habitat along land-use gradients from natural (such as ancient woodland and natural wetlands) to environments heavily used by humans (such as intensive agricultural and urban landscapes). They also used this database to search for records of wildlife hosts of known disease and parasites capable of infecting humans. As a check against error, they also cross-referenced these zoonotic species against another database containing 20,382 associations between 3,883 animal (non-human) host species and 5,694 pathogen species.

What they found was striking. As human-dominated land use increased, so did the total number of zoonotic hosts, whereas the total number of non-host species declined. In more intensively human-used areas, both the number of host species and, more so, the number of individuals of those species increased. Their results clearly suggest that the intensity of land use and human encroachment on natural systems (such as tropical deforestation and bush meat harvesting) is creating hazardous interfaces between people, livestock, and wildlife reservoirs of zoonotic disease (*69*).

Given that 2020 saw the World Health Organisation declare SARS-CoV-2 a pandemic, this research was most prescient. Previous research had shown that outbreaks of diseases such as severe acute respiratory syndrome (SARS), and bird influenza that cross over from animals to humans, have increased in the past few decades (*70, 71*), while the Rory Gibb and David Redding collaboration demonstrated that the human encroachment on natural systems was 'a'—if not 'the'—driver of this increase (*69*).

This research looked at human encroachment on more pristine ecosystems with population growth and development: it was a land-use change approach. But, as we have noted, we are in a warming world. In a warming world species' range—that is the area over which species are found—shifts. Generally, with global warming, species move towards the poles as well as to higher altitudes (they move up mountains). However, they do not do this uniformly. This is for a variety of reasons, the main of which include that different species have different degrees of climate tolerance and that the climate has a number of dimensions: it is not just temperature, rainfall, for example, is another and both are important to life (*32*). With global warming, we would expect some different species to meet who had not commonly before, in this interglacial, to cohabit the same ecosystem. Such novel species encounters in turn mean that any viruses associated with one species has an increased opportunity to jump to another.

The number of such novel encounters due to species' range shift with climate change has been estimated. In 2022, a small research collaboration, led by Colin Carlson and Gregory Albery of Georgetown University in Washington, DC, reported on their analysis of projections of geographical range shifts for 3,139 mammal species under climate-change and land-use scenarios to the year 2070. They found in particular that there will be new species encounters at higher altitudes and in areas of existing high biodiversity in Asia and Africa. Worryingly, these areas additionally tended to be of high human population density. Also worryingly, they found that even modest warming—such as keeping warming below 2°C in line with the 2015 Paris Accord's more modest target—would see the vast majority of mammal species overlap with at least one unfamiliar species somewhere in their potential future range, regardless of emissions scenario, and that the majority of first encounters would likely occur by 2040. Taking out land-use scenarios, and looking just at climate change, they found that climate change scenarios, even scenarios in line with the Paris Accord, would essentially lead to a doubling of species-to-species contacts. Finally, their work speaks to our experiencing a virus, SARS-CoV-2, originating in bats, and evolving through an intermediary host species to be able to infect humans, was not simply due to chance. They also discovered that, owing to their unique dispersal ability, bats account for the majority of novel viral sharing and are likely to share viruses along evolutionary pathways that will possibly be able to infect humans (*72*, *73*).

So, what to do?

It has been proposed that globally curbing deforestation and wildlife trade, as well as monitoring livestock and wildlife for new virus outbreaks, would cost over a score or so billion US dollars a year. However, given the global cost of the 2020 SARS-CoV-2/CoVID-19 was somewhere around several trillion US dollars (the lost GDP for 2020 globally was around 5 trillion), annually the cost of preventing further encroachment of human activity on natural systems and viral monitoring is two magnitudes less. Novel virus outbreaks are increasing and, as the last, previous serious global pandemic was that of the so-called Spanish flu roughly a century ago (1918), we might expect the next to be sooner than that in the future (*74*). This makes an annual expenditure of a score or more billion US dollars cost-effective, notwithstanding other gains such as biodiversity conservation: indeed, add these in makes the exercise positively profitable.

Here's the thing. If we are going to manage the current critical transition of the Earth's state, so that the forthcoming state is a sustainable one, we are going to have to adapt our economic perceptions, and possibly our economic system, so that our managed transition is seen as cost-effective and profitable as well as stable and secure: that such adaption is not viewed as an *unnecessary* cost but as a *necessary* investment!

We will briefly touch on transition management solutions later, but, building on the economics of nature conservation for pandemic prevention, for now I give one and that is the need to properly value natural systems.

For example, in 1996, New York City invested between US$1 billion and US$1.5 billion in natural capital, in the expectation of producing cost savings of US$6 billion to US$8 billion over 10 years, giving an internal rate of return of 90%–170% in a payback period of 4–7 years. This return is an order of magnitude higher than is usually available on relatively risk-free investments!

This may seem surprising but it is true. You see, New York's water comes from a watershed in the Catskill Mountains. Until recently, water purification processes by root systems and soil microorganisms, together with filtration in its flow through the soil, were sufficient to cleanse the water to legal, potable standards. But then sewage, fertiliser and pesticides levels in the soil rose and stopped it reaching the legal standard required. The city had the choice of either restoring the integrity of the Catskill ecosystems or building a filtration plant at a capital cost of US$6 billion to US$8 billion, not including annual running costs of around US$300 million. In other words, New York had to invest in either natural or monetary capital (*75*).

The Catskill Mountains are but one set of regional ecosystems, and New York but one city, but what of the planet as a whole: what is the value of services provided by ecological systems and the natural capital stocks that produce them that are critical to the functioning of the Earth's life-support system?

In 1997, a landmark study addressed this question. It was by a collaboration of over a dozen, primarily US-based, economists, geographers, and environmental scientists. What they did was to identify a suite of 17 environmental services and functions and associate these to 19 biome types (with urban as being classified as one of these).

The environmental services they considered included: gas (atmosphere) regulation (such as the generation of oxygen by marine algae), climate regulation (such as the drawdown of carbon dioxide by trees), genetic resources (such as relatives of crop cultivars found in tropical forests), and nutrient cycling (such as by wetlands). In 1994, the year from which this study drew its data, urban land accounted for just 0.64% of the Earth's surface but provided no discernible natural service and has no natural capital, yet in conventional economic terms it provides the largest proportion of gross national product (GNP): over half the global population lives in cities and that half earns on average more than those living in rural settings. Conversely, non-polar deserts occupied 3.7% of the global surface, yet have extremely low population and contribute a negligible amount to global GNP. But what of the global value of ecosystem services?

The 1997 study estimated the most valuable ecosystem services on a *per* area basis were provided by swamps and flood plains in controlling floods and providing a waste treatment function (for example, marshlands trap nutrients): their value was estimated at US$19,580 *per* hectare. (Again, all the US dollar values cited here pertain to the year 1994.) But, given that globally the area of swampland and flood plains is just 165 million hectares, the global value of the ecosystem services they provide was US$3,231 billion. The ecosystem types conferring the most value were, due to their covering 4,855 million hectares, forests at US$4,706 billion, tropical forests providing US$3,813 billion, and temperate boreal forests US$894 billion.

In total, they calculated that the planet then provided US$16–$54 trillion ($10^{12}$) *per* year, with an average of US$33 trillion *per* year. They say that because of the nature of the uncertainties, this must be considered a minimum estimate. This compares with global gross national product of the conventional economy in 1994 of around US$18 trillion *per* year (*76*).

If we are to preserve natural systems and the services they provide, then we need to value them. This will mean a substantive shift in the way we run the global economy.

Progress is being made. For example, in 2012, the United Nations approved a universal framework for global deployment, known as the System of Environmental–Economic Accounting. From this, a new metric called gross ecosystem product (GEP) has been developed and is beginning to be used in a number of countries including China. But, a quarter of a century on from the 1997 ecosystem services study, progress is painfully slow (*77*).

Individually, each of the problems in managing the current transition to a sustainable mode are formidable: our current economic model's ingrained dependence on fossil carbon; the limited fossil carbon budget we have left before substantive climatic change; the trajectory of humanity's world system; Beddington's 'Perfect Storm' synergy of confounding factors (*56*); the increase in pollution; our growing global population; ensuring security and stability of new technological and socio-economic systems with back-ups in place; and the need change our economic system so as to value ecosystem services. Collectively, together, they may seem insurmountable. Yet there are other benefits as informed by the example of ceasing our encroachment on the remaining natural systems so reducing the emergence of new infectious diseases. The good news is that if we really want to, we have the technical ability to make these changes; that is—as mentioned earlier—if our political leaders have the necessary skills and wisdom to undertake them.

Of course, if we (humans) are having to face these challenges on Earth then arguably it is likely that technology-wielding species elsewhere will have, or have had, to also address them.

## Fermi re-visited. Who will inherit?

As discussed earlier (Chapter 7), the co-evolution of life and planet narrative can perhaps contribute to the question of life elsewhere. Given that life on Earth arose so early suggests that it is an 'easy' evolutionary step (Chapter 2) so that life somewhat analogous to prokaryotes should be common on Earth-like worlds. On Earth photosynthesis, at least anaerobic photosynthesis, also evolved independently a number of times and so it too seems to be an 'easy' evolutionary step. Only oxygenic photosynthesis arose once (though then was refined a number of times) and so its origin appears to be a 'difficult' step, or alternatively was so successful that after it first arose it out-competed subsequent potentially putative rivals (Chapter 3). Symbiogenesis also took place a number of times and so that too seems to be an 'easy' evolutionary step (Chapter 4) as does multicellularity (Chapter 4 again). Finally, 'intelligence' has arisen independently a number times and so it too seems to be a reasonably 'easy' step even if only one species (our own) has sufficiently developed sentience to manifest a complex, technological society (Chapter 8). This suggests (on an Earth-like world) that the rise of technology-wielding intelligence is only a matter of time once biological intelligence becomes repeatedly manifest, albeit earlier in a less-developed manner.

The biological route to technology-wielding species therefore consists largely of 'easy' evolutionary steps as is evidenced by most of them either occurring quickly (in the geological sense) or independently a number of times. The reason why it took over 3 billion years for a technology-wielding species to evolve and dominate the planet is because many evolutionary steps—life arising, photosynthesis, oxygenic photosynthesis, endosymbiosis, and multicellularity, simple sentience, and then full-blown sentience—all had to take place first and in strict order, with the entire planet flipping its state essentially three (or four if one counts the abiotic-to-biotic Earth change) times. All this takes geological time and a significant part of a G-type star's main sequence lifetime.

Nonetheless, given an Earth-like planet in the right part of a G- or K-type star's habitable zone (so as to allow for stellar evolution and habitable zone shift) and given a few billion years, the evolution of a technology-wielding species is not unlikely.

So, as with the Fermi Paradox (Chapter 7), the question to address is 'if there are technology-wielding alien species where are they?'

One answer is that they are rare both in time (they take billions of years to evolve) and space (Earthlike planets need to be in the right place within the habitable zone—so as to allow for habitable zone shift with the star's main sequence evolution—as well as need to have a large, planetary axis stabilising moon, and these are few). Indeed, even G- and K-type stars are minority types in our galactic disk (again see Chapter 7).

However, this is not an entirely satisfactory solution to the Fermi Paradox given the Galaxy's size—while large with a diameter of roughly 200,000 light years, assuming Einsteinian limitations, even at 1% let alone 10% of the speed of light, it could be traversed a score or so times over its main disk's life to date. And its size also begets opportunity: there are a couple of hundred billion or so stars (of which several to a score of billion will be G- and K-type). So another answer to Fermi would be helpful.

One solution, I contend, is that any star-faring species—should they exist—would have first had to confront the technological (hence population growth) success in harnessing resources that springs from science. They would have had their own fossil-fuelled boom and dominated their planet, but in doing so—and in facing the issues raised in this chapter—would have learned to live sustainably. In short, they would have abandoned the economic model of ever-increasing resource use that we have been employing to date and which we, currently, take to be the norm: they would have a new norm of living within resource constraints and sustainably.

This does not mean that they would stagnate: it is always possible to innovate, do more with less up to thermodynamic limitations, improving efficiency. It is not difficult to think of plenty of sustainability and efficiency examples, including: lighting, renewable energy, crop cultivars, *etc.* For example, your home computer is both more powerful and uses less resources (in this case metal) than a home computer of only a decade ago. It is also possible to generate new intellectual value (factual knowledge, designs, fictional stories, songs, and so forth) without excessive resource use. Any species that managed its transition to a sustainable mode would also have brought their planetary population down to sustainable levels and this would reduce competition for real estate: they would not need to colonise the entire Galaxy.

Make no mistake, adapting to a sustainable mode would have a lasting societal impact. World War II lasted half a decade, but markedly shaped socioeconomics and politics of the subsequent half-century. Arguably, more recent major events less so: here, recent exemplars might include President Trump's 'America first' and Brexit. Yet considering that the transition to managing our global system sustainably and bringing global population down to sustainable levels will take a couple of centuries, then the transition's impact on the global culture will be far, far greater and more deeply embedded in our social consciousness and heritage.

If this had happened with a technology-wielding species elsewhere, and did so over some generations, their culture would cease to be one based on continued growth in the conventional sense. Given this, why would they spend considerable effort going to the stars in any major way unless they really had to? They may have some interstellar probes, but as with us, much of their exploration of interstellar space would be observational from their home world.

Of course, there are drivers for interstellar travel (assuming it is possible, and here even today we can at least envisage the necessary sub-light-speed technology). Exploration is one. There are even drivers for interstellar colonisation such as an insurance policy against existential threats, be it nuclear war or an asteroid strike, such as envisioned by the renowned physicist Stephen Hawking. He has called for humanity to move to the stars (*78*). But one would only need to establish a very few colonies on other worlds to ensure a species' survival: an endless move through the Galaxy simply would not be required.

Removing the need for endless growth undermines one of the assumptions inherent in the Fermi Paradox: relentless expansion through the Galaxy. In short, other technology-wielding species may well exist but, given the comparative rarity of suitably positioned, large-moon-partnered, Earthlike planets around K- and G-type stars compared to all the stars in the Galaxy, they will not necessarily—indeed, they would be unlikely to—be close to us.

Much depends on how many of such civilisations exist today in our galaxy. (We might safely ignore the question of technology civilisations in other galaxies as they are too far away: the nearest, Andromeda, is roughly 2 million light years away.) Here, though, there has been considerable debate, I do not want to spend much time on this (if only because any answer would be so highly speculative as to be next to useless), but I do address civilisations in our galaxy briefly so as to indicate how the co-evolution of life and planet narrative can contribute.

One significant strand of the debate began with astronomer Frank Drake. In 1960, the National Academy of Sciences approached the Drake asking him to help convene a small, informal SETI (Search for Extra-Terrestrial Intelligence) conference at the Green Bank observatory. This went ahead in 1961, and its participants included three Nobel laureates. To spark discussion, Drake suggested that they break down the problem into key factors that would determine the number of technological extraterrestrials in the Galaxy and came up with this, now famous, equation.

$$N = R^{*} f_p n_e f_l f_i f_c L$$

Where:

N = the number of such civilisations in our galaxy
$R^{*}$ = the average rate of star formation in our galaxy *per* year
$f_p$ = the fraction of those stars that have planets
$n_e$ = the average number of planets that can potentially support life *per* such star
$f_l$ = the fraction of planets that could support life that actually develop life
$f_i$ = the fraction of planets with life that go on to develop intelligent life
$f_c$ = the fraction of civilisations that develop communication technology
L = the length of time for which such civilisations exist in years

The first factors in this equation are reasonably well defined. $R^{*}$ at the time of the Greenbank meeting was thought to be around 10. $f_p$ was considered to be between 2 and 0.5. Today we are beginning to get a better handle on this with NASA's Kepler and Transiting Exoplanet Survey Satellite (TESS) missions along with other research programmes. Where the co-evolution of life and planet narrative can contribute is in estimating the values of $f_l$, $f_i$, and $f_c$.

Of course, as briefly mentioned in Chapter 7, much depends on the longevity of technological civilisations. The 1961 Greenbank meeting assumed L = 10,000 years and concluded that N = 10. If this was so then—assuming a uniform distribution through the spiral arms—the nearest extant civilisation to us would be very roughly 50,000 light years away.

Subsequently, Drake became more optimistic and he created a car plate '*N* = *L*' (*79*). But even with N being a thousand times greater, with 10,000 extant civilisations in our galaxy, the closest would on average be a thousand light years or so away. Indeed even with the Chapter 7, fairly large pinch of salt, guestimate of 42,500 civilisations then—assuming random spacing—on average they would still be several hundred light years apart.

Assuming that such a civilisation more advanced than ours did want to colonise two, three, or four worlds for Hawking's long-term, hedging-bet-for-survival reasons, they would only need to colonise worlds less than several light years from their home planet to be safe from a near-home-world supernova (assuming this to be the most likely existential astronomical threat to their world and planetary system): they would not need to go to the tremendous effort and expense of further colonisation. Such a multi-star civilisation would therefore still be around a thousand light years, or even several hundred light years, away from us.

Indeed, for a star-faring species, spreading throughout the Galaxy would be arguably a dangerous venture to undertake from that species' own perspective. For instance, suppose it left its home system migrating around the spiral arm in both directions at a hundredth or even a tenth the speed of light, then between over 150,000 and 1,500,000 years' time the two waves would meet on the other side of the Galaxy. At the very least, what one migration wave would encounter meeting the other would be a different culture. Indeed, possibly (quite likely) over that time the species itself would have biologically diverged so that when they finally met they would encounter a different species to their own: they would become their own aliens!

It should be pointed out that it would not just be our *succeeding* being able to manage our current transition (to comparatively smooth the critical transition to a globally more sustainable mode), and our managing the Earth system, that would turn our species away from expansive growth. The alternative possibility, *failing*, would also turn us away from endless population growth and ever-increasing resource use. That possibility being that we *fail* to manage the transition, would see serious climate change and undermine many of the natural systems on which we rely. That alternative future would be one of considerable human suffering, if not outright war, and a lowering of our global population in a less than benign manner. So, even in this dystopic scenario, generations later, there would be survivors and they too would have learned the sustainability lesson; albeit learned it the hard way. For them the need for relentless expansion to the stars would appear just as unnecessary.

I therefore contend that the inherent assumption within the Fermi Paradox of a single, technology-wielding species travelling *everywhere* in the Galaxy is unlikely. Given this, the Fermi Paradox ceases to be a paradox.

Of course, this is only my conjecture; others may disagree and no doubt many will.

Yet what is more likely, from both the Drake equation and the co-evolution of life and planet narrative, is the proximity of non-technology-wielding life. Indeed, the greatest likelihood is that extraterrestrial life in its simplest form arises; something vaguely analogous to prokaryotic life. So great is this likelihood that we may not even have to leave the Solar System to encounter it (see Chapter 7). Here, potential discoveries on Mars are most enticing. Given the number of Mars missions currently taking, as well as slated to take place, such discoveries may not be that far off in the future.

Before leaving the question of putative, technological extraterrestrial civilisations, there is another—admittedly more fanciful, science fictional—explanation for the Fermi Paradox arising from the co-evolution of life and planet narrative developed in this book. If this narrative broadly has some substance (albeit that new scientific discoveries add detail or change some of the narrative's nuance, though do note the 'if'), then it will also apply to an extraterrestrial intelligence's world. They too would be aware of the problems of life arising around long-lived but cool red dwarf stars as well as the difficulty of there being enough time for multicellular life to arise around hotter, but short-lived stars (O-, B-, A-, and F-types) than our G-type Sun.

As discussed in Chapter 7, if they were more advanced than us, it would not be difficult to imagine their having a more developed sense of morality and ethical code. It would not be hard to envisage that they might want to prevent our being culturally contaminated and remain hidden. In popular culture terms, think of it as a form of *Star Trek's* 'Prime Directive'. This explanation for the Fermi Paradox is known as the zoo hypothesis (*80*).

If they were also a very long-lived technological species then they might outlive their star's main sequence stage and relocate. Knowing that complex life arises around G- (and possibly K-) type stars, and wishing to keep any biospheres around them pristine, they might instead move to an Earthlike planet around warmer, short-lived stars, seeding the world with their native oxygenic photosynthesisers as well as multicellular species (so skipping lengthy evolutionary development), secure in the knowledge that complex, multicellular life would never evolve there naturally. Here, I contend (very speculatively), planets around F-type stars would be ideal relocation candidates: F-types spend a much shorter time in their main sequence phase than our G-type Sun, but still last a couple or so billion years on their main-sequence phase. If there were any long-lived extraterrestrial civilisation capable of interstellar travel, who wanted to conserve biodiversity on a galactic scale, then around short-lived F-type stars is likely where they will be found, even if they would need to move on every couple of billion years or so. Indeed, an Earth-sized or larger planet in the habitable zone around an F-type star with oxygen and methane in its atmosphere (that might be detected by CHEOPS or a successor mission) could be evidence of extraterrestrial intelligence.

The one problem here is the possible ease, or difficulty in which oxygenic photosynthesis (OP) first evolved. The original evolution of OP on Earth might have taken so long because it was a 'difficult step': though its subsequent refinement a number of times speaks to its further development being 'easy steps'. On the other hand, if we find OP had evolved on Mars (see Chapter 7) possibly early when the planet had a sea then this speaks to OP being an 'easy' evolutionary step, facilitated on Mars by nutrient run-off from its large proportion of land draining into a small-volumed sea with limited buffering capacity: the exact reverse situation on Earth (little primordial land and a large, voluminous ocean). OP might therefore evolve quickly on a Mars-like world around an F-type star but not likely a technological civilisation (other than one arriving from elsewhere as wildly speculated above). In this regard, the current and forthcoming Martian space missions are likely to be informative.

Meanwhile, here on Earth solutions are desperately needed to overcome the many difficulties in managing the transition to a sustainable planetary state.

## Possible solutions

The obvious tasks required to combat human-induced climate change have been detailed countless times before. They include the aforementioned need to transition from a fossil-fuelled powering of our global society to that using non-fossil carbon energy: renewables and nuclear. Also required is the need to use energy resources efficiently. Again, much has been written about this and there is no need to re-cap these arguments here. However, from the existing dialogue, there are a few things that have been either omitted or understated.

For instance, not only do we need to improve the efficiency of agricultural production in a sustainable way (here soil conservation is a major failing) but we urgently need to tackle waste.

An often quoted estimate is that 'as much as half of all food grown is lost or wasted before and after it reaches the consumer'. The idea that food waste is an issue is not new. The first World Food Conference in 1974 identified reduction of post-harvest losses as part of the solution in addressing world hunger. As a result, the Food and Agriculture Organization (FAO) of the United Nations established the Special Action Programme for the Prevention of Food Losses. But progress has been poor. It has been estimated that approximately one-third of all fresh fruits and vegetables produced worldwide is lost before it reaches consumers. Farm-to-shop losses in the United States are estimated to be between 2% to 23%, depending on the commodity, with an overall average of 12%. And all this excludes waste once it reaches households. For instance, it has been estimated that food waste solely within UK households is on average 25% of that purchased (by weight); this is in addition to waste earlier in the food supply chain (*81*). A recent (2021) estimate for total food waste in China is of 27% (some 349 million tonnes annually). That study also suggested that the land, water, carbon, nitrogen, and phosphorus footprints associated with this waste are similar to those of a medium-sized country (such as the United Kingdom's in the case of carbon footprint) (*82*).

Global, annual greenhouse gas emissions from agricultural production between 2000 to 2010 were estimated at 5.0–5.8 gigatonnes of carbon dioxide equivalent or about an eighth of all global emissions (*83*). This may seem a small amount, but it is a significant one: reducing total food waste (plough to plate) by a half might cut global emissions by around 6%.

Even less a greenhouse-related issue is that of reducing waste in less relevant areas of the economy/society. Human activity, of whatever sort, involves human-power, skill/expertise, time, investment, *etc.* If it is not efficient then there is a waste of resources that could go towards addressing greenhouse and sustainability issues. Here I will provide just a couple of examples; there are many, but someone else can write the book on this topic.

The United States (at 5.0294 Gt $CO_2$ $yr^{-1}$) is currently (2019, the pre-CoVID-19 lockdown year) the second largest emitter of carbon dioxide through fossil fuel use (so excluding things like cement production) after China (9.9052 Gt $CO_2$ $yr^{-1}$) but US citizens are on average the largest emitters (at 15.01 tonnes C $yr^{-1}$ *per capita*): this compares with China's (at 7.075 tonnes $CO_2$ $yr^{-1}$ *per capita*). (See Appendix 2 data and sources.) So the question arises as to whether there are significant savings to be made in the US economy that could be redirected to adapting the nation to one of a zero fossil carbon economy?*

Here, one often ignored sector is US healthcare, which is the most economically inefficient on the planet. Plot the world's nations' citizens' average longevity against *per capita* healthcare spend and you get a curve. Figure 9.7 is based on

* British readers might be interested to note for comparison that the United Kingdom's carbon dioxide emissions from fossil fuel in 2019 was 0.3802 Gt $CO_2$ and emissions *per capita* were 5.66 tonnes $CO_2$: the UK imports manufactured goods and has considerable economic carbon leakage.

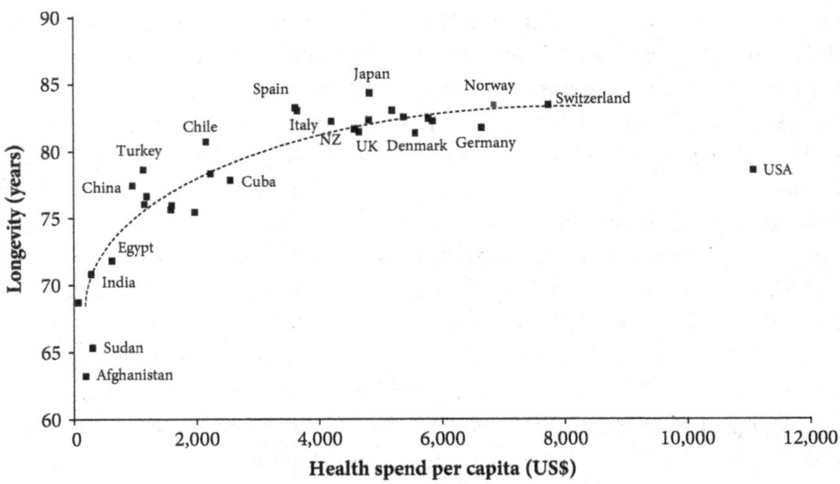


**Figure 9.7** Nations' health spend *per capita* against longevity. 2019 data from the Global Health Expenditure Database of the UN's World Health Organisation and nation citizens' average longevity from the *CIA The World Factbook*, both accessed in 2021.

2019 data and depicts 32 of these countries including the most industrial and the most populous. All the world's nations, save one, lie on this curve with *per capita* annual health care spend up to US$8,000. (Note: this is healthcare spend and does not include biomedical research which is considerable in a number of western European Union nations, the UK as well as the United States. There are OECD rules as to how nations' health care and biomedical research spend are calculated: so here we are talking like-for-like.) It can be seen that that those individuals with the least health care spend need only a modest increase in spending to make a substantive improvement in longevity. But the more that is spent on citizens' health care, so the harder it becomes to improve longevity. Citizens of the wealthiest nations can expect an average longevity of between 81 and 84 years.

The one exception is the United States, whose citizens, while they can expect to live to over 78 years of age, spend twice as much as many western European nations, Japan, New Zealand, and the like. Not only this, US citizens cannot expect to live as long. So, for example, if the United States were to adopt a system similar to the UK's National Health Service, they could halve their health spend and not only maintain their average longevity but likely see it increase!

A simple calculation, assuming the US population to be a round 300 million, shows that, adopting the UK system, the annual saving in 2019 terms would be around US$1.8 trillion. This would not be an insignificant sum to throw at that nation's fossil decarbonising efforts and investment in transforming its economy to one that is sustainable and green.

Another measure is The Global Peace Dividend initiative that was launched at the end of 2021 by over 50 Nobel laureates and the presidents of five science academies. It calls on all countries to jointly reduce military spending by 2% each year and instead contribute to a global fund to tackle climate change, pandemics, and extreme poverty. World military spend increased by more than 2% in 2020. If countries agreed to a common reduction of 2% starting in 2025, US$1.3 trillion will have been saved by 2030. For comparison, global investment in clean energy and energy efficiency amounted to US$700 billion in 2020 (*84*).

Yet as worthy and desirous as The Global Peace Dividend initiative is—and as possible as it might have seemed over the three decades since the fall of the Iron Curtain in 1989/1990—the events of 24th February 2022, with Russia's invasion of Ukraine, let alone China's activities in the Sea of Japan and around Taiwan, put paid to any possibility of reductions in global military spend being redirected towards creating a non-fossil carbon global economy. Though ironically, the Russia-Ukraine conflict did highlight—through soaring fossil fuel prices—how dependent the global economy still was on fossil carbon (let alone western Europe on Russian fossil fuel) and the need to move away from this (*85*). Nonetheless, we are going to have to actively invest—provide new spend—in non-fossil carbon technology as opposed to relying solely on cuts or efficiency savings elsewhere to cover the cost. Whether there is the genuine political will for this, as opposed to purely virtue-signalling, remains to be seen. (If only there had been someone back in the 1990s saying that one of the 'opportunities' tackling global warming affords us would be to wean ourselves off the need for fossil fuel imports and the dependency on potentially hostile nations [*86*]).

Irrespectively, no one mechanism will get us on track to keep warming to below 1.5°C (or even 2°C): instead we need a portfolio of several We already know how to do this, and it is common knowledge that we need to replace fossil fuels with non-fossil energy sources, improve energy efficiency, stop deforestation, *etc., etc.* Indeed, in 2004 Robert Socolow and Stephen Pacala of Princeton University came up with several ways that if implemented would bring down the current trajectory of increasing emissions in a series of wedges (*87*). That up till now (2025) global emissions have continued rise (see also Figure 8.3) speaks not to a failing of technology but of political will. This brings us on to one measure, one that is rarely discussed in climate debates, and that is to reduce corruption.

Be it politicians spreading misinformation such as pandering to climate-change deniers, fraud, or plain thieving, corruption, in the broadest sense of the term, is arguably the greatest barrier to the successful transition to a non-fossil carbon global economy. A 2013 survey of nearly 70,000 people across 69 countries by WIN/Gallup International confirmed corruption as the world's top problem (*88*). And we really do not need the 2008 global financial crisis to see that.

Again, there is not the space here to go into too much detail. The dangers of politicians lying or being 'economical with the truth' are obvious. But it is the scale of corruption that is the threat to our successfully navigating the current critical transition to an environmentally sustainable, non-fossil-fuelled world. To take just one example, in 2020 an international collaboration of researchers led by Emma Witbooi, of Nelson Mandela University, using 2009 data, estimated that 18% of the global fish catch, valued at US$10–$23.5 billion, between 2000 and 2003 was lost to illegal or unreported fishing. This while figures of the UN Food and Agriculture Organization (2015 data) estimate that 59.9% of the world's commercial fish stocks are now fully fished and a third of the global fish stocks are overexploited (*89*).

The European Union's first report into corruption was in 2014 (when the United Kingdom was still a member), and it gives us an idea of the scale of the problem. It estimated that the annual cost of corruption to the Union's nations was some 120 billion Euros (about £102 billion or US$138 billion). This is of a similar magnitude to that the African Union estimated in 2002, that indicated that corruption costs its fifty-three nations US$150 billion (*88*). This compares with the failed UN climate Conference of Parties international goal for developed nations to deliver to less developed countries US$100 billion a year of climate finance by 2020. Or the estimated deforestation prevention requirement estimate of US$200 billion a year (*90*).

Suffice to say that tackling corruption in the broadest sense of the term needs to be a far greater priority in all countries but, if successful, it would release the funds to facilitate our traversing the current critical transition. As outlined in this section, there are the means to finance the solutions we can implement if we want to and if our political leaders have the vision and will. What we lesser folk can do is tell them.

## Looking at information-processing in the future

If increased information-processing is one of the fundamental hallmarks of each successive state of the evolving Earth system (see Chapter 8) then what in the future will it, information-processing technology, be like? Of course, if we are about to be—or are in the process of—crossing a critical transition to this new state, then clearly we have no hard evidence as to what it will be like. Indeed, because of this, this transition will be a 'singularity' with an 'event horizon' beyond which we humans cannot currently see or predict. This is not just because the future is unknowable with any certainty but because up to now human actions and human technological advancements have shaped our world. With the rise of artificial intelligence (AI) then as it gets increasingly sophisticated, so there will come a point when AI itself—should we let it (and someone probably will)—will design its own improvements: intellectual human agency will not play a part in this development even if human physical agency does. How AI develops from here on in will (even if only in part) be unknowable to humans, hence is an 'event horizon'. This point in time is known as the 'technological singularity', a term popularised by the mathematician and writer Vernor Vinge (*91*).

However, what we can see is that just as eukaryotes and then multicellular species respectively came to dominate the semi- and fully-oxygenated states of the Earth, so our species combined with technological information processing (the human-techno complex, if you will) will dominate the forthcoming state; indeed, arguably this complex is already doing this and undertaking niche construction at a near global level. However, just as we can see from some Archaea and their prokaryotic symbionts precursor exemplars to eukaryotic cells (Chapter 4), or see from quasi-structured colonial species and coenocytial species, how they might be precursor stages to multicellularity, so analogously we might get an inkling of what might be done with information processing by looking at where we are and current trends.

Equally, of course, how painful a time we will have crossing the current critical transition to an environmentally sustainable, lower population and fossil-carbon-free future influenced and ultimately 'managed' by increased information processing, remains to be seen. We can but hope we will gain the wisdom, and will, to implement the necessary, sustainable stewardship of the planet. Though with this transition, while global climate change in the form of a Snowball Earth is (most) unlikely, a global hyperthermal is very much on the cards. But it is not as if there are no solutions, there are: some were discussed in the previous section. It is the to-date lack of sufficient movement, political deftness and the willingness of our leaders to look sufficiently beyond their power bubbles, and not least their lack of skill set, that will result in a painful transition.

But what of the future beyond the current critical transition?

First, just to recap a bit, James Lovelock, the originator of the idea of the Earth as a Gaian self-regulating system (see Chapter 2) (*92*), recognised that, while the Earth system was self-regulating, the Earth has previously existed in different self-regulating states (notably before and after initial oxygenation). As with this thesis presented here, his final book, *Novacene: The Coming Age of Hyperintelligence* (*93*), notes that as life evolved, so did its ability to process information: first genetically, with nucleic acids, through symbiogenesis and the rise of eukaryotes; then with multicellularity which in turn gave rise to even more information processing with the evolution of animal brains; and finally with humanity's creation of electronic information processing. He envisions a future in which humanity lives in symbiosis with 'artificial intelligence'. Actually, he refers to 'cyborgs', but that term relates to something part-human, part-machine as used by the *New York Times* in 1960 before being brought into the science fiction lexicon, notably by Frank Herbert in 1966, and greatly popularised in the 1970s with the TV show *The Six Million Dollar Man* that was in turn based on Martin Caidin's novel *Cyborg* (*94*, *95*). (Though James Lovelock broadens that definition of cyborg in *Novacene* to encompass artificial intelligence [*93*].) Given the current trend in information processing and, indeed, Moore's Law (that integrated circuits' number of transistors—hence processing capability—doubles roughly once every two years) (*96*), that has so far held for over six decades, suggests that one day true artificial intelligence will be possible.

Before continuing, it is worth noting what is meant by 'artificial intelligence'?

We currently have AI systems that run household washing machines, control traction in cars, fly planes, and so forth. We even have AI that can discern verbal commands, such as asking Alexa to play music and play games such as chess and Go. However, though we are coming tantalisingly close, we do not (quite yet) have AI that truly thinks like a person, with whom we could have a conversation that would be absolutely indistinguishable from that with another human (or pass what is called the 'Turing Test'). Such an advanced AI is known by those working in the field as 'Artificial General Intelligence' (AGI), as distinguished from the intelligence responding to a specific stimulus to undertake a specific task (*97*). Though, as Arthur C. Clarke anticipated in *2001: A Space Odyssey* (*98*), there is debate in the community as to whether AGI as true intelligence is possible or whether it would just mimic intelligence. In this book, I mostly use the term AI to mean AGI or even artificial sentience.

In science fiction, though, there are some great AI stories, but AI often gets a bit of a raw deal: often AIs go rogue—*2001: A Space Odyssey* (*98*) and *Colossus* (also known as *The Forbin Project*) (*99*)—and sometimes they seek to become human—*The Demon Seed* (*100*) or Data in *Star Trek: The Next Generation* (*101*)—or even embark on war with humanity—*The Terminator* (*102*) and *The Matrix* (*103*). James Lovelock is of the opinion that such conflict is unlikely as we are heading for a true, mutualistic symbiosis with machines: they need us as much as we need them (*93*). Certainly, science fiction has given us instances of machine revolt wiping out humanity, but it sometimes it doesn't go well for the machines: in Robert Cargill's *Sea of Rust* (*104*) without humans, robots take to scavenging each other for spare parts while super-AIs embark on a civil war that sweeps up lesser AIs.

Yet science fiction is fiction. What might this future human–AGI relationship in reality look like? Well, nobody knows. However, the notion of a cyborg hive-mind civilisation as in *Star Trek*'s (*101*) alien borg is most likely pure fantasy even if intriguing as a device in fiction. Having said that, human–machine interface technology is increasing all the time. For example, one recent (2021) development is a neural interface that directly translates the thinking of letters into text. Two small probes are each attached to a small (4 mm square) microelectrode array previously implanted onto the surface of the brain. The subject then thinks about handwriting a letter and a computer decodes the brain's neural impulses to recreate the shape of the letter. This brain–computer interface is not a step towards creating a borg, but has a serious purpose in that it could help paralysed stroke victims communicate.(*105, 106*) Nonetheless, it is unlikely that our relationship, as biological beings, with technological information processing will generally be surgically invasive.

If I had to go for a science fictional instance of such a future world, then it might perhaps be a little more like the relationship portrayed in Iain Banks 'Culture' series of novels, such as in *Excession* (*107*) in which there is a true partnership: it could be over-summarised as with the AIs doing a greater share of the thinking/computing and number-crunching, with the Culture's biological citizens undertaking a little more of their share of the doing (as well as generally enjoying life).

A complicating factor is that, with increased information processing, not only are we heading towards full-blown artificial intelligence but—with genomics and gene-editing techniques such as CRISPR—we have the potential (ethical concerns aside) to edit our own genes, possibly enhancing our own biological ability to process information. If we did then *Homo sapiens sapiens* would be leaving the strictures of 'natural' Darwinian evolution for what Harvey Fineberg (President of Washington's Institute of Medicine) calls 'neo-evolution' (*108*). If we did go down this ethically questionable route of editing our own genes for anything other than inheritable genetic-disorder clinical reasons, then most likely artificial intelligence would be used to guide the process. But this is arguably too speculative to dwell on for now, though some are seriously considering the gene-editing option. Indeed, if it becomes a possibility, even if illegal in many nations, some will be tempted to pursue it (*109*).

Returning to developing full-blown AI, we are already some way towards achieving this goal: it is arguably closer than many think even though AI tools such as ChatGPT (Chat Generative Pre-trained Transformer) and Midjourney (an art/image developer), that were both launched in 2022, have garnered some public attention. One recent (2021) development—Project Debater—is a fully autonomous computer system that can take part in live debates with people on a hundred different topics. So, while it cannot converse at will about anything, this is a start (*110, 111*) and since then (unsurprisingly) the technology has improved.

One other area in which AI development is making great strides is that of driverless cars. Here there has been a recent (2022) major advance. It is one thing to drive a car on a civilian road and avoid other cars and pedestrians as well as obey the highway code. It is quite another to drive on a racetrack and aim to win, competing against rivals out to do the same. Recently, researchers—from Sony AI based in New York, Tokyo, and Zurich—have developed an artificial intelligence (AI), GT Sophy, based on a modelled neural network and let it loose on the motor racing video game *Gran Turismo*. *Gran Turismo* is very realistic: among much else, it includes modelling the friction on front and back wheels and so, for example, it can replicate over- and under-steer going into and out of corners. It is so realistic that professional racing drivers use it. Kazunori Yamauchi even used the video game to find ways of tweaking his real racing car to overcome a recurring problem that he was having when taking a corner at the Nurburgring, a Grand Prix track in Germany that has the nickname 'The Green Hell'.

Now, there have already been AIs that can play *Gran Turismo* before GT Sophy. Indeed, the AI Audi TTS Shelley can do lap times comparable with human champions. However, such AIs rely on a computer model of racing physics. GT Sophy is different, its neural network enables it to be taught through practice and failure and more practice. It learns! (Technically, this is called deep reinforced learning.) Through this training process, GT Sophy learns to take different lines through the corners in response to different conditions. In one case, two human drivers attempted to block the preferred path of two GT Sophy cars, yet the AI succeeded in finding two different trajectories that overcame this block and allowed the AI's cars to pass. The researchers say that such neural network AI capable of deep reinforced learning will have real world impacts on collaborative robotics, swarms of aerial drones, as well as, obviously, autonomous vehicles (*112*, *113*).

Indeed, developments in AI are such that we are already facing complex ethical questions: this is as much about the ethics of AI development as the ethics we should instil in AIs.

Now, this last is trickier than you may think, in no small part because both society's ethics evolve and that ethics differ between societies. This has also been exemplified by those developing AI-controlled, driverless cars. Take the following 'trolley bus'-type problem. An autonomous vehicle is self-driving along with its AI minding its own business when, suddenly, someone steps out into the road. The AI could swerve but in doing so possibly mount the pavement with potentially serious risk to pedestrians. What should it do?

Here a research team largely based at the Massachusetts Institute of Technology reported in 2018 on their developing a Moral Machine, an online experimental platform designed to explore the moral dilemmas faced by autonomous vehicles. This platform gathered nearly 40 million decisions in ten languages from millions of people in 233 countries and territories in an attempt to find out a society's ethical stance.

The actual question they posed was this. An autonomous vehicle experiences a sudden brake failure. Staying on course would result in the death of two elderly men and an elderly woman who are crossing on a 'do not cross the road' signal. Swerving, under this scenario, would result in the death of three passengers: an adult man, an adult woman, and a boy. What should the AI do?

Here answers differed and also came in geographical cultural clusters. One cluster—the western cluster—contained North America as well as many European countries of Protestant, Catholic, and Orthodox Christian cultural groups. Another, the eastern cluster, contains many far eastern countries such as Japan and Taiwan that belong to the Confucianist cultural group, and Islamic countries such as Indonesia, Pakistan, and Saudi Arabia. The third (the southern) cluster consisted of the Latin American countries of Central and South America, in addition to some countries that are characterised in part by French influence, such as French overseas territories.

The Moral Machine found the preference to spare younger characters rather than older characters was much less pronounced for countries in the Eastern cluster, and much stronger for countries in the Southern cluster. The same was true for the preference for sparing higher status characters. Similarly, countries in the Southern cluster exhibited a much weaker preference for sparing humans over pets, compared to the other two clusters. Only the (weak) preference for sparing pedestrians over passengers and the (moderate) preference for sparing the lawful over the unlawful appear to be shared to the same extent in all clusters.

This demonstrates that not only will we have to create a suite of ethics over developing AI, but that we will have to imbue AIs with the means to make decisions that reflect those ethics of the cultures in which they operate (*114*).

Yet despite soft hurdles, such as the afore ethical issues, and hard technological difficulties that have yet to be overcome, it is very clear that for the foreseeable (long-term) future technological information processing will increase bringing with it more automation and in all likelihood the eventual creation of full-blown artificial intelligence. Let's hope the benefits far outweigh any disadvantages: I for one do not relish political rivals each having speeches written for them, and debate points provided them, by AIs.

Irrespective of this, if our future, highly information-processed world is to be sustainable, and for us to see a much lower populated (hence environmentally sustainable) world—of youth, fewer-than-now economically active in traditional sectors as they will be in education (not just academia but in trade, craft, and vocational sectors), and more-than-now economically active in older age (retiring later and part-time and even voluntary work)—then AI and automation will be key to freeing up human resources for such a demographic transition to a smaller, more elderly population. This, though, will mean that we will have to reformulate the global economic model (in addition to taking into account natural capital discussed earlier): some form of taxation of certain types of information-processing may well have to be considered (though expect the wealthy, information processing commerce to complain bitterly). Whatever transpires, the middle of the 21st century is likely to be as different to the early 21st century as the early 21st century was to the pre–World War I world at the start of the 20th.

## Testability

Any scientific narrative being postulated needs to be capable of making predictions so as to be tested. Even if the co-evolution of life and planet narrative being presented here needs nuance tweaking in the light of new discoveries, its core essence—for it to be scientific—needs to be able to make a number of predictions so it can be tested.

To get the ball rolling, I now suggest a few.

This narrative begins with life arising very early on in Earth's primordial existence (Chapter 2). If this is so then we should see the discovery of more primordial strata with carbon inclusions rich in carbon-12 (depleted in carbon-13) that are structurally less ambiguous than those we have now. Less reliably, but providing some tentative corroboration, we might expect new evidence from molecular clocks.

Also, if there was the evidence for early life in the early, water-rich Mars then this too would argue for an early start to life on Earth.

Resolving the current Martian atmosphere's seasonal oxygen source detected by landers is of particular interest. At the moment, there is no satisfactory explanation for this that does not involve biology (see Chapter 7). If it is due to oxygenic photosynthesis (OP) then that would be most revealing as OP would have evolved far sooner on Mars than it (currently) seems on Earth. This would speak to a lack of nutrients in the Earth's oceans as well as the size of the anaerobic Earth system to be transformed as being the delaying factors: the primordial Mars had a far higher land-area-to-ocean-volume ratio with slightly acidic rain releasing nutrients from the land both into oceans and crater lakes. The latter, especially if some included hydrothermal vents, would be ideal sites for life's arising and evolving OP.

The earlier life got going on Earth means that there is greater time in which for oxygenic photosynthesis to evolve as well as the rise of eukaryotes (Chapters 3 and 4): we might have to revaluate when these first arose. If they evolved earlier, it would be corroborating (not firm) evidence for these being the drivers of Snowball I (and not arising in response to that event): there would also be more time for them to overcome the anaerobic Earth system.

OP discovered on Mars, especially if it evolved on an early Mars and extant OP was genetically different to that on Earth, would also mean that it was a comparatively 'easy' evolutionary step: it would have evolved independently twice. That OP only evolved once (even if a number of variations subsequently became manifest) on Earth, and if it was an 'easy' step, then it would speak to it being such a neat trick that it trumped any subsequent putative rivals.

Here, the current and future Martian rover explorations will be most fascinating.

The idea that a young Venus might never have been able to have oceans could be tested by building direct-imaging space telescopes that can observe similar exoplanets (exo-Venuses) around young stars and take spectra of their atmospheres. This area of research would also help determine how star systems' habitable zones evolve. Here, the James Webb Space Telescope is making a start.

Another way to test (or disprove) this early Venusian ocean hypothesis would be to measure the composition of Venus's surface. If it was found that it contained metamorphic rocks that necessitate liquid water to form then this would argue for the past presence of surface water and oceans (Chapter 7) (*115*).

Analysing the atmospheres of exoplanets, such as through the European Space Agency's CHEOPS (CHaracterising ExOPlanet Satellite) mission, could reveal if there was both oxygen and methane in worlds around a red dwarf. Could it be possible that with enough time, three photosystems could be bolted together (not just two as on Earth) to enable OP in a red dwarf star system? Remember, such stars have a much longer main sequence lifespan than our own Sun. (See Chapters 3 and 7.)

Similarly, such atmospheres associated with planets around very short-lived K-type stars (too short-lived for OP to evolve) might speak to them having been colonised and 'terraformed' by a colonising species who brought OP with them. This might even be the first sign we get of extraterrestrial intelligence?

Of course, here there is a confounding factor. What if we found that OP had evolved on a primordial Mars aided by its large land-to-primordial-ocean ratio so providing adequate nutrients. This would not only explain the seasonal oxygen detected on Mars (see Chapter 7), it would also indicate that OP might also be able to evolve naturally on planets around short-lived K-type (orange dwarf) stars. Alas it will be some time before we know how early OP arose on Mars (if it ever did). And if we do detect oxygen and methane in the atmosphere of a planet around a K-type star then perhaps we should also be on the lookout for techno-signatures (such as radio waves) from planetary systems around such stars.

Finally, should we find firmer and less ambiguous evidence for multicellular life before Snowball II (Chapters 4 and 5), especially if evidence were found for terrestrial multicellular life (especially Tim Lenton and Andrew Watson's proto-lichens) (*116*) would speak to multicellular life causing Snowball II and not arising, or booming, as a consequence of it.

All of the above (and no doubt many other ways of testing the narrative presented here) could be undertaken by researchers with today's theoretical technological capability. In short, we may get some answers within a few years or decades (not centuries). The future of research in Earth system science, with special regard to the co-evolution of life and planet, will be interesting.

Aside from testability, the other benefit of any scientific hypothesis or narrative that has a degree of merit is that it might speak to, and inform, other areas of science and philosophy. Here, it would be easy to speculate wildly, but that would most likely only give us highly debatable results. However, I cannot leave you without an example (or a teaser) purely for illustrative purposes only.

In 1990, the theoretical physicist John Wheeler (who is best known for popularising the terms 'black hole' and 'wormhole') suggested that information is fundamental to the physics of the universe. According to this 'it from bit' doctrine, all things physical are information-theoretic in origin (*117*). He included this idea in developing what is called the Participatory Principle of the Universe. Here there is no 'real' reality (reality is not quite what we think it is) but that the Universe is the result of intelligence observing of, and experimenting in, the Universe itself. It is this observation that determines the precise nature of the Universe. This notion may seem outlandish but it arises from what physicists call 'Copenhagen Interpretation' (*118*) and there is even, let's say 'corroborating', evidence for this in what is called 'Wheeler's delayed-choice experiment' (*119*). I will not go into it in detail here, but readers can look up the references or even check out the 2022 *PBS Space Time* episode 'Does the Universe Create Itself?' on YouTube (*120*).

The co-evolution of life and planet narrative developed in this book includes increasing information processing with each evolutionary leap: the rise of life sees information encoded in genetic molecules (RNA and then DNA); the rise of eukaryotes sees the incorporation of prokaryotic genetic molecules into the proto-eukaryote in the form of mitochondrial and chloroplast DNA; the rise of multicellularity sees brains; and the species with the most developed brain has created writing and, more recently, non-biological computational information processing. If the idea that we are living in a participatory universe is correct, then it would be of no surprise that increased information processing would be at the heart of biological evolution as suggested by this co-evolution of life and planet narrative.

And that is enough of whimsical speculation.

## In conclusion

Humanity in the 21st century faces many challenges, not least will it move into a fossil carbon free and environmentally sustainable future?

Life and the Earth system has come a long way in the arm's-length journey from the shoulder to fingertip. Along the way, the system has critically transitioned between key global states with each state seeing a new form of life (prokaryotic, eukaryotic, and multicellular) dominate. Are we about to see a new state, as significantly distinct a state as the others, with a single species (ourselves) and its techno-bio-complex actively managing the planet?

While the narrative presented here is evidence-based, as you can see there is still some speculation. Irrespectively, if we are going to overcome the sustainability challenges we are going to have to have a better understanding of life's interaction with, and life itself as a part of, the Earth system. Knowing how our planet and life co-evolved across deep time is a key part of this understanding. Such knowledge is not just fascinating—it is that too—it is of import to navigating this century and the next.

Welcome to the Singularity.

## References

1. Intergovernmental Panel on Climate Change, *Climate Change 2001: Synthesis Report.* IPCC: Geneva, Switzerland (2001).
2. Gutjahr, M., Ridgwell, A., Sexton, P. F., *et al.*, Very large release of mostly volcanic carbon during the Palaeocene–Eocene Thermal Maximum. *Nature*, **548**, 573–577 (2017).
3. Intergovernmental Panel on Climate Change, *Climate Change 2001: Mitigation—A Report of Working Group III.* Cambridge: Cambridge University Press, Great Britain (2001).
4. Intergovernmental Panel on Climate Change, *Climate Change 2013: The Physical Science Basis. Contribution of Working Group I to the Fifth Assessment Report of the Intergovernmental Panel on Climate Change.* Cambridge University Press: Cambridge, Great Britain (2013).
5. Dickens, G. R., O'Neil, J. R., Rea, D. K., & Owen, R. M., Dissociation of oceanic methane hydrate as a cause of the carbon-isotope excursion at the end of the Paleocene. *Paleoceanography and Paleoclimatology*, **10**, 965–971 (1995).
6. Foster, G. L., Hull, P., Lunt, D. J., & Zachos, J. C., Placing our current 'hyperthermal' in the context of rapid climate change in our geological past. Philosophical Transactions of the Royal Society A, **376**, 20170086 (2018).
7. Zeebe, R. E., Ridgwell, A., & Zachos, J. C., Anthropogenic carbon release rate unprecedented during the past 66 million years. *Nature Geoscience*, **9** (4), 325–329 (2016).
8. Steffen, W., Rockström, J., & Richardson, K., *et al.*, Trajectories of the Earth system in the Anthropocene. *Proceedings of the National Academies of Sciences USA*, **15** (33), 8252–8259 (2018).
9. Gardner, A. S., Moholdt, G., Wouters, B., *et al.*, Sharply increased mass loss from glaciers and ice caps in the Canadian Arctic Archipelago. *Nature*, **473**, 357–360 (2011).
10. Broecker, W. S., The great ocean conveyor. *Oceanography*, **4** (2), 79–89 (1991).
11. Caesar, L., Rahmstorf, S., Robinson, A., Feulner, G., & Saba, V., Observed fingerprint of a weakening Atlantic Ocean overturning circulation. *Nature*, **556**, 191–196 (2018).
12 Thornalley, D. J. R., Oppo, D. W., Ortega, P., *et al.*, Anomalously weak Labrador Sea convection and Atlantic overturning during the past 150 years. *Nature*, **556**, 227–230 (2018).
13. Praetorius, S. K., North Atlantic circulation slows down. *Nature*, **556**, 180–181 (2018).
14. Intergovernmental Panel on Climate Change, *Climate Change 2021: The Physical Science Basis. Contribution of Working Group I to the Fifth Assessment Report of the Intergovernmental Panel on Climate Change.* Cambridge University Press, Cambridge, Great Britain (2021).
15. Ditlevsen, P., & Ditlevsen, S., Warning of a forthcoming collapse of the Atlantic meridional overturning circulation. *Nature Communications*, **14**, 4254 (2023).
16. Hugelius, G., Loisel, J., Chadburn, S., *et al.*, Large stocks of peatland carbon and nitrogen are vulnerable to permafrost thaw. *Proceedings of the National Academies of Sciences USA*, **117** (34), 20438–20446 (2020).
17. Nottingham, A. T., Meir, P., Velasquez, E., & Turner B. L., Soil carbon loss by experimental warming in a tropical forest. *Nature*, **584**, 234–237 (2020).
18. Barnosky, A. D., Hadly, E. A., Bascompte, J., *et al.*, Approaching a state shift in Earth's biosphere. *Nature*, **486**, 52–58 (2012).
19. Gatti, L. V., Basso, L. S., Miller, J. B., *et al.*, Amazonia as a carbon source linked to deforestation and climate change. *Nature*, **595**, 388–393 (2021).
20. Boulton, C. A., Lenton, T. M., & Boers, N., Pronounced loss of Amazon rainforest resilience since the early 2000s. *Nature Climate Change*, **12**, 271–278. (2022).
21. Denning, S., Southeast Amazonia is no longer a carbon sink. *Nature*, **595**, 354–355 (2021).
22. Forzieri, G., Dakos, V., McDowell, N. G., Ramdane, A., & Cescatti, A., Emerging signals of declining forest resilience under climate change. *Nature*, **608**, 534–539 (2022).

23. McKay, D. I. A., Staal, A., Abrams, J. F., *et al.*, Exceeding 1.5°C global warming could trigger multiple climate tipping points. *Science*, **377**, 1171–1181, eabn7950 (2022).
24. Cohen, A. S., Coe, A. L., & Kemp, D. B., The Late Palaeocene Early Eocene and Toarcian (Early Jurassic) carbon isotope excursions: A comparison of their time scales, associated environmental changes, causes and consequences. *Journal of the Geological Society*, **164**, 1093–1108 (2007).
25. Villani, L., McKee, M., Cascini, F., Ricciardi, W., & Boccia, S., Comparison of deaths rates for CoVID-19 across Europe during the first wave of the CoVID-19 pandemic. *Frontiers in Public Health*, **8**, 620416 (2020).
26. Second Johnson Ministry. https://en.wikipedia.org/wiki/Second_Johnson_ministry (accessed June 2021).
27. Brown Ministry. https://en.wikipedia.org/wiki/Brown_ministry (accessed June 2021).
28. Vallance, P., & Al-Khalili, J., *The Patrick Vallance Interview*. BBC Radio 4. (12 October 2021).
29. Subbaraman, N., First science advisor in US President's cabinet talks CoVID, spying and more. *Nature*, **594**, 311 (2021).
30. Mottl, O., Flantua, S. G. A., Bhatta, K. P., *et al.*, Global acceleration in rates of vegetation change over the past 18,000 years. *Science*, **372**, 860–864 (2021).
31. Intergovernmental Panel on Climate Change, *Climate Change 2007: The Physical Science Basis—Working Group I Contribution to the Fourth Assessment Report of the Intergovernmental Panel on Climate Change*. Cambridge University Press: Cambridge, Great Britain (2007).
32. Cowie, J., *Climate Change: Biological and Human Aspects*, 2nd edition. Cambridge University Press: Cambridge, Great Britain (2013).
33. Nolan, C., Overpeck, J. T., Allen, J. R. M., *et al.*, Past and future global transformation of terrestrial ecosystems under climate change. *Science*, **361**, 920–923 (2018).
34. New, M., Liverman, D., Betts, R., Anderson, K., & West, C., eds., Four degrees and beyond: The potential for a global temperature increase of four degrees and its implications (Thematic issue). *Philosophical Transactions of the Royal Society A*, **369**, 1–195 (2011).
35. Nicholls, R. J., Marinova, N., Lowe, J. A., *et al.*, Sea-level rise and its possible impacts given a 'beyond 4°C world' in the twenty-first century. *Philosophical Transactions of the Royal Society A*, **369**, 161–181 (2011).
36. IMBIE Team, Mass balance of the Greenland Ice Sheet from 1992 to 2018. *Nature*, **579**, 233–239 (2020).
37. Paolo, F. S., Fricker, H. A., & Padman, L., Volume loss from Antarctic ice shelves is accelerating. *Science*, **348**, 327–331 (2015).
38. Garbe, J., Albrecht, T., Levermann, A., Donges, J. F., & Winkelmann, R., The hysteresis of the Antarctic Ice Sheet. *Nature*, **585**, 538–544 (2020).
39. Mora, C., Dousset, B., Caldwell, I. R., *et al.*, Global risk of deadly heat. *Nature Climate Change*, **7**, 501–506 (2017).
40. Im, E-S., Pal, J. S., & Eltahir, E. A. B., Deadly heat waves projected in the densely populated agricultural regions of South Asia. *Science Advances*, **3**, e1603322 (2017).
41. Xu, C., Kohler, T. A., Lenton, T. M., Svenning J.-C., & Scheffer, M., Future of the human climate niche. *Proceedings of the National Academies of Sciences USA*, **117** (21), 11350–11355 (2020).
42. Philip, S. Y., Kew, S. F., van Oldenborgh, G. J., *et al.*, Rapid attribution analysis of the extraordinary heatwave on the Pacific Coast of the US and Canada June 2021. World Weather Attribution (2021). https://www.worldweatherattribution.org/wp-content/uploads/NW-US-extreme-heat-2021-scientific-report-WWA.pdf (accessed August 2022).
43. Boivin, N. L., Zeder, M. A., Fuller, D. Q., *et al.*, Ecological consequences of human niche construction: Examining long-term anthropogenic shaping of global species distributions. *Proceedings of the National Academies of Sciences USA*, **113** (23), 6388–6396 (2016).

44. Allen, M. R., Frame, D. J., Huntingford, C., *et al.*, Warming caused by cumulative carbon emissions towards the trillionth tonne. *Nature*, **458**, 1163–1166 (2009).
45. Malte Meinshausen, M., Meinshausen, N., Hare, W., *et al.*, Greenhouse-gas emission targets for limiting global warming to 2°C. *Nature*, **458**, 1158–1162 (2009).
46. Schmidt, G., & Archer, D., Too much of a bad thing. *Nature*, **458**, 1117–1118 (2009).
47. Tong, D., Zhang, Q., Zheng, Y., *et al.*, Committed emissions from existing energy infrastructure jeopardize 1.5°C climate target. *Nature*, **572**, 373–377 (2019).
48. Welsby, D., Price, J., Pye, S., & Ekins, P., Unextractable fossil fuels in a 1.5 °C world. *Nature*, **597**, 230–234 (2021).
49. Nogrady, B., Fossil-fuel reserves must remain untapped to hit 1.5 °C goal. *Nature*, **597**, 316–317 (2021).
50. Meadows, D. H., Meadows, D. L., Randers, J., & Behrens, W. W., III, *The Limits to Growth: A Report for the Club of Rome's Project on the Predicament of Mankind.* Pan: London, Great Britain (1972).
51. Anon., Limits to growth? It's time to end a 50-year argument. *Nature*, **603**, 361 (2022).
52. Gribbin, J., *Future Worlds.* Abacus: London, Great Britain (1979).
53. Turner, G., *A Comparison of Limits to Growth with Thirty Years of Reality.* CISIRO: Canberra, Australia (2008).
54. Hausfather, Z., Drake, H. F., Abbott, T., & Schmidt, G. A., Evaluating the performance of past climate model projections. *Geophysical Research Letters*, **47**, e2019GL085378 (2019).
55. Kay, J. E., Early models successfully predicted global warming. *Nature*, **578**, 45–46 (2020).
56. Beddington, J., *Food, Energy, Water and the Climate: A Perfect Storm of Global Events?* Government Office for Science: London, Great Britain (2009).
57. Morales-Caselles, C., Viejo, J., Martí, E., *et al.*, An inshore–offshore sorting system revealed from global classification of ocean litter. *Nature Sustainability*, **4**, 484–493 (2021).
58. Smith, J., & Vignieri, S., A devil's bargain. *Science*, **373**, 34–35 (2021).
59. Geyer, R., Jambeck, J. R., & Law, K. L., Production, use, and fate of all plastics ever made. *Science Advances*, **3**, e1700782 (2017).
60. Bradshaw, C. J. A., & Brook, B. W., Human population reduction is not a quick fix for environmental problems. *Proceedings of the National Academies of Sciences USA*, **111** (46), 16610–16615 (2020).
61. Coleman, L., How a town's internet was sabotaged by a tiny creature linked to one of the world's richest families. *Metro.* 2nd August 2023. https://metro.co.uk/2023/08/02/how-a-villages-internet-outage-could-be-down-to-a-tiny-creature-linked-to-famous-family-19235779/ (accessed August 2023).
62. Letwin, O., *Apocalypse How? Technology and the Threat of Disaster.* Corvus: London, Great Britain (2020).
63. Bentley, A., Avert global wheat crisis due to invasion of Ukraine. *Nature*, **603**, 551 (2022).
64. Ghanem, M. E., Correspondence: Africa's forgotten crops could offset food insecurity. *Nature*, **605**, 30 (2022).
65. Ostfeld, R. S., & Keesing, F., Species that can make us ill thrive in human habitats. *Nature*, **584**, 346–347 (2021).
66. Hu, B., Zeng, L-P., Yang, X-L., *et al.*, Discovery of a rich gene pool of bat SARS-related coronaviruses provides new insights into the origin of SARS coronavirus. *PLoS Pathogens*, **13** (11), e1006698 (2017).
67. Mallapaty, S., Coronaviruses closely related to the pandemic virus discovered in Japan and Cambodia. *Nature*, **588**, 15–16 (2020).
68. Lam, T. T-Y., Jia, N., Zhang Y-W., *et al.*, Identifying SARS-CoV-2-related coronaviruses in Malayan pangolins. *Nature*, **583**, 282–285 (2020).
69. Gibb, R., Redding, D. W., Chin, K. Q., *et al.*, Zoonotic host diversity increases in human-dominated ecosystems. *Nature*, **584**, 398–402 (2020).

70. Jones, K. E., Patel, N. G., Levy, M. A., *et al.*, Global trends in emerging infectious diseases. *Nature*, **451**, 990–993 (2008).
71. Smith, K. F., Goldberg, M., Rosenthal, S., *et al.*, Global rise in human infectious disease outbreaks. *Journal of the Royal Society Interface*, **11**, 20140950 (2014).
72. Carlson, C. J., Albery, G. F., Merow, C., *et al.*, Climate change increases cross-species viral transmission risk. *Nature*, **607**, 555–562 (2022).
73. Baker, R. E., & Metcalf, C. J. E., Heating and stirring the global viral soup *Nature*, **607**, 455–456 (2022).
74. Dobson, A. P., Pimm, S. L., Hannah, L., *et al.*, Ecology and economics for pandemic prevention. *Science*, **369**, 379–381 (2020).
75. Chichilnisky, G., & Heal, G., Economic returns from the biosphere. *Nature*, **391**, 629–630 (1998).
76. Costanza, R., d'Arge, R., de Groot, R., *et al.*, The value of the world's ecosystem services and natural capital. *Nature*, **387**, 253–260 (1997).
77. Daily, G. C., & Ruckelshaus, M., 25 years of valuing ecosystems in decisions. *Nature*, **606**, 465–466 (2022).
78. Srinivas, S., Interstellar explorations: Stephen Hawing shows the way. *Science Reporter*, **50** (5), 54–55 (2018).
79. Tarter, J., & Impey, M., If you want to talk to ET, you must first find ET. In Impey, C., Lunine, J., & Funes, J., eds., *Frontiers of Astrobiology*, Cambridge University Press: Cambridge, Great Britain, Pages 286–305 (2012).
80. Ball, J. A., The zoo hypothesis. *Icarus*, **19**, 347–349 (1973).
81. Parfitt, J., Barthe, M., & Macnaughton, S., Food waste within food supply chains: Quantification and potential for change to 2050. *Philosophical Transactions of the Royal Society B.*, **365**, 3065–3081 (2010).
82. Xue, L., Liu, X., Lu, S., *et al.*, China's food loss and waste embodies increasing environmental impacts. *Nature Food*, **2**, 519–528 (2021).
83. Intergovernmental Panel on Climate Change, *Climate Change 2014: Mitigation of Climate Change—A Report of Working Group III*. Cambridge University Press: Cambridge, Great Britain (2014).
84. Smerlak, M., & Rovelli, C., Correspondence: Put defence money into planetary emergencies, urge Nobel winners. *Nature*, **601**, 308 (2022).
85. Thompson, H., The geopolitics of energy is reshaping the world again. *Nature*, **603**, 364 (2022).
86. Cowie, J., *Climate of Human Change: Disaster or Opportunity?* Parthenon Publishing: London, Great Britain (1998).
87. Pacala S., & Socolow R., Stabilization wedges: Solving the climate problem for the next 50 years with current technologies. *Science* **305**, 968–972 (2004).
88. Witbooi, E., Ali, K-D., Santosa, M. A., *et al.*, Organized crime in the fisheries sector threatens a sustainable ocean economy. *Nature*, **588**, 48–56 (2020).
89. Holmes, H., *Corruption: A Very Short Introduction*. Oxford University Press: Oxford, Great Britain (2015).
90. Climate Change Committee, *COP26: Key Outcomes and Next Steps for the UK*. Climate Change Committee: London, Great Britain (2021).
91. Vinge, V., The coming technological singularity: How to survive in the post human era. *Whole Earth Review*, **81**, 88–95 (1993).
92. Lovelock, J. E., *Gaia: A New Look at Life on Earth*. Oxford University Press: Oxford, Great Britain (1979).
93. Lovelock, J., *Novacene: The Coming Age of Hyperintelligence*. Allen Lane: London, Great Britain (2019).
94. Prucher, J., *Brave New Words: The Oxford Dictionary of Science Fiction*. Oxford University Press: Oxford, Great Britain (2007).
95. Caidin, M., *Cyborg*. Del Rey: London, Great Britain (1972).

96. Moore, G. E., Cramming more components onto integrated circuits. *Electronics*, **38**, 114–117(1965).
97. Boden, M. A., *Artificial Intelligence: A Very Short Introduction.* Oxford University Press: Oxford, Great Britain (2016).
98. Clarke, A. C., *2001: A Space Odyssey*. Hutchinson & Co.: London, Great Britain (1968).
99. Jones, D. F., *Colossus.* Rupert Hart-Davis: London, Great Britain (1966).
100. Koontz, D, *Demon Seed.* Bantam, New York, USA (1973).
101. Rodenberry, G., *Star Trek: The Next Generation.* Paramount Domestic Television: Los Angeles, California, USA (1987–1991).
102. Cameron, J., *The Terminator.* Herndale & Pacific Western: Los Angeles, California, USA (1984).
103. Wachowski, L., & Wachowski, A., *The Matrix.* Warner Brothers & Village Roadshow: Los Angeles, California, USA (1999).
104. Cargill, R., *Sea of Rust.* Gollancz: London, Great Britain (2017).
105. Willett, F. R., Avansino, D. T., Hochberg, L. R., *et al.*, High-performance brain-to-text communication via handwriting. *Nature*, **583**, 249–254 (2021).
106. Rajeswaran, P., & Orsborn, A. L., Neural interface translates thoughts into type *Nature*, **583**, 197–198 (2021).
107. Banks, I. M., *Excession.* Orbit: London, Great Britain (1996).
108. Fineburg, H., Neo-evolution: Is *Homo sapiens* ready? *British Medical Journal*, **343**, d7649 (2011).
109. Stock, G., *Redesigning Humans: Choosing Our Children's Genes.* Profile Books: London, Great Britain (2002).
110. Slonim, N., Bilu, Y., Alzate, C., *et al.*, An autonomous debating system. *Nature*, **591**, 379–384 (2021).
111. Reed, C., Argument technology for debating with humans. *Nature*, **591**, 373–374 (2021).
112. Wurman, P. R., Barrett, S., Kenta Kawamoto, K., *et al.*, Outracing champion Gran Turismo drivers with deep reinforcement learning. *Nature*, **602**, 223–228 (2022).
113. Gerdes, J. C., Neural networks overtake humans in a racing game. *Nature*, **602**, 213–214 (2022).
114. Awad, E., Dsouza, S., Kim, R., *et al.*, The Moral Machine experiment. *Nature*, **563**, 59–64 (2018).
115. Kasting, J. F., & Harman, C. E., Venus might never have been habitable. *Nature*, **598**, 259–260 (2021).
116. Lenton, T., & Watson, A., *Revolutions That Made the Earth*, Oxford University Press: Oxford, Great Britain (2011).
117. Wheeler, J. A. Information, physics, quantum: The search for links, in Zurek, W. H., ed., *Complexity, Entropy, and the Physics of Information.* Addison-Wesley: Redwood City, California, USA (1990).
118. Howard, D., Who invented the Copenhagen Interpretation? A study in mythology. *Philosophy of Science*, **71** (5), 669–682 (2004).
119. Wheeler, J. A., The 'past' and the 'delayed-choice double-slit experiment. In Marlow, A. R., ed., *Mathematical Foundations of Quantum Theory.* Academic Press: London, Great Britain, pages 9–48 (1978).
120. O'Dowd, M., Does the Universe create itself? *PBS Space Time*, **8**, 15 (2022). https://www.youtube.com/watch?v=I8p1yqnuk8Y (accessed December 2022).

APPENDIX 1

# Abbreviations and symbols

These abbreviations are not only found in this book but also are commonly used in the literature referenced. This list should therefore help not only your understanding this book but also should you decide to undertake further reading elsewhere.

| | |
|---|---|
| **δ** | The Greek letter lower case delta is used in mathematics to symbolise a small change or a change from. Used in this book to describe a change in carbon isotopes (typically a change away from the **VDPB** value), a change in deuterium isotopes (typically a change away from the **VSMOW** value), and also a change in energy in a chemical reaction. |
| **μ** | The Greek letter lower case mu is used in mathematics to symbolise an order of magnitude $10^{-6}$ (a millionth). Hence a micrometre (μm) is a millionth of a metre. It relates to the Greek '*micros*' meaning small. |
| **‰** | *Per* thousand |
| $^{39}$**Ar** | Argon-39 (the argon isotope with an atomic weight of 39—radioactive) |
| $^{40}$**Ar** | Argon-40 (the argon isotope with an atomic weight of 40—stable) |
| $^{12}$**C** | Carbon-12 (the carbon isotope with an atomic weight of 12) |
| $^{13}$**C** | Carbon-13 (carbon with an atomic weight of 13) |
| $^{16}$**O** | Oxygen-16 (the oxygen isotope with an atomic weight of 16) |
| $^{18}$**O** | Oxygen-18 (oxygen with an atomic weight of 16) |
| $^{32}$**S** | Sulphur-32 (the sulphur isotope with an atomic weight of 32) |
| $^{33}$**S** | Sulphur-33 (with an atomic weight of 33) |
| **ADP** | **A**denosine **D**i**p**hosphate |
| **AGI** | **A**rtificial **g**eneral **i**ntelligence |
| **AI** | **A**rtificial **i**ntelligence |
| **AMOC** | **A**tlantic **m**eridional **o**verturning **c**irculation |
| **ATP** | **A**denosine **t**ri**p**hosphate |
| **BIFs** | **B**anded **i**ron **f**ormations |
| **bya** | **B**illion **y**ears **a**go |
| **C** | Atomic carbon |
| **CIE** | **C**arbon **i**sotope **e**xcursion |
| **CFC** | **C**hloro**f**luoro**c**arbon |
| **ChatGPT** | **Chat** **G**enerative **P**re-trained **T**ransformer |
| **CHEOPS** | **CH**aracterising **ExOP**lanet **S**atellite |

| | |
|---|---|
| **CRISPR** | **C**lustered **R**egularly **I**nterspaced **S**hort **P**alindromic **R**epeats |
| **CoVID-19** | **C**orona**V**irus **I**nfectious **D**isease 2019 |
| **CSIRO** | **C**ommonwealth **S**cientific and **I**ndustrial **R**esearch **O**rganisation (Australia's national science agency) |
| **D** | Atomic deuterium (the heavier isotope of hydrogen) |
| **DNA** | **D**eoxyribose **n**ucleic **a**cid |
| **EARS** | **E**ast **A**frican **R**ift **S**ystem |
| **EJ** | Exajoules = $10^{18}$ J—see also Appendix 2 |
| **FAO** | **F**ood and **A**gricultural **O**rganization |
| $Fe^{2+}$ | Ferrous iron |
| $Fe^{3+}$ | Ferric iron |
| **GEP** | **G**ross **e**cosystem **p**roduct |
| **GNP** | **G**ross **n**ational **p**roduct |
| **GtC** | **G**iga**t**onnes **c**arbon (1 gigatonne = 1 petagram) |
| **Gt** $CO_2$ | Gigatonnes of carbon dioxide |
| **Gt** $CO_2\ yr^{-1}$ | Gigatonnes of carbon dioxide *per year* |
| $Gtyr^{-1}$ | Gigatonnes *per* year |
| **Gyr** | **G**iga-**y**ea**r**s (billions of years) |
| **Gya** | **G**iga-**y**ears (billions of years) **a**go |
| **H** | Atomic hydrogen |
| **Hg** | The element mercury (atomic mercury) |
| **IETM** | **I**nitial **E**ocene **t**hermal **m**aximum—old common nomenclature (see also **PETM**) |
| **IPCC** | **I**ntergovernmental **P**anel on **C**limate **C**hange |
| **J** | Joule (a unit of energy which in household electrical terms is equivalent to that of one watt of power expended for just one second). See also Appendix 2. |
| **K** | The symbol of the kelvin scale of absolute temperature with each kelvin being equivalent in temperature difference to that of a degree Celsius, but with 0 K being approximate to -273°C, and 273 K approximately equalling 0°C. (To be precise, 273.15 K equals 0°C as the kelvin is officially defined in relation to the triple point of water.) A container of a substance at 0 K sees all motion of the substance's atoms, and all atomic vibration within molecules, cease. |
| $km^2$ | Square kilometres |
| $km^3$ | Cubic kilometres |
| **kyr** | Thousand (kilo) years—see also Appendix 2. |
| **K/Pg** | Cretaceous-Paleogene boundary (new geological nomenclature—see **K/T**) |
| **K/T** | Cretaceous-Tertiary boundary (old geological nomenclature—see **K/Pg**) |

| | |
|---|---|
| **LGM** | Last Glacial Maximum |
| **LHB** | Late Heavy Bombardment |
| **LIP** | Large igneous province. Plural, LIPs. |
| **MIF** | Mass independent fractionation |
| **MIF of S** | Mass independent fractionation of sulphur isotopes |
| **mol** | Mole (see Appendix 2) |
| **mtoe** | Million tonnes of oil equivalent |
| **mya** | Million years ago (synonymous with mega years ago, but not strictly considered as such in nomenclature terms, as 'mega' abbreviates to 'M' upper case). (Similarly **Gya**—a thousand million years ago—'giga' abbreviates to 'G' [upper case] under the International System of Units—see also Appendix 2) |
| **myr** | Million years (equivalent to mega years, but not in strict nomenclature terms, as 'mega' abbreviates to 'M' [upper case])—see also Appendix 2. |
| **mt DNA** | Mitochondrial DNA |
| **$N_2O$** | Nitrous oxide |
| **O** | Atomic oxygen |
| **$O_2$** | Two oxygen atoms bonded to form a molecule (the common form of gaseous oxygen) |
| **$O_3$** | Ozone |
| **OP** | Oxygenic photosynthesis |
| **PAL** | Present atmospheric level (here commonly used in reference to oxygen). Oxygen currently composes nearly 21% of the atmosphere, so that equals 100% PAL. Conversely, the 4% PAL of oxygen thought needed for multicellular life equates to an atmospheric composition 0.84%. |
| **Pb** | Lead |
| **PETM** | Palaeocene Eocene thermal maximum (new common nomenclature)—see also **IETM**) |
| **Pg** | Petagram (1 petagram = 1 gigatonne)—see also Appendix 2 |
| **PgC $yr^{-1}$** | Petagram carbon *per* year |
| **ppbv** | Parts per billion by volume |
| **PREDICTS** | Projecting Responses of Ecological Diversity In Changing Terrestrial Systems |
| **RCP** | Representative Concentration Pathway |
| **RNA** | Ribonucleic Acid |
| **S** | Sulphur |
| **SARS-CoV-2** | Severe Acute Respiratory Syndrome Corona Virus 2 |
| **SF** | Science fiction (or speculative fiction that includes fantasy) |
| **Si** | Silicon |
| **Se** | Selenium |
| **SETI** | Search for Extra-Terrestrial Intelligence |

| | |
|---|---|
| **SMOW** | Standard **m**ean **o**cean **w**ater. See **VSMOW** |
| **sp** | Species (singular) |
| **spp** | Species (plural) |
| **SSP** | **S**hared **S**ocio-economic **P**athway |
| **STEM** | **S**cience, **T**echnology, **E**ngineering, **M**edicine/**M**athematics |
| **Te** | **Te**llurium |
| **Th** | **Th**orium |
| **U** | **U**ranium |
| **UN** | **U**nited **N**ations |
| **UNEP** | **UN E**nvironment **P**rogramme |
| **UV** | **U**ltra**v**iolet |
| **VPDB** | The **V**ienna **P**ee**D**ee **b**elemnite. The Vienna PeeDee belemnite standard is a standard $^{13}C/^{12}C$ ratio against which geoscientists commonly relate their $^{13}C/^{12}C$ ratio measurements. The belemnite here is fossil remains of a cephalopod (a class of mollusc, a phylum that includes species such as the octopus) and it is found in the PeeDee formation in South Carolina (US). |
| **VSMOW** | **V**ienna **s**tandard **m**ean **o**cean **w**ater. VSMOW is a water standard defining the isotopic composition of ocean water and is similar to (but not the same as) **SMOW**. VSMOW is often used for technical and pragmatic reasons. |
| **WMO** | **W**orld **M**eteorological **O**rganization |
| **yr** | Year |
| **yr$^{-1}$** | *Per* year |
| **ZJ** | **Z**etta**j**oules = $10^{21}$ J. See also Appendix 2. |

APPENDIX 2

# Nomenclature, Data, and Terms

## Nomenclature

**The biological nomenclature** used follows the Institute of Biology's *Biological Nomenclature: Recommendations on Terms, Units and Symbols* (1989 & 1997) which is recognised by all leading UK examination boards and academic publishers. As part of this, the five-kingdom classification of organisms is principally used. The five-kingdom classification was proposed by Margulis and Schwartz in 1988 and builds on an earlier framework by Whittaker in 1959.

**Geological dating** (geological eons, periods, epochs, and ages) follows that of the International Commission on Stratigraphy (2018 and 2023). See www.stratigraphy.org.

**Key scientific terms used** are defined later in this Appendix.

## Data: Assumptions and sources

**Energy statistics and data** are presented using the following criteria.

Where statistics are given in millions of tonnes of oil equivalent (mtoe) for quantities of nuclear and renewable energy (such as hydroelectric power) they are expressed in terms of the oil equivalent *as if* the energy were generated by an oil-fired electricity-generating plant with the equivalent of a 1965–2000 assumed efficiency of 36%; between 2000 and 2017 there is an assumed a linear increase in efficiency from 36% to 40%; and the assumption thereafter is that power station efficiency will linearly increase to 45% by 2050. Consequently, it is possible to make some comparison and to see how these non-fossil energy sources (could) displace oil sources.

The principal data source for various fuels contributing to global primary energy production was the *BP Statistical Review of World Energy* (2022 ed.). For the year 2023, increases in energy production/consumption in various primary energy sources over 2021 were taken from the *Energy Institute Statistical Review of World Energy 2023.*

Historic proportional energy-supply estimates prior to 1950 are taken from Morgan and Murray (1976), and early carbon emissions were assessed from a variety of sources, including from the Carbon Dioxide Information Analysis Center (CDIAC) of the US Department of Energy and also the Woods Hole Research Center, Massachusetts, USA. A check for carbon emissions was made using the International Energy Agency's *$CO_2$ Emissions from Fossil Fuel Combustion Annual Data.* Estimates for fossil fuel use before 1950 were made after 1950 and were published in the 1980s and 1990s cross-checked against the US Government's Energy Information Administration www.eia.gov, the *BP Statistical Review of World Energy* (2022 ed.), and a 2023 update from the *Energy Institute Statistical Review of World Energy 2023.* These were adapted where necessary, as *per* previous power-station efficiency assumptions. The Worldwatch Institute's *Vital Signs* series (2003), which in turn draws on additional sources, was used as a further check on statistics post-1950. The UK Department of Trade and Industry's *Energy Trends* (also known as

Digest of UK Energy Statistics, or DUKES) was similarly used as a check for UK data. (It is important to note that there is some variation, fortunately mainly minor, between these data sources, hence the need for cross-checking.)

Key websites that have energy and/or carbon data:

- British Petroleum—http://www.bp.com (with statistical reviews up to 2022)
- Carbon Dioxide Information Analysis Center—http://cdiac.ornl.gov
- Energy Institute—http://energyinst.org (with statistical reviews 2023 onwards)
- International Energy Agency—http://www.iea.org
- US Energy Information Administration (EIA)—http://www.eia.gov

Data from 1950 onwards strictly relates to commercially traded fuels and does not include locally gathered or produced domestic fuels in less-developed nations, such as burning of rural wood or animal dung.

- 1 t (metric) of oil = 1.1023 short tons = 256 gallons (imperial) = 308 US gallons
- 1 mtoe (million tonnes of oil equivalent) produces around 4 TWh = $4 \times 10^{12}$ Wh in a power station of about 33% efficiency.
- 1 mtoe, when burned, releases approximately $4.25 \times 10^{13}$ Btu (British thermal units) or $4.48 \times 10^{16}$ J.
- A machine doing work consuming energy at the rate of 1 $Js^{-1}$ consumes 1 W each second. If this continues for an hour (60 s × 60 min), 1 kWh of energy is consumed, which is equivalent to 3,600 J.

## Data: Orders of magnitude

Other than fossil-fuel energy content (above), other non-biological units and nomenclature are based on Baron (1988).

| *Factor* | *Prefix* | *Symbol* |
|---|---|---|
| $10^{-6}$ | micro | μ |
| $10^{1}$ | deca | da |
| $10^{3}$ | kilo | k |
| $10^{6}$ | mega | M |
| $10^{9}$ | giga | G |
| $10^{12}$ | tera | T |
| $10^{15}$ | peta | P |
| $10^{18}$ | exa | E |
| $10^{21}$ | zetta | Z |
| $10^{24}$ | yotta | Y |

## Terms

| | |
|---|---|
| **abiotic** | Without, or in the absence of, life. Pertaining to non-living things. |
| **adenosine diphosphate** | Fundamental to life as we know it, this is a molecule (biochemical) that can absorb energy when it gains a phosphate group to become **adenosine triphosphate.** The energy to do this either comes from photosynthesis (such as in plants) or chemical reactions such as respiration (such as in animals). Adenosine diphosphate is in all living cells. |
| **adenosine triphosphate** | A molecule (biochemical) that releases energy when it loses a phosphate group to become **adenosine diphosphate.** This energy release drives virtually all the metabolic processes of life as we know it. As such, it is fundamental to life on Earth. Adenosine triphosphate is in all living cells. |
| **aerobic** | With oxygen. |
| **algae** | Algae are a number of eukaryotic (see **eukaryote**) species that can be unicellular, multicellular, or colonial organisms; they are usually photosynthetic and commonly belong to the plant kingdom (but, confusingly, some species belong in the kingdom **Protoctista** [Protists or Protista]). |
| **anaerobic** | (In the) absence of oxygen. |
| **Angiospermae** | Angiosperms, or Angiospermophyta, are species of a major phylum of plants: the plants that have flowers. |
| **Anthropocene** | A geological epoch that was (early in the 21st century) proposed to begin in 1950, it being the start of the time humanity began to make a discernible mark on the geological record. At the time of writing, it has not been approved by the International Commission on Stratigraphy and is not on its chronostratigraphic chart. However, it is increasingly being used by geoscientists and others, and may be considered an informal term for the period of time in which humans had a discernible impact on the Earth system. (Previously Ernst Haeckel, in 1868, introduced the idea of an 'Anthropozoic age' to highlight humans' profound impact on the environment, but this included humanity's impact on forests, which do not have a noticeable impact on the geological record let alone the Earth system as a whole.) |
| **apatite** | A group of phosphate minerals that contain calcium and phosphate. Among other locations, it can be found in hydrothermal vents. |
| **Archaea** | A **domain, superkingdom**, kingdom, or class, depending on which taxonomic classification system is used, of single-celled **prokaryotes** or prokaryotic microorganisms that were among the first living celled life on Earth and which contains the archaebacteria. (These days considered as distinct from **Monera.** The classification of **Archaea** has seen a number of changes over the years and might even be considered by some a contentious area of systematics [species classification].) |

| | |
|---|---|
| **archaeon** | Of Archaea. |
| **Archaean** | The geological eon spanning 2.5 bya–4 bya. |
| **atmosphere** | Other than its meaning as a noun referring to the covering of air around the Earth or another planet/moon, it is also a unit of pressure. Atmospheric pressure is defined as a present-day standard value of $1.01325 \times 10^5$ N/m$^2$ or 101,325 pascals, and 1.01325 bar at sea level. |
| **autocorrelation** | The mathematical cross-correlation of a signal with itself that can be used as a mathematical method of finding and analysing repeating patterns. Autocorrelation is found to change a little before a critical transition (or threshold) is crossed and so can be used as a predictive tool. |
| **autotrophs** | Species such as plants able to elaborate all their chemical constituents from simple inorganic compounds such as carbon dioxide. The term comes from the Greek '*autos*' meaning 'self' and the Greek '*trophe*' from '*trephein*' meaning 'to nourish'. See also **heterotrophs** and **chemolithoautotrophs.** |
| **Avalon explosion** | The explosion or 'boom' of new species early in the **Ediacaran**; as such it is sometimes referred to as the 'Avalon boom'. |
| **bacteriochlorophyll** | The photosynthetic pigment in photosynthetic bacteria. See also **chlorophyll.** |
| **biome** | Large, landscape-level assemblages of ecosystems. For example: tundra, savannah, temperate forest, and tropical rain forest. |
| **biotic** | Relating to life. |
| **bikont** | A taxonomic term for a supergroup of eukaryotic species that include plants. Many single-celled members of the group, and the presumed ancestor, have two flagella. Bikonts and **unikonts** arose or became distinct somewhere at the very least between 1.2 billion years ago and 900 million years ago but possibly earlier. Though this term is still used by some biologists, these days it rarely is. |
| **bleb** | A biomedical term meaning a small vesicle (saclike space) containing fluid. |
| **blue-green algae** | This is a term now discouraged, as it suggests that such species are algae (which would make them **eukaryotes**). It is an old name for **cyanobacteria** (which are **prokaryotes**). |
| **boreal** | The northern hemisphere biogeographic region around and just below the Arctic circle characterised by short summers of long days and long harsh winters of short days. |
| **bya** | 'Billion(s) of years ago'. See also **gyr.** |
| **byte** | An information and computer science term corresponding to the amount of information (number of 'bits', in computer science) required to define a single character (letter, number, or symbol). |
| **$C_3$ plants** | Plants that use the older photosynthetic pathway with three carbon atoms in the first product during carbon dioxide assimilation. |

| | |
|---|---|
| **$C_4$ plants** | Plants that use the more recently evolved photosynthetic pathway capable in intense sunlight (hence tropical low latitudes) of taking more carbon dioxide out of the atmosphere than $C_3$ photosynthesis. |
| **Cambrian** | The geological period spanning 541 mya–485 mya. |
| **Cambrian boom** | These days many refer to this as the 'Cambrian explosion', it being the explosion, or 'boom', of new, multicellular species including the **Eumetazoa** that took place early in the **Cambrian.** (Also known as the 'Cambrian diversification'.) |
| **carbon leakage (ecological)** | This can take place when ecosystems that store carbon are disrupted. Ploughing a field can release carbon back into the atmosphere, as can burning woodlands and forests. |
| **carbon leakage (economic)** | A nation releases greenhouse gas carbon, but some of these emissions may come from burning fossil fuels whose energy is used to produce goods that are exported to another country. These emissions may not be counted in the importing nations' greenhouse emissions statistics. |
| **Cephalopoda** | A class of marine animals. Cephalopods include squid and octopus species. The term comes from the Greek '*kephale*' meaning 'head' and '*podos*' meaning 'foot'. |
| **chaos theory** | A field within mathematics relating to the behaviour of dynamic systems in which outcomes can vary considerably depending on the system's initial conditions and when systems are affected by a new, or a change in a, factor. Chaos exists when the present determines the future, but the approximate present does not approximately determine the future. Hit a billiard ball hard into a grouping of other balls and it is virtually impossible to replicate exactly where all the balls end up as all attempts to replicate the initial shot are 'approximate' (not exactly the same). The smallest of differences will result in a greatly different outcome of pattern of balls. |
| **chemolithoautotrophs** | Species that use inorganic sources of carbon, nitrogen, and other chemicals found in, or emanating from, stone or rock. 'Auto' comes from the Greek '*autos*' meaning 'self', 'chem' relates to chemicals, and the 'litho' relates to stone or rock from the Greek '*lithos*'. See also '**autotrophs**' and '**heterotrophs**'. |
| **chemosynthesis** | The use by some bacteria of energy derived from chemical reactions (such as involving sulphur or ammonia). |
| **cherts** | Siliceous geological strata often including the remains of siliceous organisms. |
| **chimera** | An organism composed of two different cell types (commonly such as through grafting or inserting a cell into an embryo). |
| **chlorophyll** | The principal photosynthetic pigment of plants. It is found in the **chloroplasts** of plant cells. See also **bacteriochlorophyll.** |
| **chloroplast** | The photosynthetic **organelle** of plant cells. See also **photosynthesis.** |
| **citric acid cycle** | See the **tricarboxylic acid** (TCA) **cycle**, also known as the **Krebs cycle.** |

| | |
|---|---|
| **clathrates** | See **methane clathrates** below. |
| **coenocyte** | A single cell with many nuclei arising from the nucleus dividing a number of times without the cell itself dividing. (Of note to biology students, this is quite distinct from a 'syncytium', which is also a single cell with many nuclei but here there were originally many cells that aggregated whose cell walls then disappeared leaving a single cell.) |
| ***corpora pedunculata*** | The most conspicuous part of the protocerebral lobes of the insect brain. Also known as 'mushroom bodies', they have been shown to the source of cognitive and innate behaviour. In cockroaches, flies, bees, and wasps, among other insect species they are particularly large. |
| **Cretaceous** | The geological period between 145 mya and 66 mya. (In the academic literature, it is abbreviated to 'K'.) |
| **CRISPR** | Stands for **C**lustered **R**egularly **I**nterspaced **S**hort **P**alindromic **R**epeats in genetic acids; these are the basis for a gene-editing tool that can remove or even replace genes in a genome. For the purposes of this book, you need not worry about the details, just know that we are developing tools (such as one based on CRISPR) to ever more precisely edit genetic sequences. |
| **Cryogenian** | The period of time between 720 and 635 mya (which encompasses much of the time of the Earth's second big flip). |
| **cyanobacteria** | A group of **prokaryotes** capable of **oxygenic** (oxygen generating) **photosynthesis.** The Greek '*kyanos*' means 'dark blue'. |
| **deoxyribonucleic acid** | **DNA**, the large-molecule nucleic acid of genetic material as used by most organisms. |
| **domain** | The highest taxonomic rank of organisms in the three-domain system of taxonomy. Also known as '**superkingdom**'. According to this system, the tree of life consists of three domains: **Archaea,** Bacteria, and Eukarya or **Eukaryota(e).** The first two are all prokaryotes and the latter, of course, are eukaryotes. (Derived from 'an area under one rule', from the Latin '*dominus*' meaning 'lord'.) |
| **detritovore** | An organism that feeds on detritus—predominantly small pieces of dead and decomposing plants and animals. |
| **eclogitic** | Relating to a coarse-grained rock consisting of bright green omphacite and deep red almandine-pryope garnet. There may be trace amounts of other minerals such as quartz present too. (Search the internet if you want to find out more; meanwhile just note 'eclogitic' relates to a specific rock type.) |
| **ecological carbon leakage** | The release of carbon from an ecosystem to the atmosphere. For example, through deforestation or through soils releasing their carbon. |
| **economic carbon leakage** | The exporting of fossil carbon emissions by one country through another country creating goods in part by fossil fuel use. These goods are then imported. |
| **Ediacaran** | The geological period of time, 635 mya–541 mya, that followed the **Cryogenian** Snowball Earth II. This period saw the **Avalon explosion.** |

**endocytosis** The process of a cell engulfing molecules, liquids, or solids and then 'ingesting' them. (See also **phagocytosis.**) In terms of its Greek roots, 'endocytosis' comes from the word '*endo*' meaning 'within', '*cyt*' meaning 'cell', and '-*osis*' meaning 'process'.

**endosymbiosis** The living together of two species with one inside the other. Its Greek roots come from '*endo*' meaning 'within', '*syn*' meaning 'together', and '*biosis*' meaning 'living'. 'Symbiosis' itself is when two species live together with each partner conferring some benefit to the other.

**environmental proxies** Also known as **palaeo-indicators**—such as tree rings, isotopes, fossil species, and so forth—that indicate past environmental conditions such as climate, pH, land erosion, *etc.*

**Eocene** The geological epoch spanning 56 mya–34 mya.

**Eukaryota(e)** The Eukaryota are a biological **domain**. Eukaryotes are 'higher' organisms than **prokaryotes/Prokaryota(e)**. The cells of eukaryotes have a distinct, membrane-bound nucleus as well as membrane-bound organelles such as mitochondria and chloroplasts. The term relates to the Greek '*eu*' meaning 'good' and '*karyon*' meaning 'nut' or 'kernel': 'nut' and 'kernel' standing in for the biological term 'nucleus'. (Not to be confused with the Latin word for 'kernel'—see **nucleus.**) Hence 'eukaryote' in terms of its Greek roots means 'good nucleus'.

**eukaryote** A type of cell of species belonging to the **Eukaryota(e).**

**Eumetazoa** These are a subkingdom of the **Metazoa** kingdom (multicelled animals) within the domain **Eukaryota(e)**. They may be considered to be more evolutionary developed than other metazoans such as the simpler sponges (Porifera) that under some species classification systems belong to the **Parazoa** subkingdom. The prefix term relates to the Greek '*eu*' meaning 'good'.

**exo-Earth** An Earthlike planet located outside the Solar System.

**exoplanet** A planet located outside the Solar System.

**extremophiles** Meaning 'liking extremes'. Prokaryotic extremophile species inhabit environments we humans consider extreme: they 'like extreme' conditions. Such habitats might be very hot (thermal springs), very salty (ponds on salt pans), have high concentrations of elements and compounds poisonous to humans, or combinations thereof.

**forams (Foraminifera)** Foraminifera (or forams) are single-celled marine animals possessing shells usually made of calcium carbonate. They are an order within Sarcodina. (Sarcodina are a superclass within the kingdom of **Protoctista.**) (Older biologists, and many textbooks, still refer to forams as an order within 'protozoa', which is a phylum no longer recognised under the five-kingdom system of species classification. Nonetheless, 'protozoa' is a term that still has a certain traction within biology, so don't be surprised if you come across it in your own further reading.) The carbon in forams' carbonate shells is used for isotopic analysis that helps illuminate carbon cycling. The oxygen in their carbonate shells is also used for isotopic analysis that informs either about sea temperature or (if there is a lot of ice on the planet, as there has been the past 35 million years in Antarctica) how much ice there is on the planet.

| | |
|---|---|
| **galaxy** | An assemblage of many, gravitationally bound stars (common noun) typically isolated from other galaxies by millions of light years. |
| **Galaxy, The** | The galaxy in which the Earth is located (proper noun), also known as the Milky Way galaxy. |
| **Gaskiers** | The final, and milder, cold phase some of Snowball Earth II that took place 580 mya. |
| **genomics** | The science of a species' cellular nucleic acid—its complement of genes, or genome. |
| **glacial** | A cold part of an **ice age**. (See also **interglacial**.). (It is also an adjective meaning pertaining to a glacier, *e.g.* **loess**.) |
| **Gyr** | **Giga years** ($10^9$ years) or billions of years. (See 'Orders of magnitude' in the preceding subsection.) See also **bya**, which is commonly used in this text, but some of the reference titles and their contents employ Gyr. |
| **heterotrophs** | Species that require carbon in organic form (such as sugars and protein). These include all animals and fungi. The term comes from the Greek '*heteros*' meaning 'other' and '*trophe*' from '*trephein*' meaning 'to nourish'. See also '**autotrophs**' and '**chemolithoautotrophs**'. |
| **Holocene** | The epoch covering the past 11,700 years. See **interglacial**. It is the most recent epoch within the **Quaternary** period. |
| **hominid** | Any member of the primate family Hominidae, including modern humans and their ancestor species. It *includes* the genus *Gorilla* (gorillas). (Not to be confused with **hominin** in the literature; such is the similarity of terms.) |
| **hominin** | Any member of the *Homo* species genus, the taxonomic tribe Hominini. It *excludes* the genus *Gorilla* (gorillas). Some say that it should also exclude the genus *Pan* (chimpanzees and bonobos), though commonly they are included. (Not to be confused with **hominid** in the literature; such is the similarity of terms.) |
| ***Homo sapiens*** | Anatomically modern man. The species interbred with some other archaic human species. The term comes from the Latin '*homo*' meaning 'man' and '*sapiens*' meaning 'wise'. |
| ***Homo sapiens sapiens*** | The present as well as ancestral population of humans that gave rise to the extant generation of anatomically modern humans. There is some debate as to the usage of this term. It is used in this book to underline recent generations of our species having a major impact on the biosphere. This arguably had its roots with the megafauna extinctions outside of Africa where megafauna had not co-evolved with humanity. Today our species is having a global impact with the altering of the planet's atmosphere and ongoing additions of new mineral species. It translates from the Latin as 'wise, wise man' or 'very wise man'. Given the environmental damage our species is doing to the planet, the term might be considered as oxymoronic. |

**horizontal gene transfer** A way bacteria can exchange genetic material. (This is different from sexual reproduction, which sees a mixing of genes from both parents within a strand of nucleic acid.)

**hyperthermal** A warm period a few degrees Celsius above the temperature before and after. The event typically lasts a couple of hundred thousand years or so. The PETM (previously known as the IETM) is an example of a hyperthermal.

**ice age** Confusingly, this is a term with two meanings. In the strict geological sense, an ice age is made up of cold **glacial** times interspersed with warmer **interglacials.** In a looser sense the term 'ice age' can also be used as a synonym for '**glacial**'. Geologists, palaeo-environmental scientists and Earth system scientists usually have little problem discerning which sense is meant from the context of use. However, those new to exploring these disciplines need to be wary of possible confusion.

**igneous rock** Rock formed from high temperature, molten silicate (sometimes rarely carbonate) and generally rock solidified from magma that has reached the surface.

**interglacial** The warm parts of an **ice age.** (See also **glacial.**) The past 11,700 years has seen the Earth in the **Holocene** interglacial epoch which itself is within the **Quaternary** period.

**joule** The joule is a fundamental unit of energy denoted by the symbol 'J'. The application of energy to do work necessitates a rate of energy consumption, or power consumption. A rate of power consumption of 1 joule *per* second is equivalent to one watt. To give you a household example of what this means, a typical one-bar electric fire is rated at about a thousand watts, or a kilowatt.

**Krebs cycle** See **tricarboxylic acid** (TCA) **cycle.** Also known as the **citric acid cycle.**

**large igneous province** This is a large area, typically thousands of square miles or more, of rock that when formed is effectively a sea of molten lava. The formation of large igneous provinces (LIPs) has been associated with a number of extinction events in the geological past.

**loess** A fine, permeable material composed of minerals of the consistency of very fine sand or course flour. Once mixed with organic matter it makes for good soil. Most comes from wind-blown glacial rock flour, though some can be of desert origin.

**Lorenz attractor** A set of **chaotic** solutions of the Lorenz system; when plotted they often resemble a butterfly or figure eight. In **chaos theory** they can represent islands of comparative stability surrounded by instability, with the system moving about the attractor.

**magnetosphere** The volume around planets whose core generates a magnetic field. They deflect Solar wind and so protect small planets' atmospheres from ablation (loss). The International Space Station orbits within the Earth's magnetosphere, and Saturn's moon Titan largely orbits within Saturn's.

**main sequence stars** Stars are born and ultimately die mostly ending up as either a supernova and then supernova remnant, neutron star, or black hole, depending on the star's size. Between birth and death, most stars are said to lie on the 'main sequence' (of the Hertzsprung-Russell diagram). This time on the 'main sequence' sees the star primarily convert hydrogen into helium. As the hydrogen is used up, other forms of fusion play a significant part in the star's energy production and it leaves the 'main sequence'. While on the 'main sequence', the star's energy output will very gradually increase. The Sun is a 'main sequence' star.

**mantle** Part of the Earth beneath the crust and the core roughly 30–2,900 km down. It is under pressure and effectively solid but can slowly deform and so moves. It has a slow circulation that drives plates tectonically.

**Marinoan** The second major cold phase of Snowball Earth II (sometimes called the **Varanger**) that ended around 635 mya.

**metamorphic rocks** Rocks that have been changed from their original form by heat and pressure.

**Metazoa** (Animalia) The kingdom of species consisting of multicelled animals that contain the subkingdom of **Parazoa** (which evolved first and some say, and it is argued here, prior to Snowball II) and **Eumetazoa** (which are considered to have evolved after Snowball II).

**megafauna** Large animals. Other than domesticated spp., many now extinct outside of Africa, where megafauna co-evolved with humans.

**methane clathrate** A clathrate is a form of a chemical compound in which one component is enclosed by a cage of another. In this case methane is enclosed by a cage of water molecules. This soft, icy-like substance is created at depth in the sea above certain pressures and below certain temperatures. However, it easily decomposes, releasing the methane either when the pressure is reduced or the temperature increases.

**methanogens** Methane-generating **prokaryotes.**

**mineral species** A mineral with *both* an unique chemical composition and an unique molecular/crystal structure.

**Miocene** The geological epoch spanning 23 mya–5 mya.

**mitochondria** An **organelle** found within **eukaryotic** cells, responsible for energy production.

**mole** An amount of an element or a pure chemical compound (a chemical combination of elements) that contains as many atoms or molecules as does 12 grams of carbon-12. (Using moles is a way of relating atomic and molecular reactions to workable amounts of material in the lab.) It is abbreviated as 'mol'.

**Monera** A kingdom of single-celled **prokaryotic** microorganisms that were among (or at least similar to) the first living celled life on Earth. (Technically, they are distinct from **Archaea**, but this is a consideration that only those studying biology at undergraduate level and above need seriously worry about: both Monera and Archaea are **Prokaryotes**). Monera comes from the Greek '*moneres*' meaning 'solitary', from '*monos*' meaning 'single' or 'alone'.

| | |
|---|---|
| **moon** | The natural satellite of a planet (common noun). |
| **Moon, The** | Earth's natural satellite (proper noun). |
| **nano** | From the Greek '*nanus*' meaning dwarf. Hence 'nanoplankton' meaning small plankton, which are typically of a size of 0.2 to 20 μm. (Not to be confused with the Greek '*micros*' meaning small.) |
| **Neogene** | The geological period spanning 23–5 mya. This was a warm period in the Earth's history. |
| **net primary production** | The amount of (dry) biomass produced *after* the biomass used in respiration has been taken into account. Abbreviated as NPP. Conversely, gross primary production *includes* the biomass consumed by respiration. Meanwhile secondary production is the biomass generated by species feeding on primary producers such as plants. See also (biological) **productivity**, with which it should not be confused. |
| **niche construction** | The process by which a species alters its own (or another species') local environment. |
| **nucleus** (biology) | A membrane-bound **organelle** within **eukaryotic** cells that contains the cell's genetic material. (The word is based on the Latin word '*nuc*' (the stem of '*nux*') meaning 'kernel' or 'nut'. '*Nuc*' + '*ula*' therefore means 'little nut'. Do not confuse it with the ancient Greek for 'kernel'—see **eukaryote**.) |
| **nucleus** (chemistry) | The centre of an atom that contains **protons** and (other than the most common form of hydrogen) neutrons. The nucleus is surrounded by electron orbitals. Together, the nucleus and their surrounding electrons make up atoms. |
| **nunatak** | A rock peak sticking out of the surface of an ice sheet. |
| **organelle** | A distinct structure within a **eukaryotic** cell responsible for carrying out specific functions. Examples include **chloroplasts** and **mitochondria**. |
| **olivine** | A group of minerals all consisting of silicate ($SiO_4$) combined with the elements of either magnesium, (ferrous) iron, manganese, or calcium or mixtures thereof. Often yellow-green, green, or brown-green in colour. Found on both Earth and Mars. The action of water on olivine at low temperatures releases hydrogen that can combine with carbon to form methane. |
| **oxygenic** | Oxygen generating. |
| **palaeo-** | A prefix that relates the meaning of whatever follows it (the suffix) to the past. It comes from the Greek word '*palaios*' meaning 'ancient' or 'very old'. For example, 'palaeoecology' is the study of fossils so as to elucidate how plants and animals related to each other in the distant past. |
| **palaeoclimate** | The climate of the past. |
| **palaeo-indicators** | Also known as **environmental proxies**. |
| **palaeosols** | Fossil soils. |

| | |
|---|---|
| **Paleogene** | The geological period spanning 66–23 mya. This period saw the Earth begin to cool from the warmer **Neogene** towards the cold **Quaternary**. |
| **Parazoa** | The subkingdom of **Metazoa** (Animalia) multicelled animals that evolved before **Eumetazoa**. Parazoans include sponges (of the phylum Porifera). 'Para' comes from the Greek meaning '*beside*', and '*zoa*' comes from the Greek '*zoion*' or '*zoon*' meaning 'living being' or 'animal'. The term has become deprecated in a number of recent classification systems, being replaced by the phylum **Porifera** (but it is still often found in the literature, hence it is a term of which students will need to be aware). |
| **perchlorates** | Salts of perchloric acid ($HClO_4$). They include $NaClO_4$, $Mg(ClO_4)_2$, and $Mg(ClO_3)_2$, all of which have been detected in brines seeping from the sides of slopes on Mars. |
| **peridotitic** | Refers to rocks consisting of wholly or largely olivine together with other ferromagnesian minerals but devoid of feldspar. (Feel free to search-engine 'olivine' and 'feldspar' if you wish, but for purposes of this book just note that 'peridotitic' refers to a specific rock type.) |
| **phagocytosis** | The process of a cell engulfing particles/solids and then 'ingesting' them. (See also **endocytosis**: phagocytosis is a specialised type of endocytosis.) 'Phagocytosis' comes from the Greek word '*phagein*' meaning 'to devour', '*kytos*' meaning 'cell', and '*-osis*' meaning 'process'; and so from its Greek roots, 'phagocytosis' means the 'process by which cells devour'. |
| **photolysis** | Meaning 'light breaking', photolysis is the splitting of a molecule by light, typically ultraviolet. This process can often result in **radicals.** A synonym for 'photodissociation'. |
| **photosynthesis** | The use of energy from light (or in some cases infrared heat) to drive a chemical reaction, notably the reduction of carbon dioxide to carbohydrates. In the process an oxidation reaction has to occur. With plants this is the oxidation of water to oxygen. Photosynthesis in bacteria involves the oxidation of hydrogen sulphide to sulphur. |
| **planulae** | The larval, reproductive stage of some corals. |
| **Porifera** | A phylum of sessile, aquatic, filter-feeding creatures that include sponges. The name derives from the Greek meaning 'pore bearer'. |
| **primary energy** | Primary energy comprises commercially traded fuels, including modern renewables used to generate electricity. Electricity, when generated from fossil fuels, is a 'secondary' energy resource, as the consumption of a primary energy resource is needed for its production. |
| **production** | A biological term. See **net primary production.** |
| **productivity** | This term is not used in this book but is included for the benefit of young, life science undergraduates as it can be confused with the term 'production' (see '**net primary production**'). 'Productivity' is a term commonly used by agricultural biologists meaning 'productivity *per* unit area'. (Agriculturalists tend to talk about productivity as they consider matters *per* hectare or acre.) It is easy to confuse 'productivity' with 'production'. |

**Prokaryota(e)** Prokaryotes constitute a biological kingdom of simple, prokaryotic organisms such as bacteria and Archaea. (Sometimes considered a **superkingdom**, or a **domain**, depending on which taxonomic classification system is used.) All prokaryotic organisms (or prokaryotes) lack membrane-bound nuclei and envelope-bound organelles. (These contrast with more advanced cellular life—**eukaryotes (the Eukaryota[e])**—that do have such features.) Prokaryotae include **cyanobacteria**, which in turn used to be called 'blue-green algae' (but, as '**algae**' are '**eukaryotes**', taxonomist biologists (those who engage in the 'systematic' classification of species) are discouraging the term 'blue-green algae'). '*Pro*' relates to the Greek word for 'before'. 'Kayron' relates to the Greek meaning 'nut' or 'kernel', and is the equivalent of the Latin-based term 'nucleus'. Therefore, in terms of its Greek roots, 'prokaryote' means 'before a nucleus'.

**prokaryote** A type of cell or species of the **Prokaryota(e)**.

**prokaryotic** Of **prokaryotes** or the **Prokaryota(e)**.

**Protoctista** (A term effectively synonymous with '**Protista**') are eukaryotic organisms (that have a membrane-bound nucleus in their cells) which are neither animals, plants, nor fungi, and which are often unicellular or assemblages of similar cells. Protoctistans (Protists) were the first **eukaryotes** to evolve. (Specialist taxonomists would, though, make a subtle—but important to them—distinction between Protoctistans and Protists. However, the average reader, or non-biological scientist, need not worry about this technicality.)

**proton** A key subatomic particle found in the nuclei of atoms. Hydrogen, the simplest element, consists of a single proton in its **nucleus** surrounded by an electron orbital consisting of a single electron. Take away the electron and you are left with the proton. Some other elements have a greater affinity to electrons than hydrogen, and so hydrogen can be a proton donor. Early life most likely relied on hydrogen, acting as a proton donor, as an energy source.

**Protista** See **Protoctista**. From the Greek '*protiston*' meaning 'the (most) first of all ones'. The terms 'Protista' and 'protist' are still used by biologists but in a more informal way referring to simple, single-celled organisms, the first **eukaryotes**. Protista are today not considered a proper phylum, and so it is known as a paraphyletic term.

**Quaternary** The current geological period covering the past 2.59 million years characterised by an ice age of cold **glacials** and shorter warmer **interglacial** times such as the current **Holocene** epoch.

**radical** A term in chemistry (it means something else in biology and again in mathematics) for a molecule or atom that presents an odd number of electrons: the use in this text is purely in the chemical sense. They can be formed by **photolysis** (*e.g.* $OH^{\bullet}$ from water). They are often very short-lived and highly reactive.

**regolith** Weathered material between the surface and bedrock. It can be a few tens of metres deep. Either it is formed *in situ* by the weathering of the bedrock or it can be transported by water wind or ice.

**REPAIR** — **R**NA **E**diting for **P**rogrammable **A** to **I** **R**eplacement is an RNA editing tool.

**siderophile** — Elements that have an affinity to—like to bond with—iron.

**Signor–Lipps effect** — Since the discovered fossil record of organisms is unlikely ever to be complete, neither the first nor the last organism of a given species will be identified.

**stromatolites** — The fossilised remains of algal mats. Often, they are laminated with calcium carbonate. Today's algal mats often contain cyanobacteria.

**Sturtian** — The first major cold phase of Snowball Earth II that ended some 700 mya.

**superkingdom** — The highest taxonomic rank of organisms in the three-domain system of taxonomy. Also known as '**domain**'—see that entry for further detail.

**symbiogenesis** — The process by which **endosymbiosis** arose. This took place multiple times and, though true-eukaryotes arose only once, this suggests that the evolution of eukaryotes was an easy evolutionary step.

**syncytium** — Adjective: syncytial. See **coenocyte.**

**Tertiary** — A term for the geological period from 66 to 2.58 million years ago that since the 2000s is rarely used (students researching past geological papers will invariably see this term). It covers two periods now used: the Palaeogene and Neogene. Its abbreviation is 'T'.

**Theia** — (Sometimes spelled Thea.) The protoplanet thought to have collided with the Earth to form the Moon.

**thermohaline** — '*Thermo*' from the Greek '*Thermos*' meaning heat and '*haline*' from the Greek '*halós*' meaning salt. The term is use in 'thermohaline circulation' whereby warm, surface ocean water moving to higher latitudes becomes both saltier (due to evaporation) and colder (due to the higher latitude), therefore becomes more dense and so sinks. This thermohaline circulation driver allows more water to be transported from the tropics.

**tholins** — Organic compounds of hydrogen and carbon formed from the breakdown products of methane (having been subjected to ultraviolet light) reacting with each other. They are found today in the atmosphere of Saturn's moon Titan and likely existed in the early, anaerobic, Earth's atmosphere. After the Greek '*tholós*' meaning 'not clear' or 'muddy'.

**tricarboxylic acidcycle** — Sometimes abbreviated to the TCA cycle, it is also known as the **citric acid cycle** and the **Krebs cycle.** It is a fundamental process involved in the metabolic breakdown of carbohydrates to release energy which life can then use. It is not necessary in terms of this book to note the detail, but in essence it involves condensing acetate with oxyalacetate to form citrate. Four oxidation reactions and two decarboxylations follow to ultimately regenerate the oxyalacetate and the start of a new round of the cycle. It is so fundamental that it is taught to older teenagers at school who go on to study for a bioscience degree at university.

| | |
|---|---|
| **trophic level** | A division within a food chain represented by a broad category of species within an ecosystem characterised by mode of food supply. For example: plants (primary producers); herbivores (the secondary producers that eat the plants); and the carnivores (that eat the herbivores). |
| **unikont** | A taxonomic term for a supergroup of eukaryotic species that include animals and fungi. Many single-celled members of the group, and the presumed ancestor, have a single flagellum. **Bikonts** and unikonts arose or became defined somewhere at the very least between 1.2 billion years ago and 900 million years ago but possibly earlier. (These days some protistologists prefer the term 'Amorphea' to 'Unikont'.) |
| **Varanger** | A rarely used alternative name for the **Marinoan**. |
| **vascular plants** | Advanced plants with specialist conducting tissue for the transport of water, metabolites, and minerals. They include club mosses (seedless), ferns, cycads (one of five extant divisions of seed-bearing plants), gymnosperms (one of five extant divisions of seed-bearing plants), and angiosperms (flowering plants). |
| **water activity** | The water activity of a solution is the ratio between the water-vapour pressures of that solution and of pure water at the same temperature and pressure. |
| **zoonotic** | A human disease is said to be zoonotic when it passes from an animal to our own species, the transference process being known as zoonosis. |

# Nomenclature and data sources for appendices 1 and 2


Allaby, M. *Dictionary of the Environment.* Macmillan Press: London, Great Britain (1977).

Baron, D. N. *Units, Symbols and Abbreviations: A Guide for Biological and Medical Editors and Authors.* Royal Society of Medicine Services: London, Great Britain (1988).

BP Economics Unit (eds.). *BP Statistical Review of World Energy.* British Petroleum Corporate Communications Services: London, Great Britain (2019 and 2020).

Daintith, J., Martin, E., Hine, R. S., *et al. Oxford Dictionary of Science* (7th ed.). Oxford University Press: Oxford, Great Britain (2017).

Daintith, J. (ed.). *A Dictionary of Physical Sciences.* Pan: London, Great Britain (1978).

Institute of Biology and the Association for Science Education (eds.). *Biological Nomenclature: Recommendations on Terms, Units and Symbols.* Institute of Biology: London, Great Britain (1989, 1997, and 2009).

Lawrence, E. (ed.). *Henderson's Dictionary of Biological Terms.* Longman: Harlow, Great Britain (1989).

Margulis, L., & Schwartz, K. V. *Five Kingdoms: An Illustrated Guide to the Phyla of Life on Earth.* W. H. Freeman and Co: New York, USA (1988).

Morgan, R., & Murray, R. B. *Energy Resources and Supply.* Wiley Interscience: London, Great Britain (1976).

Stiegeler, S. E. *A Dictionary of Earth Sciences.* Pan: London, Great Britain (1976).

Walker, P. M. B. (ed.). *Chambers Science & Technology Dictionary.* W & R Chambers and Cambridge University Press: Edinburgh & Cambridge, Great Britain (1988).

Whittaker, R. H., On the broad classification of organisms. *Quarterly Review of Biology,* **34**, 210–226 (1959).

Worldwatch Institute, *Vital Signs 2003–2004: The Trends That Are Shaping Our Future.* Earthscan: London, Great Britain (2003).

APPENDIX 3

# Bio-geological Timescale

| AGE (mya) | EON | ERA | | PERIOD | EPOCH | MAJOR EVENTS | TIME BEGINNING (mya) |
|---|---|---|---|---|---|---|---|
| 0.0117 | Phanerozoic | Cenozoic | Neogene | Quaternary | Holocene | Current interglacial (part of which is sometimes called the Anthropocene) | 0.0117 mya (11,700 ya) |
| 2.58 | | | | | Pleistocene | Series of inter-glacials and glacials. Rise of *Homo* spp. | 2.58 |
| 5.333 | | | | Tertiary | Pliocene | Rise and decline of the australopithecines | 5.333 |
| 23.03 | | | | | Miocene | 35–15 mya temps 3°C–4°C warmer than present but cooling | 23.03 |
| 33.9 | | | Palaeogene | | Oligocene | $C_4$ grasses begin to expand. Gets cooler. Antarctic sheet ice growing. | 33.9 |
| 56 | | | | | Eocene | The Eocene is warm. It begins with the IETM/PETM hyperthermal. | 56 |
| 66 | | | | | Palaeocene | Extinction recovery and the Earth begins to warm | 66 |
| 100.5 | | Mesozoic | | Cretaceous | Upper | End-Cretaceous extinction(and the dinosaur extinction) | 100.5 |
| 145 | | | | | Lower | | 145 |
| 201.4 | | | | Jurassic | | $CO_2$ levels begin a slow, variable trend down to Quaternary levels | 201.4 |
| 251.9 | | | | Triassic | | $CO_2$ levels rise to ~5x present level. Dinosaurs begin to rise. | 251.9 |

| | | | | | | |
|---|---|---|---|---|---|---|
| 298.9 | | Palaeozoic | Permian | | End-Permian extinction | 298.9 |
| 323.2 | | | Carboniferous | Pennsylvanian | $CO_2$ levels low. Rise of woody plants. | 323.2 |
| 358.9 | | | | Mississippian | | 358.9 |
| 419.2 | | | Devonian | | $CO_2$ levels peak at over 22x present level following a period of low levels and glacials | 419.2 |
| 443.8 | | | Silurian | | | 443.8 |
| 485.4 | | | Ordovician | | | 485.4 |
| 538.8 | | | Cambrian | | 'Age ofinvertebrates' | 538.8 |
| 635 | Precambrian | Proterozoic | Neoproterozoic | Ediacaran | Avalon boom of Ediacaran (or Vendobiota) spp. | 635 |
| 720 | | | | Cryogenian | Snowball Earth II | 720 |
| 1,000 | | | | Tonian | The 'boring billion' | 1,000 |
| 1,200 | | | Meso-proterozoic | Stenian | | 1,200 |
| 1,400 | | | | Ectasian | | 1,400 |
| 1,600 | | | | Calymmian | | 1,600 |
| 2,500 | | | Palaeo-proterozoic | | Snowball Earth I (from 2.4 bya) $O_2$ photosynthesis begins to increase | 2,500 |
| 4,000 | | Archaean | | | Fossil evidence of bacteria and cyanobacteria. | 4,000 |
| ~4,567 | | Hadean | | | | ~4,567 |

The dates given are those designated by the International Commission on Stratigraphy (ICS, 2023). The spellings are English. The breakdown of nomenclature includes commonly used terms such as 'quaternary' which the ICS has moved (though not yet implemented) to discontinue. Yet because of common use, and for student ease (especially by non-geologists), such terms are retained in this chart.

# Index

*For the benefit of digital users, indexed terms that span two pages (e.g., 52–53) may, on occasion, appear on only one of those pages.*

*Note. Nations are only listed here in relation to aspects of the Earth system or economy. Mentions of nations where researchers are based are not included.*


Africa 75, 99, 112–113, 171–172, 174–175, 181–182, 263–269, 271–279, 284–285, 301, 312, 315–316, 326–327, 329–330, 342

agriculture (*includes* fishing) 131, 242, 246–247, 257–258, 273–279, 284–285, 308–309, 312–313*f*, 315, 317, 321–324, 327–328, 331, 334, 338–339, 342

albedo 52, 54–55, 58–59*f*, 61–64, 98–99, 148, 152, 163, 164, 166–167, 181–182, 204, 223, 263–264, 275, 298–299

algae 20–21*f*, 56*f*, 80, 110–111, 114, 122–123, 135–136, 139, 141, 144, 160–161, 177–178, 331

Amazon 63–64, 210, 298–299*f*, 301, 305–306, 315–316

America(s) 99, 112–113, 171, 177–178, 181–182, 276, 295*b*, 297, 304, 311–312, 316, 347

amino acids 20, 22–27, 281

  handedness 24–26, 102, 221–222

AMOC (*see* Broecker conveyor)

animals (*see* Metazoa)

Antarctica 57*f*, 57–58, 63–64, 91–93, 171–172, 181–182, 190–191, 243*f*, 251, 263–265, 272–273*f*, 299*f*–299, 309, 311–312, 314

Anthropocene 254–255, 259–260, 296*f*

antibiotic resistance 77, 87, 158, 216

Archaea 26–27, 35, 71, 74, 79, 116, 223–224, 343 (*see also* prokaryotes)

Archaean 17, 20–23, 29–32, 35–38, 44, 47, 51, 72–73, 74–76, 79, 83–86, 101

Arctic 116, 120, 175, 177, 179, 182*f*, 183–184, 298–300, 311–312

artificial intelligence 284, 342–348

Asia 181–182, 267, 273, 276, 297, 315–316, 326, 329–330, 347

asteroid(s) 5–6, 18–19, 27–29, 50–51, 58–59, 67, 151, 171, 173–174, 177–180, 241, 334

Atlantic 63–64, 117, 171–172*f*, 184–185*f*, 188–189, 191, 209–210, 250, 254, 300–303, 305

atmosphere 16–17, 19–24, 29, 31–32, 35–36, 38–42, 43–44, 47, 56*f*, 76–79, 82–91, 94–98, 101, 103–104, 107–110, 121–122, 136–141, 144, 146–149, 152–154, 159, 162*f*, 164, 167, 171–174, 175, 178, 183–191, 194, 199–202, 204–205, 208–217, 218–221, 223, 225–230, 242, 244–247, 248–251, 253, 257–259, 261–262, 263–264, 283–284, 285*f*, 292–294, 295*b*, 297, 299–301, 303–306, 317–319, 331, 337–338, 348–349

Australia 16–17, 121–122, 160, 181–182, 256*f*, 301, 311–312, 315–316

autocorrelation 65–66, 302–303, 305–306

autotrophs 26–27, 35, 39 (*see also* chemolithoautotroph)

Avalon boom (explosion) 150, 170, 256–257

bacteria 2, 20–22, 26–28, 35–37, 43, 46–47, 70–72, 72*f*, 74–78, 79–80, 82, 83–84, 86–87, 103, 110–111, 114–116, 120–122, 127, 132, 141, 144, 216 (*see also* prokaryotes; cyanobacteria)

banded iron formations (BIFs) 19–22, 44, 47, 67, 72–73, 83, 260, 262–263, 285*f*

bikonts 123, 135

biomass 35–41, 45–47, 89–90, 103, 107, 132–133, 140, 142, 148–149, 179–180, 225–226, 254, 257, 259, 282*f*

boring billion, the 108–110, 120–123, 125–126, 128–129, 131, 143, 148–149, 194

Broecker conveyor (*also includes* AMOC) 63–64, 254, 300–303, 311–312

$C_3$ plants *and* $C_4$ plants 156–157, 187, 227, 265–266, 269, 295*b* (*see also* plants)

Cambrian boom (explosion or diversification) 137, 143, 151, 157, 170, 256–257
Canada 19, 22, 28–30, 175, 299, 301, 311–312, 316
carbon 16–17, 20–22, 35–43, 69, 72–73, 88–89, 121, 133, 140, 142, 143–145, 149–151, 162*f*, 167, 174, 176, 186–191, 196, 199–200, 242, 254, 258, 262, 297–300, 303–304, 317–319, 326, 338–339, 348 (*see also* carbonate, carbon dioxide, carbon isotope excursion, methane, permafrost, soil carbon)
  $^{12}$C *and* $^{13}$C 20–21, 21*f*, 28–29, 35–42, 37*f*, 50, 72–73, 83, 97–98, 141, 144–145, 148, 176, 183–190, 185*f*, 242–244, 253, 269, 292–294, 348
carbonate 21*f*, 35–36, 37*f*, 40–42, 97–98, 104, 141, 144–145, 148, 152–153, 174, 178, 184, 185*f*–186–188, 189–190, 204, 212–213, 227, 269, 297
carbon dioxide 16–17, 19, 21*f*, 22–24, 35–47, 56*f*, 71–72, 76, 80, 81*f*, 82–83, 85, 90–92, 94–95, 97–99, 101, 103–104, 120–121, 140–142, 146–149, 152–154, 156–157, 159–160, 172, 174–176, 178–179, 181–182, 187–189, 200–201–201*t*, 202, 204, 206, 210, 211–213–214*t*, 215, 219, 223, 226–227, 241, 242–244, 245–249, 251, 253, 263–264, 274–275, 283–284, 292–293, 295–295*b*, 297–301, 303, 305–306, 317–318, 331, 339
Carboniferous 149–150, 195
carbon isotope excursion (CIE) 173, 177, 180–184, 186–187, 189–191, 242–244, 253–255, 284, 292–294, 297, 307–308, 310
carbon leakage
  economic 255–256*f*, 339
  ecological 39, 41, 122 (*see also* soil carbon)
chaos theory 61–65
  Lorenz attractors 61–63, 65, 295*b*
chemolithoautotrophs 26–27, 43, 71, 73, 211, 215, 219, 221, 223–224
Chile 216–217, 311–312, 340*f*
China 136–137, 139–140, 151, 255–256*f*, 267, 285*f*, 315–316, 319, 328, 332, 338–341
chloroplasts 111, 113–120, 261, 350
climate change 3–4, 50–51, 53–58, 60–61, 64–66, 89, 135, 142–143, 167, 173, 176, 178–179, 191–192, 192*t*, 199, 203, 242–253, 254–255, 263–265, 268, 272–275, 283*t*–284, 292–293, 294–301, 302–304, 306–308, 310–318, 320, 323–324, 329–330, 338, 343 (*see also* global warming; Snowball Earth)
continental drift 16, 29–30, 99, 108, 112–113, 152, 171 (*see also* plate tectonics; Wilson Cycle)
convergent evolution 129–130
coral 11–15, 101, 299*f*
CoVID-19 254–255, 308, 327–330, 339
Cretaceous 17–18, 151, 172, 176–181, 191–192–192*t*, 283, 307–308
critical transitions (*see* transitions)
cyanobacteria 74, 78, 80–83, 86–87, 89, 102–103, 110–111, 114–117, 120–123, 125–128, 135, 159, 160–161, 164–165, 191–192, 213, 217, 225, 229, 261, 280, 283–284, 285*f* (*see also* bacteria)

deep sea vents (*see* thermal vents)
Denmark 189, 340*f*
deuterium 20, 57–58, 75, 201–202, 210, 272 (*see also* hydrogen)
dinosaurs 17–19, 67, 92, 121, 151, 172, 173, 176–182, 283
DNA 20, 24–25, 43, 70, 102, 109, 111–117, 120, 129, 192*t*, 219, 224, 229, 241–242, 260–261, 280–281, 282*f*, 283*t*, 350 (*see also* nucleic acid; RNA)

Earth-like planets 1–3, 50, 225–226, 227–228, 230, 232, 332–333, 337–338
economics 250, 254–255, 308–309, 315–320, 326–327, 330–334, 339–342, 347–348
ecosystem function 150, 176, 306
ecosystem services 257–258, 330–332
Ediacaran(s) 121, 136–137, 139, 143, 150–151, 157, 170, 256–257, 261
Egypt 254, 256*f*, 276–279, 340*f*
Enceladus 201*t*, 223–224
endosymbiosis 113, 115–120, 123, 125–126, 128, 229–230, 241–242, 261, 332–333, 343–344 (*see also* symbiosis)
enzymes 20–21, 23, 149–150, 157–158
eukaryotes 71, 74, 91, 97, 102–103, 111–130, 135–137, 149, 154, 161, 163–167, 191–192, 192*t*, 199, 211–212, 222–223, 229–230, 232, 234, 241–242, 256–257, 260–263, 280–284, 286, 343–344, 348, 350
Europa 201*t*, 219–223
Europe 8, 171, 174, 188, 189, 267–268, 277–278, 301, 308, 311–312, 319, 327, 339–342, 347
exobiology 199, 205–207, 210–217, 219–231, 234–235, 332–333, 337, 348–349

exoplanet(s) 2–3, 27, 50, 199, 224–226, 227–228, 230, 234, 349
extinction 58–59, 70, 127, 170, 174, 241, 284–286, 307–308 (*see also* mass extinction)
extraterrestrial civilisations 219–220, 232–236, 332–338, 349

faint Sun (*see* Sun)
fecundity 257, 326
feedback(s) 52–56, 58, 60, 66–67, 90, 104, 109–110, 150, 152, 160, 186, 226–227, 241, 251, 271–272, 294–295, 297–300, 310, 314
Fermi Paradox 233–236, 332–338
fertility (*see* fecundity)
flips of state (*see* transitions, critical)
forest(s) (*includes* woodland) 37–38, 40–41, 63–64, 131, 190, 258, 263, 265, 269–271, 277, 293, 299–301, 304–306, 310–311, 315–316, 327–331, 341
fossil fuel 21*f*, 38, 40–41, 69, 79, 104, 144, 162*f*, 189, 195, 244, 246, 248–249, 249*f*, 254–255, 262, 284, 293–294, 305–306, 317–320, 332, 333–334, 338–339, 341–342
France 255–256*f*, 273–274, 301, 347
Fungi 123, 129–130, 135, 141, 145, 147, 149–150, 159, 173, 177–178, 205, 217, 259

Gaia 53–55, 58–59, 213–214, 343–344
Galaxy 1–2, 7–8, 27, 224–225, 231–232, 235–236, 333–337
Gaskiers (*see* Snowball Earth, Snowball II)
Germany 255–256*f*, 340*f*, 345–346
glacial(s) 8–9, 50–51, 55–56, 57*f*–58, 60–61, 67, 90, 93, 96–98, 143–148, 156, 160, 163–164, 170, 227, 243*f*, 245–248, 263–265, 272–275, 294–298 (*see also* last glacial maximum)
global (human) population 2–3, 246–247*f*, 255, 257, 259–260, 281, 284–285, 307–308, 312, 315, 320–323, 327, 331–332, 333–334, 336–337, 343, 347–348
global warming 50–51, 63–64, 69–70, 104, 176, 178, 183–190, 211–212, 242, 244–253, 254–255, 284–285, 299–301, 307–308, 314–318, 329–330 (*see also* climate change)
grass 40, 131, 156–157, 187, 227, 257–258, 265–266, 269–270, 275, 297, 327–328
greenhouse (*both* gas *and* effect) 40, 52, 55–56, 76, 90–91, 96–97, 104, 120–121, 146, 148, 164, 176, 178–179, 183–186, 189, 202–204, 206, 209, 211–212, 220, 223, 226–227, 242–253, 274–275, 277, 283–284, 292, 294–295, 297–298, 301, 303–307, 310, 315, 317–319, 339
Greenland 16, 21–22, 57–58, 63–64, 171, 188–189, 251, 264–265, 272, 298–299, 301

Hadean 17–19, 22, 27–28, 41
heterotrophs 39
Holocene 50–51, 57–58, 60–61, 254–255, 260, 274–276, 294–298, 296*f*, 309–311
Hominids 263, 265–266, 268–269, 271, 273*f*
*Homo spp.* 50–51, 69, 170, 244, 253, 254–255, 256–259, 262–263, 265–275, 279–281, 285*f*, 342–345
hot springs (*see* springs (hot))
Huronian glaciation (*see* Snowball Earth, Snowball I)
hydrogen 2, 16–17, 20, 22–27, 43–46, 57–58, 71–103, 109, 120, 157, 201–202, 210, 215–216, 219, 221, 223–224, 226–227 (*see also* deuterium)
hydrogen peroxide 157–159, 163–164, 221
hyperthermal 67, 183–184, 191, 292–298, 303–304, 307–308, 310, 342
hydrothermal vent(s) (*see* thermal vents)

ice age(s) 8–9, 50–51, 55–58, 59*f*, 60–61, 67, 76, 90, 154–157, 246–247, 263–264, 272, 294–295, 297–298 (*see also* Snowball Earth)
Iceland 29–30, 156, 171, 175
ice sheet(s) 55, 57–58, 63–64, 91, 93, 158–160, 163–164, 166, 251, 263–264, 272, 274–275, 294–295, 298–299, 312, 314
IETM (*see* PETM)
India 172, 179, 181–182, 255–256*f*, 297
information (*and* information processing) 67, 70–71, 120, 147, 191–192–192*t*, 194, 199, 241–242, 260–262, 268, 269–270, 273–274, 277–282, 292, 342–348, 350
  byte(s) 278–279, 281–282
infrared (light) 6, 20, 45–47, 75–78, 80–82, 86–87, 89, 166, 174, 213, 222
interglacial(s) 8–9, 50–51, 55–58, 60–61, 67, 156, 170, 246–247, 263–265, 272, 274, 294–298, 309–311
Intergovernmental Panel on Climate Change (IPCC) 60–61, 162, 185*f*, 245–246, 248–253, 292–294, 302–304, 310, 315, 317–318, 323
Ireland 277

iron 5–7, 16–17, 18–20, 23–24, 28–29, 44, 46–47, 67, 72–73, 83–84 (*see also* banded iron formations)
isotope(s) 4–6, 7–8, 20, 21–22, 36–37, 75, 84–86, 96–98, 100, 108–109, 144–145, 174, 178, 186, 187, 189, 201–202 (*see also separately* carbon, deuterium; sulphur)

Jupiter 6, 218–221
Jurassic 17–18, 176, 178–181, 307–308

Kenorland (*see* supercontinent)
Krebs cycle (*see* tricarboxylic acid cycle)

large igneous province(s) (LIPs) 151, 153–154, 156, 167, 171–176, 177–181, 188, 203–204, 241
last glacial maximum (LGM) 243*f*, 272, 275, 309–311
late heavy bombardment (LHB) 18–20, 21–22, 27–30, 72–73, 85–86, 201–202, 211, 234, 284–285*f*
lichens 141–147, 149, 159–161, 164, 195, 349–350
Lokiarchaeota 116–120, 123, 127
Lorenz attractors (*see* chaos theory)

magnetic field (*also* magnetosphere) 98, 202, 204, 209–210, 218, 221
Makganyene Snowball (*see* Snowball Earth, Snowball I)
mammals 129, 180, 186–187, 261–262, 264, 273–274, 329–330
Marinoan (*see* Snowball Earth, Snowball II)
Mars 6, 8–9, 72–73, 92, 201*t*, 207–218, 221, 223–225, 226–227, 337, 348–349
mass extinction 93, 151, 170, 172–181, 307–308 (*see also* extinction)
mercury (element) 179
Mercury (planet) 199–201, 225–226
Metazoa 10–11, 23, 37–38, 45, 70, 91, 107–108, 110–111, 114, 121–124, 129–130, 132, 135–136, 138–140, 148–150, 158–159, 173, 176, 180, 181–182, 186–187, 190, 192*t*, 195–196, 230, 241–242, 254, 258–259, 261–262, 268–269, 273–274, 284–285, 328, 343–344
methane 17–23, 25–27, 41, 76–77, 82–88, 90–91, 94–98, 102–103, 109–110, 117–119, 157, 176, 178, 179, 184–187, 190, 202, 206, 213–215, 217–219, 223–224, 226–227, 230, 242, 245–246, 251–253, 283–284, 296*f*, 298, 299*f*, 303, 306–307, 337–338, 349
Mexico 31, 177–178
Middle East 254, 273, 275–276, 278–279, 347
Milankovitch 170, 294–295
mineral species (*see* species, mineral)
mitochondria 111–117, 118*f*, 120, 241–242, 261, 350
molecular clocks 27, 109, 116–117, 122–126, 128*f*, 140, 195, 348
moon(s) 8–9, 77, 200–201*t*, 218–225, 235, 333–335
Moon, the 4, 6–16, 17–19, 128, 165, 220, 236, 277, 285*f*
mountains 29–31, 39, 65–66, 90–91, 93, 99–102, 107, 147, 152–153, 164–165, 181–182, 295*b*, 297, 329–331
multicellularity (*also* multicellular) 89–90, 103, 107, 121, 123–124, 127, 129–133, 135–141, 143–149, 154, 160–161, 163–164, 167, 170, 176, 191–192, 192*t*, 194–196, 199, 211–212, 222–225, 227–228, 230–235, 241–242, 256–257, 260–263, 279–281, 283*t*, 285*f*, 286, 333, 337–338, 343–344, 349–351

natural capital (*see* ecosystem services)
niche construction 285–286, 310, 315, 317, 343
nitrogen 23, 25–26, 76, 109, 200–201, 204, 214*t*, 218, 223, 226–227, 338–339
Norway, 171, 188–189, 340*f*
nucleic acid 24–27, 70–71, 102, 120, 192*t*, 204, 212–213, 221–222, 224, 241–242, 260, 280–281, 282*f* (*see also* DNA; RNA)

ocean(s) (*includes* seas) 17, 19–20, 25, 29–30, 35, 43, 47, 56*f*, 63–64, 71, 75, 91–93, 99–100, 104, 120, 127–128, 131–133, 146–147, 149, 152–153, 161, 165–166, 174, 176, 180, 184, 188, 201–205, 210–212, 217, 219–227, 244, 259, 298–299, 300–303, 314, 338, 348, 349
orbit(s) 6–7, 9–10, 14*f*, 15–16, 18, 58–59*f*, 191, 203, 215, 218, 220–221, 223, 228, 230–231, 263–265
oxygen (*includes* oxygen isotopes) 7–8, 16, 17–18, 26–27, 29, 43–47, 74–75, 78–86, 87–91, 94–104, 107–110, 113–114, 116–117, 120–126, 128*f*, 128, 132, 135–141, 143–150, 154, 157–161, 163–165, 167, 174, 178, 186–187, 191–192, 194–195, 199, 201–202, 204, 211, 213–217, 219–231, 234, 241–242, 256–263, 279,

280–281, 283–286, 331–333, 337–338, 343–344, 348–349

Pacific 156, 184–185*f*, 316
Paris Accord 297–298, 306–307, 314, 319–320, 329–330
Perfect Storm 324*f*–325, 332
permafrost 38, 41, 91, 190–191, 210, 215, 244, 251, 298, 299*f*, 303–304, 306–307
Permian 151, 173–176, 177–181, 189, 191–192–192*t*, 307–308
PETM 171, 181–192, 292–294, 297–298, 303–304, 307
phosphine 205
phosphorous 101–103, 107, 146, 160–161, 205, 212–213
photolysis (*also* photo-dissociation *and* radiolysis) 16–17, 83–84, 95–96, 103–104, 109, 157, 211–212, 213–214, 216, 218, 220, 221–222, 227
photosynthesis 20–22, 27, 35–39, 45–47, 56*f*, 71–73, 120–121, 144–145, 149, 161–162, 174, 213, 217, 222, 227, 244, 257, 305–306, 332–333
  anaerobic 20, 47, 74, 76–79, 80–82, 86–87, 110–111, 174, 262–263, 332–333
  oxygenic 45–46, 79–84, 86–91, 94–99, 102–104, 107–111, 114, 116–117, 120–127, 131–132, 135–136, 138–141, 144–145, 147–149, 154, 156–165, 167, 174, 177–178, 187, 191–192, 194–195, 199, 204, 211, 213, 215–217, 219–220, 222, 224–225, 227, 229–231, 234, 241–242, 256–258, 260–263, 279–280, 283–284, 285*f*–286, 332–333, 337–338, 348–349
plants 10*f*, 20, 21*f*, 37–39, 44–46, 70–71, 74, 80–81, 107–108, 110–111, 114, 121–123, 129–130, 132–133, 135, 140–141, 143–144, 150, 156–157, 159–161, 173, 176–178, 180, 186–187, 190, 195–196, 227, 241–243, 259, 265–266, 269, 271, 277, 295*b*, 304, 305–306, 309–310 (*see also* $C_3$ and $C_4$ plants)
plastic pollution 255, 259, 325
plate tectonics 29–31, 38–39, 74–75, 99–100, 102, 112–113, 167, 171, 194, 199, 202–204, 209–212, 221, 253, 285*f*, 297 (*see also* continental drift and Wilson cycle)
Pluto 8, 200
politics 244, 250–251, 253, 307–309, 318–319, 324, 326–327, 330–332, 334, 339–342, 347–348
Pongola glaciation 82–83, 88, 96–97, 102, 123, 128*f*
population, human (*see* global population)
Porifera (*see* sponges)
productivity (agricultural *and* biological) 35, 222, 246–247, 257–258, 268 276–278, 306, 320–323, 324–325, 326–327, 338–339
prokaryotes 2, 20–21, 25–28, 35–39, 45–47, 71, 76–78, 80–82, 86–87, 91, 103, 110–120, 123, 125–128, 131–132, 135–136, 149, 158–159, 161–167, 174, 192*t*–201, 205, 211–213, 215–217, 221–226, 228–230, 232, 234, 241–242, 259–261 280, 283–284, 332–333, 337, 343, 350–351 (*see also* Archaea; bacteria)

Quaternary (ice age) 50–51, 55–58, 60–61, 67, 156–157, 263–265, 272, 294–295

Rain 40–41, 56*f*, 75–76, 91, 100–103, 107, 120–121, 127, 140, 142, 143–145, 152–153, 159, 165, 178, 186, 190, 200–202, 210, 212, 216–217, 227, 264, 272, 274–275, 301–302, 311–314, 329, 348
refugia 27–28, 78, 93, 103, 147
respiration 23, 39–40, 45, 79–82, 87, 121, 140, 148–149, 162*f*, 219, 221–222
RNA 24–26, 43, 74, 102, 219, 224, 229, 260, 281, 350 (*see also* DNA; nucleic acid)
Russia 174, 311–312, 326–327, 341

Saturn 77, 218, 223
science fiction/fantasy 2–3, 17–18, 43, 51–52, 76, 78, 116, 151, 178–179, 199–200, 206–209, 217, 219–220, 233, 334, 344–345
sea-level 17, 44, 93, 107, 182, 189–190, 201, 208–209, 212, 251, 272–273, 294–295, 296*f*, 299, 312, 314, 338
seas (*see* oceans)
Signor–Lipps effect 126–127, 193, 266–269, 278–279
silicate drawdown (of carbon) 141–142, 146, 149, 153–154, 156, 159–160, 162, 227, 295, 297
singularity (technological) 342–343, 351
Snowball Earths 69, 94, 99, 143, 148, 151–162, 163–167, 192*t*–192, 199, 260, 262–263, 282*f*, 343
  Snowball I 88–93, 96–98, 100–105, 110, 120–121, 125–128, 131, 133, 136, 138, 146–147, 153–154, 164, 194, 280, 348
  Snowball II 91, 135, 137–150, 170, 195, 259, 280–281, 282*f*, 349–350

soil(s) 62–63, 84, 132, 143–144, 147, 174–175, 186, 242, 257–258, 330–331, 338 (*see also* soil carbon)
soil carbon 21*f*, 38–39, 145, 162*f*, 187–188, 190, 251, 269, 303–304 (*see also* soils)
Solar System 1–2, 4–8, 18, 199–200–201*t*, 218–220, 223, 225
species
  biological 9, 10*f*, 10–12, 21*f*, 26–27, 35–36, 40–41, 47, 50–51, 58, 67, 70, 71, 74, 78, 80–81, 86, 89, 93, 107–108, 109–110, 114–116, 117–119, 119*f*, 121–124, 126–127, 129–133, 135–137, 138–144, 146, 148–151, 159–161, 163–167, 170, 173–174, 177, 178–179, 180–181, 187, 191–192, 195–196, 205, 216–217, 221, 223–224, 227–228, 230, 232–233, 234–236, 241, 253, 256–259, 260–272, 274–276, 278–279, 280–281, 283–286, 307–309, 311, 317, 326, 327–330, 332–333, 334–335, 336–337, 343, 349
  mineral 192–196, 199, 210, 242, 258–260, 262, 279, 284, 325
sponges 129–130, 138–139, 148–149, 161
springs-hot 19, 23, 25, 27–28, 35, 43, 75, 87, 91, 93, 102, 147, 192*t*, 205, 212–213, 219, 224, 242 (*see also* thermal vents)
stability 8–9, 16, 18, 38, 42–43, 48–57*f*, 59–60, 62–63, 66–67, 226–227, 232, 235, 274–275, 297, 332
star(s) 1–3, 6, 7–8, 12–13, 50, 92–93, 202, 224–225, 226–228, 334–335 (*see also* stellar type; Sun)
stellar type 228–232, 234–235, 333, 337–338, 349 (*see also* star(s); Sun)
sterols 121–127, 160
stromatolites 21–22, 27, 87, 131–132, 142, 176, 285*f*
Sturtian (*see* Snowball Earth, Snowball II)
sulphur 18–19, 22–23, 26–27, 30, 44, 46*t*–47, 72*f*, 74, 84–86, 87–88, 96–97, 101, 153–154, 159, 176, 199–201, 205
Sun, the 4–9, 10–11, 15–16, 18, 28, 51, 58, 76–77, 85, 148, 174, 200–201*t*, 203, 208–211, 215–218, 225–226, 228–230, 231–232, 234–236, 241, 244, 274–275, 337, 349 (*see also* star; stellar type)
  faint Sun 76, 202, 206, 226–228
supercontinent 30, 39, 90–92, 97–99, 101–103, 107–108, 112–113, 116, 120–121, 127, 151–158, 161–167, 171, 193–196, 199, 212, 222, 242, 258–259
  Columbia 194–195
  Gondwanaland 112–113
  Kenorland 99, 108, 152, 154–155, 164–165, 194
  Nena-Columbia 108, 154–155
  Pangaea 30, 112–113, 149–150, 152, 154–155, 195–196
  Rodinia 108, 153–155, 164, 195
symbiosis 114, 128, 139, 141, 343–344 (*see also* endosymbiosis)
Sweden 160

tectonics (*see* plate tectonics)
thermal vents 19, 22, 25–27, 39, 43–47, 75–78, 80–82, 86–87, 102–103, 117–120, 127, 165–166, 174, 188–189, 204, 219, 221–222, 242, 348 (*see also* springs-hot; volcanoes)
tholins 77, 82–83, 85, 95–96, 202, 206, 218–219, 227
threshold(s) (*see* transitions)
tides 10–16, 77, 128, 165–166, 220–223, 228, 230, 278
Titan 77, 85, 201*t*, 218–219
transitions 3–4, 30, 48, 61, 75–76, 108, 347–348
  critical 31–32, 42–43, 47, 49–51, 55–58, 60–67, 69, 83–86, 87–90, 93–96, 102–104, 107–110, 122–123, 126, 128, 132, 135–137, 142–143, 148–149, 157, 159, 161–165, 167, 170, 173, 180, 191–192, 194–196, 199, 204–205, 226, 236, 241–242, 256–263, 279, 282–286, 292–308, 317, 326, 330, 334, 342–343, 351
trees 9, 11–12, 36–37, 41, 129, 149–150, 195, 227, 257–258, 265, 269, 331 (*see also* forests)
tricarboxylic acid (TCA) cycle 23–24, 26

Ukraine 301, 326–327, 341
ultraviolet (light) 19–20, 25, 43, 77, 83–86, 94–96, 102–103, 109, 157–158, 204, 212–213, 216–217, 218–219, 221–222, 224, 229
UNEP 250
unikonts 123, 129–130, 135
United Kingdom 250, 255–256*f*, 277, 301, 308, 334, 338–341
Uranus 218
USA 160, 177–178, 180, 186, 255–256*f*, 309, 311–312, 316, 319, 330–331, 334, 339–341

Venus 200–206, 225–226, 228, 241
volcanoes 5, 19, 22–23, 30, 39, 43–44, 50, 75, 97–99, 104, 127, 147–148, 153–154, 156, 159, 167, 171, 173–176, 179, 189, 194,

202–205, 244, 253, 292–293 (*see also* large igneous province; thermal vents)

Wilson cycle 30, 99, 108, 152, 167, 171–172, 193–194, 199, 258–259 (*see also* continental drift; plate tectonics)

woodland (*see* forest(s); trees)
World Meteorological Organization 250
World Model 320–323

zircon(s) 17–18